The Dynamics of Innovation in Eastern Europe

NEW HORIZONS IN THE ECONOMICS OF INNOVATION

Series Editor: Christopher Freeman, *Emeritus Professor of Science Policy, SPRU – Science and Technology Policy Research, University of Sussex, UK*

Technical innovation is vital to the competitive performance of firms and of nations and for the sustained growth of the world economy. The economics of innovation is an area that has expanded dramatically in recent years and this major series, edited by one of the most distinguished scholars in the field, contributes to the debate and advances in research in this most important area.

The main emphasis is on the development and application of new ideas. The series provides a forum for original research in technology, innovation systems and management, industrial organization, technological collaboration, knowledge and innovation, research and development, evolutionary theory and industrial strategy. International in its approach, the series includes some of the best theoretical and empirical work from both well-established researchers and the new generation of scholars.

Titles in the series include:

Innovation Networks
Theory and Practice
Edited by Andreas Pyka and Günter Küppers

Systems of Innovation and Development
Evidence from Brazil
Edited by José E. Cassiolato, Helena M.M. Lastres and Maria Lucia Maciel

Innovation, Competence Building and Social Cohesion in Europe
Towards a Learning Society
Edited by Pedro Conceição, Manuel V. Heitor and Bengt-Åke Lundvall

The Dynamics of Innovation Clusters
A Study of the Food Industry
Magnus Lagnevik, Ingegerd Sjöholm, Anders Lareke and Jacob Östberg

Technological Systems and Intersectoral Innovation Flows
Riccardo Leoncini and Sandro Montresor

Inside the Virtual Product
How Organisations Create Knowledge Through Software
Luciana D'Adderio

Embracing the Knowledge Economy
The Dynamic Transformation of the Finnish Innovation System
Edited by Gerd Schienstock

Innovation and Institutions
A Programmatic Study
Edited by Steven Casper and Frans van Waarden

The Dynamics of Innovation in Eastern Europe
Lessons from Estonia
Per Högselius

Technology and the Decline in Demand for Unskilled Labour
A Theoretical Analysis of the US and European Labour Markets
Mark Sanders

The Dynamics of Innovation in Eastern Europe

Lessons from Estonia

Per Högselius
Lund University and Royal Institute of Technology, Sweden

NEW HORIZONS IN THE ECONOMICS OF INNOVATION

Edward Elgar
Cheltenham, UK • Northampton, MA, USA

Published by
Edward Elgar Publishing Limited
Glensanda House
Montpellier Parade
Cheltenham
Glos GL50 1UA
UK

Edward Elgar Publishing, Inc.
136 West Street
Suite 202
Northampton
Massachusetts 01060
USA

A catalogue record for this book
is available from the British Library

ISBN 1 84542 367 4

Printed and bound in Great Britain by MPG Books Ltd, Bodmin, Cornwall

Contents

Tables

Figures

Abbreviations and Acronyms

ASO	Andmeside Osakond (Data Communications Department)
ATEKB	Arvutustehnika Erikonstrueerimisbüroo (Computer Engineering Special Design Bureau, a Soviet-era Estonian implementation-oriented R&D organization)
CBSS	Council of the Baltic Sea States
CEE	Central and Eastern Europe
CEO	Chief Executive Officer
CoCom	Coordinating Committee for Multilateral Export Controls
COMECON	Council for Mutual Economic Assistance
COPERNICUS	Cooperation in Science and Technology with Central and East European Countries (a European research program)
DSL	Digital Subscriber Loop
EAS	Ettevõtluse Arendamise Sihtasutus (Enterprise Estonia)
EDACS	An Ericsson radio communications system for police cars
EEK	Eesti kroon (the Estonian currency from 1992)
EENet	The Estonian parallel to SUNET and FUNET
ELIKO	Competence Center for Electronics, Information and Communications Technologies (at TTÜ)
Eripax	An Ericsson X.25-based data communications platform
ESTAG	Estonian Technology Agency
EU	The European Union
FDI	Foreign Direct Investment
FUNET	Finnish University Computer Network
FUUG	Finnish Unix Users Group
GDP	Gross Domestic Product
GPRS	General Packet Radio Service
GPS	Global Positioning System
GSM	Global System Mobile
HTF	Helsingfors Telefonförening (a local operator in Helsinki)
ICT	Information and Communications Technologies
IP	Internet Protocol
ISDN	Integrated Services Digital Network
IT	Information Technology

IVR	Interactive Voice Response (a system for handling and recording telephone calls in large organizations)
KBFI	Keemiline ja Biofüüsika Instituut (Institute of Chemical Physics and Biophysics, an Estonian academic institute)
kbps	kilobytes per second (data transmission speed)
KTH	Kungliga Tekniska Högskolan (The Royal Institute of Technology in Stockholm)
LTH	Lunds Tekniska Högskola (Lund Institute of Technology)
Mbps	Megabytes per second (data transmission speed)
Minsviaz	Ministerstvo Sviazi (the Soviet-era Ministry of Communications, largely identical to the network operator)
MPS	Mobile Positioning System (an Ericsson service platform)
MPSS	or Minpromsredsviaz, i.e., the Soviet Ministry for the Telecommunications Equipment Industry (largely identical to the equipment industry itself)
NMT	Nordic Mobile Telephony
OECD	Organization for Economic Cooperation and Development
PBX	Private Branch Exchange
PTO	Public Telecommunications Operator (the commonly used notion for the incumbent operator in fixed voice services)
PSTN	Public Switched Telephone Network (the traditional, copper wire-based telecom infrastructure)
RAS	Radio Access System (an Ericsson solution for the wireless local loop)
RET	Raadio-Elektroonikatehas (an Estonian radio engineering enterprise)
RISO	Riigi Infosüsteemide Osakond (Department of State Information Systems)
R&D	Research and Development
RSTI	Raadio- ja Sidetehnika Instituut (Department of Radio and Communications Engineering, TTÜ)
SPKB	Sideministeeriumi Projekteerimis-Konstrueerimis-Büroo (a Soviet-era Estonian implementation-related R&D organization belonging under the Ministry of Communications)
SKB	Spetsiaalkonstrueerimisbüroo (Special Design Bureau)
SMS	Short Messaging Service
SSR	Soviet Socialist Republic
SUNET	Swedish University Computer Network
SWIFT	Society for Worldwide Interbank Financial Telecommunications

TEMPUS	Trans-European Cooperation Scheme for Higher Education between Central and Eastern Europe and the European Community
TPI	Tallinna Polütehniline Instituut (Tallinn Polytechnical Institute, Soviet name of TTÜ)
TREKB	Tallinna Raadioelektroonika Konstrueerimisbüroo (a Soviet-era Estonian radioelectronics R&D organization)
TTÜ	Tallina Tehnikaülikool (Tallinn Technical University, in Soviet times the name was TPI)
UMTS	Universal Mobile Telecommunications System
USSD	A Nokia service platform for mobile positioning
USSR	Union of Soviet Socialist Republics (Soviet Union)
VEF	A historically famous Latvian industrial enterprise
VoIP	Voice over IP
WAP	Wireless Application Protocol
WLL	Wireless Local Loop
xDSL	acronym for various 'digital subscriber loop' technologies

Acknowledgements

The work underlying this book has been carried out under often radically shifting circumstances – both organizationally and personally. I embarked on the research at the Department of Technology and Social Change at Linköping University, from where I later moved to the newly launched Division of Innovation at Lund University. Between these two home universities in Sweden I also spent half a year in Estonia, in connection with my field work there, and a semester as Marie Curie fellow at Bocconi University in Milan. In addition, I had the opportunity to take refuge for intense writing during two summers to Kraków, in the heart of Central Europe, and to Kvännakershamn on Gotland, at the historical crossroads between the Nordic and Baltic worlds.

If despite these twists and turns a book has in the end actually been produced, it is thanks to a number of people who, in different ways, have assisted me, inspired me and kept me on track. I would first of all like to thank Professor Charles Edquist, who, as I now realize, took somewhat of a risk employing me in Linköping five years ago. Working with Charles has been truly inspiring, and I owe a great deal of gratitude to his energetic guidance, advice and encouragement. Mikael Sandberg, as a political scientist with great knowledge of East European developments, has played an important complementary role, not least through his efforts to create some logic and clarity in my historical approach.

In addition to Charles and Mikael, I am greatly indebted to Christopher Palmberg, Staffan Jacobsson, Annika Rickne, Leif Hommen and Sven Lindmark, who made invaluable comments on two earlier drafts of the manuscript. I also have to thank Professor Franco Malerba and his colleagues at CESPRI in Milan, where I learned to appreciate economic approaches to innovation in what were, for me, often unexpected ways. Steve Casper gave pragmatic suggestions at the stage of research design during a series of international workshops on the 'European Political Economy of Innovation' in Greece and Italy in 2001 and 2002. Luigi Orsenigo provided inspiring suggestions at the DRUID conference in Denmark in January 2002. Thought-provoking comments were made by participants at the annual meeting of the Society for the History of Technology (SHOT) in San José in 2001, and at a workshop held at the School of Slavonic and East European Studies in London, organized by Krzysztof Piech and Slavo Radosevic, in 2003.

Needless to say, a great source of inspiration and comments on the research underlying this book have been my colleagues at Linköping and Lund. I would like to thank in particular Johanna Nählinder and Henrik Bohlin, who had the dubious pleasure of sharing an office with me during a period of two years each. Cornelis Dekker and Anja-Sofi Karhi have been provocative discussion partners with their views on interdisciplinarity, history and cultural aspects of technological change. Johanna, Henrik, Anja-Sofi and I belonged to a group of eager and quite diverse young researchers at Linköping, who were employed at about the same time. To this group belonged also Ericka Johnson, Petra Jonvallen, Anna Green and Elin Bommenel, who I hope will not mind too much being mentioned here, despite their quite different approaches to the economic and social study of science, technology and innovation. Within the Department of Technology and Social Change, I was also part of the Systems of Innovation Research Program, whose members at various stages came to influence the paths taken during the research underlying the book: Charles, Leif and Hans Sjögren (on the senior side), and François Texier, Marcus Zackrisson, Marie-Louise Eriksson, Håkan Alm, Esa Manninen, Johanna, Vasilis Galis, Mark Sellenthin and Emma Larsson (on the junior side).

At Lund University, I have greatly enjoyed the eager atmosphere of our start-up Division of Innovation. It is a somewhat amazing experience to participate in the build-up of something completely new at a 300-year-old university! In addition to Charles, Leif and Annika, I am indebted to Pierre Bitard, Fabrice Coulon, Devrim Göktepe, Jekaterina Novikova, Stian Nygaard, Max Rolfstam and Astrid Szogs, for their fresh thoughts and friendship. Margareta Ingloff made life easier by solving many practical problems. Dawn House read the last manuscript and improved the language.

In Estonia, Dr. Maie Pihlamägi at the Institute of History in Tallinn provided me with initial guidance through the world of Estonian pre-war and Soviet history. It was a pleasure to come as a foreign researcher to Estonia, and I would, of course, not have been able to do the research for this book without the warm interest, help and patience shown by about sixty interviewees, who were prepared to generously and patiently share their experiences and opinions with me. It is my hope that I have been able to use their great knowledge and stories to write a better book.

Per Högselius
Lund
October 2004

1. Introduction

1. SYSTEMS OF INNOVATION IN EASTERN EUROPE: CATCHING UP OR FALLING BEHIND?

After the accession of eight Central and East European countries to the European Union in 2004, there still remains considerable uncertainty about the long-term roles that these countries will play in the European economy. Forecasts differ not only with respect to how fast the Eastern countries will catch up with the economically more advanced Western countries, but also regarding the extent to which the 'new' countries will contribute actively and creatively to shaping Europe's future. During the 1990s, the post-socialist economies were to a great extent focused on adapting to becoming Western-style capitalist economies, with economic activities concentrated on the imitation and adoption of Western technologies, standards, rules and practices. Today, however, it is becoming increasingly recognized that economic and societal development in Central and Eastern Europe can hardly be sustained in the long term without active involvement in the creative generation and further development of new goods, services, technological processes, organizational forms, etc. – in contrast to the more passive adoption of already existing things. From this perspective, the Eastern countries face the challenge of not only responding to developments initiated elsewhere, but of taking a more creative part themselves in initiating change.

From the Soviet period, Central and Eastern Europe inherited heavily industrialized worlds in which *science* and *technology* were understood as integral and self-evident sources of socio-economic change and societal progress. Not surprisingly, it was hoped in the years around 1990 that these scientific and technological capabilities would contribute to a solid foundation for post-socialist economic progress (Hirschhausen and Bitzer, 2000: 1). In reality, the pace at which Central and East European economies are closing the economic and technological gaps with Western Europe can be seen to vary strongly across countries. However, it is a fact that even the economically most successful of the Eastern EU countries have produced economic growth that has only seldom been based on any creative exploitation of scientific and technological opportunities (e.g. Radosevic,

2003: 3). In general, current economic strengths of the Eastern countries are found in areas where competitiveness is determined more by the availability of natural resources and low-wage labor than by the ability to generate new products and processes – i.e., 'innovations'. The overall picture is one in which Eastern EU firms have only rarely – particularly in high-tech industries – been able to establish themselves with their own brands and based on their own innovative activities in Western markets. While they have been successful performers of low value-added, labor-intensive activities related to production, most high value-added activities related to the generation of *new* processes and products remain located in Western countries.

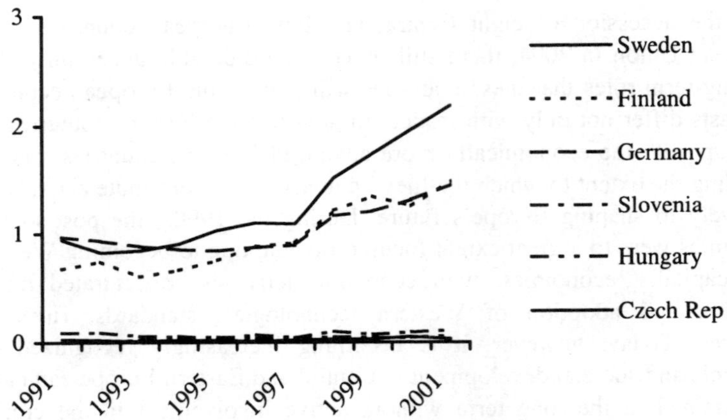

Note: The three Central and East European countries selected here are the most successful among the Eastern EU member states in terms of US patenting. For the other five Eastern EU members, the corresponding figures lie at 0.01 patents per 10,000 population, or lower.

Source: US Patent Office.

Figure 1.1 US patents granted for selected Western and Eastern EU member states, 1991-2001 (per 10,000 inhabitants)

A pessimistic interpretation of this argument would be that Central and Eastern Europe follow a similar pattern to that of many developing countries, which have often been seen to get access to the *results* of innovation, but only rarely to the *innovative process itself* (Lall, 1992: 179). At an aggregate level, this view seems to be largely confirmed by the fact that the often impressive rates of economic growth in many Central and East European countries are not paralleled by any corresponding growth in R&D investments or patenting. In terms of R&D, Western OECD firms invest on average between two and seven times as much in R&D in relation to GDP as do firms in Central and East European OECD countries, and the latter show hardly any signs of catching up in this respect (OECD, 2003: 29).[1] Moreover, R&D activities in Eastern firms have been found to be strongly skewed towards relatively simple downstream activities like testing and the implementation of new standards, as such activities are now critical for exports (Radosevic, 1997: 382, 1999: 29). In terms of patenting activities, the already very wide gap between East and West appears to be further widening, as shown in Figure 1.1; with respect to international patents, the Eastern countries can often count these on their fingers. Academies and universities, for their part, remain much more dominant R&D performers in relation to the business sector as compared to Western countries, but the negative correlation in most Central and East European countries between R&D and GDP growth (Radosevic, 2003: 3) signals that this academic R&D is often of very limited economic significance.

More qualitative evidence, to the extent it exists, similarly indicates that innovation, when it does occur, is typically of an 'imitative' rather than a genuinely 'creative' character (Sandberg, 1999: 161; Geenhuizen, 2001: 187). The technological upgrading that has taken place in the years after the fall of socialism has not been linked to the R&D system but rather to relatively simple forms of firm-based learning, particularly in the production process (Radosevic, 2003: 3). Moreover, foreign direct investment, while being of enormous importance for economic restructuring and development, has often had a *negative* rather than *positive* impact on R&D activities and innovation in Central and Eastern Europe (e.g., Farkas, 1997; Geenhuizen, 2001). With respect to *collaborative ventures* between East and West, the situation resembles that of foreign investment. Collaborative ventures have been almost exclusively based on *manufacturing*-related agreements, usually related to subcontracting. Eastern firms are predominantly incorporated into the networks of Western firms at the *lower* stages of the industry value-added chain. When cooperative agreements have been related to technology, they have not been concerned with the development of new products or processes, but rather with adjustments such as adaptations to local conditions (Sadowski, 2001).

At the same time, however, public debates in Central and East European countries contain strong arguments precisely for the establishment of advanced innovation-driven economies with effective and extensive systems of innovation and R&D bases, where education, science and technology are the main forces driving socio-economic change. Drawing as far as possible on their problematic but science- and technology-intensive Soviet-era pasts, and taking advantage of the relatively recent establishment of open capitalist economic systems, the EU's Eastern member states have set out to establish themselves among the most advanced European economies. There is often a strong belief in these countries that such a transformation can be accomplished within a foreseeable future.

Given the difficult starting point of the former socialist countries and their ambitious goal of building future economic development on the generation and exploitation of innovation, the Eastern EU economies face an immense challenge. From this perspective, the unfinished process of transition from socialism to market-based capitalist economies in Central and Eastern Europe, as well as accession to the European Union, is no more than a mere prelude to the enormous task of establishing *innovation*-driven economies. Building and transforming their innovative strengths and integrating their systems of innovation in creative ways with their neighbors in the West may thus be understood as the real challenge for Central and East European economies in the post-Soviet era.

However, despite the clear need for understanding the key mechanisms and forces in the establishment of innovation-driven economies in the EU's East, surprisingly little has been done in terms of deeper empirical research into this field. While recent studies have led to a somewhat contradictory picture of the actual innovative potential of the Eastern countries in relation to both Western countries and to each other (see in particular Radosevic, 2003), there is above all a striking absence of research into the actual processes and activities that may explain how innovation comes about in Central and Eastern Europe. As a result, the current stand of knowledge with respect to the dynamics of East European systems of innovation remains largely based on extrapolations from studies of Western countries. Similarly, views on innovation policy in Eastern Europe are typically trapped in risky transplantations of Western experiences, neglecting the peculiarities and opportunities that arise from the specific and unique history of the region.

In particular, there has been no deep research attention paid to the unexpected innovative *successes* that have been observed in a few sectors and countries and that thus challenge the aggregate, pessimistic trend as referred to above. This is the more surprising as such exceptions, to the extent that they exist, may be considered key in revealing the underlying mechanisms of the emergence of successful 'systems of innovation' in Eastern Europe. For example, it is interesting to note that the level of R&D investments in the Hungarian pharmaceuticals sector is higher than in many Western EU countries (OECD, 2003: 38). This might be a sign that creative innovation is occurring to a relatively great extent in this particular industry and country. On the more qualitative side, a study of the advanced Russian aerospace sector pointed to a case of innovative capabilities that have survived and developed further since the collapse of socialism (Shaw, 1996; cf. Cooper, 1982: 455). In the same way, it has been shown that the East European computer and software industries contain a whole array of firms engaged in innovation in its most creative sense (Dyker, 1996). It has also been pointed out that Hungary, on the eve of its EU accession, had around 10,000 software engineers, while the country also hosted Nokia's second-most important R&D hub globally in mobile switching and wireless technology (Radosevic, 2004: 120).

There is a multitude of further examples that hint of successful innovation in post-socialist Eastern Europe. Unfortunately, however, the studies referred to above have not been followed up by any deeper investigations into the specific systems of innovation in which these actual and potential cases of creative innovation occur. There has thus been little interest in delving deeper into the issue of how and why innovation *does* occur in Eastern Europe, as opposed to the focus on *non*-occurring innovation. This is unfortunate, because cases of East European innovative success may provide key insights into the problem of how innovative capabilities can be built up in East European countries. Such an understanding would also offer the opportunity to move beyond the 'convergence' debate, and thus to address the issue of how East European systems of innovation may build their success upon developmental trajectories that do not necessarily coincide with Western ones.

2. PURPOSE OF THE STUDY

The present study seeks to contribute to a better understanding of the forces and processes that account for the emergence of new – and the transformation of old – systems of innovation in Central and Eastern Europe. Building on recent theorizing on systems of innovation, and applying a case-study methodology, the study focuses on the case of the *telecommunications sectoral system of innovation in Estonia* and its evolution, especially from the late Soviet period to the country's EU accession in 2004. Earlier research has pointed to interesting characteristics of the Estonian telecommunications sector which let us hypothesize that this sector belongs to the above-mentioned cases of post-socialist innovative successes (e.g., Högselius, 2001; Kalvet, 2001; Pihl, 2001; GITR, 2002; Rannu, 2003). Such cases, according to the above argument, can be regarded as promising for studies seeking to grasp the transformation of Central and East European systems of innovation. The purpose of the study is therefore *to investigate the dynamics and evolution of the telecommunications system of innovation in Estonia from the late Soviet period to Estonia's EU accession, and to provide an in-depth explanation of how innovation has been enabled to occur in the system.*

Underlying the study is the empirical observation that the transition from socialism to capitalism in Central and Eastern Europe is a simultaneous process of, on the one hand, a *transformation* of the old Soviet-era structures into something new, and on the other hand, a *reorientation* from being deeply integrated economically with other Central and East European countries towards a new integration with the global capitalist system (and in particular Western Europe). From a systems-theoretical perspective, as will be elaborated further in Chapter 2, these two processes can be expected to be closely interrelated. In order to understand and explain the emergence of new East European systems of innovation, the study therefore takes into account both system-internal processes of change in Estonia as well as the relationships between the Estonian system and its foreign environment.

3. THEORETICAL POINT OF DEPARTURE

'Innovations', as referred to in this book, are new ways to create economic value. They may be new or improved *products* (product innovations) and *processes* (process innovations). Product innovations refer to both *goods* and *services*, while process innovations are either *technological* or *organizational* (Edquist et al., 2001: 19). 'New' does not necessarily mean that it is only a matter of 'big' changes; rather, innovations can be both 'radical' and 'incremental', and the phenomenon of innovation is therefore quite ubiquitous in any modern economy (Lundvall, 1992: 8; Freeman, 1992: 77ff.). This is even more evident if 'new' is interpreted from the perspective of a particular firm, a country or a sector of the economy rather than from the perspective of the whole world. In its broadest sense, an innovation can therefore be said to be everything from a radical world-leading innovation to a small and non-unique adaptation in a particular firm.

A distinction that will be used in this book is the one between 'creative' and 'imitative' innovation. This is a common dichotomy that has been used in research on innovation both in economics and political science (see Sandberg, 1999: 6). It goes back to authors such as Deutsch, who used the terms 'initiative' and 'imitative' innovation (Deutsch, 1949: 25f.), and Schumpeter, who suggested we distinguish between 'creative' and 'adaptive response' (Schumpeter, 1947: 150). Schumpeter associated the latter with the imitation of an already existing innovation (i.e., without improving it), while the former was seen as a 'new combination', i.e., a change never accomplished before. Schumpeter stressed that the new combination (first application) 'need not be spectacular or of historic importance', though it has to be something genuinely *novel*, i.e., must not exist anywhere else in the world (ibid.). The only exception would be two innovations that are identical but are made independently of each other. Creative innovation is thus equivalent to a (radical or incremental) innovation which is *new to the world*.

Imitative innovation, in contrast, is equivalent to a product or process that is new to a specific country or to a specific organization, but not new to the world.[2]

In Central and Eastern Europe, similar to many developing countries, one may in connection to the transition process intuitively expect to observe a higher proportion of imitative innovations than in technologically leading countries. Theoretically, this does not exclude that valuable innovative capabilities are being accumulated, since the process of imitative innovation does typically involve considerable creative efforts as well (Bell and Pavitt, 1993: 160ff.). However, the distinction between creative and imitative innovation may be helpful. The reason is that the observation of creative innovation hints in a particularly clear way at the fact that a unique development of something genuinely novel has taken place and been successfully introduced into the economy. While the present study takes into account both creative and imitative innovation, particular emphasis is therefore put on investigating and explaining how creative innovation – i.e., a product or process that is new to the world – has been enabled in Eastern Europe. Although such creative innovations may be expected to be relatively rare in the post-socialist context, this is not the same as to say that they do not occur at all.

The concept of 'system of innovation', which is central in this book, will be elaborated further in the theoretical chapter but should be clarified briefly here. Like all types of systems, a system of innovation consists of components and the relations between these. The most important *components* of a system of innovation are *organizations* and *institutions* (Edquist, 2004: 188). Organizations refer to both firms and non-firm organizations (such as universities, research institutes, government agencies, technical associations, etc.). Institutions are formal and informal rules such as laws, standards and regulations but also informal codes, routines, norms, taboos, etc. To the most important *relations* in a system of innovation belong those between organizations and institutions, and inter-organizational links such as cooperative, transactional and competitive relations among firms and other organizations. Since firms seldom innovate in isolation, the phenomenon of *inter-organizational networks* is of key importance in the analysis of systems of innovation.

The overall *function* of a system of innovation is to create, diffuse and utilize innovations (ibid.: 190). For that to happen, it is necessary that a wide array of *activities* take place in the system. The activities occurring in a system of innovation, and the relationships between them, are performed in a particular 'style' of innovation in that system. The activities that define the style are much broader than mere R&D activities; they include such diverse fields as the formulation of visions, the articulation of demand, creation of new knowledge, competence-building, new firm formation and market entry, adaptation of organizations, inter-organizational networking, provision of finance, consultancy and lobbying, creation and adaptation of formal institutions and policymaking (see further Chapter 2). A main argument of the approach proposed in this book is that the phenomenon of innovation is unlikely to be satisfactorily understood unless these broad activity areas are taken into account when analyzing the transformation and reorientation of systems of innovation.

However, a system of innovation does not include *everything*. Like all systems, it has a *boundary*, implying that some components, relations and activities belong to the system while others do not. The boundaries of systems of innovation may be defined through delimitations in terms of geography, technological field, product area, etc. One may thus speak of 'national systems of innovation', 'regional systems of innovation', 'sectoral systems of innovation', 'technological systems', etc. – all of which can be considered to be variants of the generic idea of systems of innovation (Edquist, 1997b: 12). As elaborated in Chapter 2, the present study focuses primarily on *national variants of sectoral systems*, i.e., systems that are both nationally and sectorally delimited.

When analyzing systems in a dynamic and evolutionary perspective, Myrdal (1957: 18) once argued that the 'main scientific task is . . . to analyse the causal inter-relations within the system itself as it moves under the influence of outside pushes and pulls and the momentum of its own internal processes'. The present study follows this approach by focusing on the interplay between domestic transformation and international reorientation of the Estonian telecommunications system of innovation. Thus the task set out in the purpose above is approached from two complementary angles:

1) *The momentum of the system's internal processes*. Here one may refer to evolutionary theorizing on innovation and the history of technology, where it has been argued convincingly in recent studies that the build-up of systems of innovation has to be understood in a *historical* perspective. Earlier research on innovation in Eastern Europe has not come to any clear conclusion with respect to what role the Soviet-era history plays in the build-up of post-socialist systems of innovation. There have been no in-depth empirical studies of the actual importance of inherited competencies and behaviors for enabling creative innovation in post-socialist countries. However, the critical importance of cognitive and cultural aspects for the innovative process suggests that both knowledge and informal institutions evolve in incremental ways and that they are unlikely to be radically altered or abandoned even under politically and economically disruptive changes of a kind such as in Eastern Europe around 1990. Forgetting

previous values, cognitions, rules and habits is difficult and painful, and the situation today inevitably shapes tomorrow's 'habits of thought' (Johnson, 1992; Veblen, 1925: ch. 8). Likewise, it takes time to build new knowledge, competencies and behaviors. A first major question in the present study is therefore *what role inherited competencies have played in the build-up of the post-socialist telecommunications system of innovation in Estonia, and in what ways such competencies have contributed to enabling innovation.*

2) *Pushes and pulls from the foreign environment.* Indeed, it would be difficult to find a more radical reorientation of a system's relations to the foreign environment than in the case of East European transformation. Prior to around 1990, Central and East European systems of innovation were strongly linked to other socialist countries in the East, while there were much fewer links to Western capitalist economies. After 1990, this changed dramatically, with the emergence of a variety of new connections between Eastern and Western countries. These new links have resulted in pushes and pulls from a 'new' environment. In general, these external impacts are hoped to be a powerful source for building successful post-socialist systems of innovation. In this connection the main question is how these new 'Western' pushes and pulls may contribute to building innovative strength in the Eastern countries. Given Central and Eastern Europe's deep historical and cultural relationships to the advanced economies of Western and Northern Europe, Central and East European systems of innovation may here be seen to have a potential advantage as compared to countries located farther away from advanced systems. The background for this reasoning is the widely recognized fact that innovation involves considerable interaction and interactive learning and that such interactive learning is much easier when actors share a common or similar frame of reference, with similar 'habits of thought' and common experiences of earlier interaction. Therefore, a second major question in this study is *how Estonia's political, economic and cultural similarities and links with advanced Western systems have influenced the country's ability to generate innovation.*

The roles of inherited competencies and international reorientation are both related to another important issue in the emergence of post-socialist systems of innovation, namely the problem of 'convergence' and 'divergence'. On the one hand, following up on the second question above, it appears plausible that geographical and cultural proximity to Western Europe will lead to the emergence of new East European styles of innovation that lie more or less close to those of their Western neighbors. This is so at least to the extent that it would be highly surprising if the enormous differences between the styles of capitalist and socialist systems of innovation were not to be radically reduced as a result of the transition process.

On the other hand, relating to the first major question above, the radically different histories in East and West seem to make it unlikely that East European systems of innovation will develop styles that fully converge with Western ones. Organizations and individuals in post-socialist countries live and act in an environment that for decades has been physically and mentally shaped by Soviet-style systems of innovation, and this is likely to influence the emergence of new Eastern styles of innovation. A third major question in this study is then *what the special characteristics of the Estonian post-socialist style of innovation in telecommunications actually are.* In other words, in what ways do the activities underlying innovation in Estonia differ from the activities underlying innovation in Western systems of innovation?

With respect to the mutual interplay of the above sides of the same coin – i.e., transformation and reorientation – Eastern Europe offers interesting and unusual opportunities to increase our knowledge of processes of innovation, not only in the context of Eastern Europe but generally, and thus to contribute to the advancement of innovation theory from a dynamic and evolutionary perspective. Eastern Europe provides a case of not merely the emergence of 'new' systems of innovation (as in developing countries or with respect to new emerging technologies), but of a genuine process of radical *transformation* and *reorientation* of systems that already exist. Eastern Europe therefore provides a unique setting for studying how already highly complex systems of innovation actually respond to dramatic changes in politics, economics and relations to other parts of the world.

4. OUTLINE OF THE BOOK

Methodologically, the purpose of the study, as formulated in section 2, and the major questions raised in section 3, are approached through an in-depth historical investigation of the evolution of the Estonian telecommunications system of innovation from the Soviet era to Estonia's EU accession in 2004. Far-reaching interview material and documentary data are scrutinized in order to uncover the fascinating story of the making of a system of innovation in Estonian telecommunications. It involves, among other things, a detailed analysis of the role of history in this complex process, in terms of the inherited Estonian educational system, the transformation of Soviet-era inherited organizations (firms, universities, research institutes, government agencies, etc.) and their competencies. It also involves an in-depth historical investigation of how and why pushes and pulls from the foreign environment manifest themselves in the system of innovation and its transformation, with a particular interest in the role of foreign direct investment, subcontracting, East-West joint innovation projects, cooperation in education and Western advice and consulting activities in Estonia.

The book consists of nine chapters, including this introductory chapter. Having set out the background and purpose of the study and defined the theoretical point of departure in the first chapter, Chapter 2 discusses in further depth how the dynamics and evolution of systems of innovation can be conceptualized and analyzed under the difficult conditions of simultaneous transformation and reorientation in Central and Eastern Europe.

Chapter 3 outlines the methodological approach taken for this historical investigation of the Estonian sectoral system of innovation in telecommunications. The next four chronological chapters, each dealing with a specific period of time, investigate the activities of a multitude of firms and other organizations in relation to far-reaching institutional changes and technological developments. A vast body of empirical material forms the core of the evidence, from a variety of qualitative interviews with key individuals to a massive body of documentary material.

The first of the four historical chapters – Chapter 4 – covers the 'pre-history' of Estonian telecommunications, starting in the nineteenth century and continuing into the Soviet period. It discusses in particular the dynamics of innovation in the Soviet-era Estonian system, and thus defines the starting point for the radical transformation that the republic faced from the late 1980s, when a reorientation towards the West started to become possible. Chapter 5 delineates in detail the dramatic period in which Estonia regained its independence, which proved to be a radical turning point in the evolution of the telecom sector and created a new basis for the corresponding system of innovation. Estonia's domestic transformation is here seen to be strongly related to the country's linkages to Sweden and Finland, which are investigated in detail.

Chapter 6 deals with Estonia in the mid-1990s, a period that was economically characterized by high rates of economic growth based mainly on the exploitation of the low-cost environment, while at the same time the use of new and important telecommunications technologies took off domestically: digital mobile telephony and the Internet. The last historical chapter – Chapter 7 – deals with the period from 1998 to 2004. It was a multifaceted period, starting with the Russian financial crisis in August 1998, which had a severe impact on the Estonian economy, yet coinciding with the global Internet boom and a dramatic diversification of telecommunications services, followed by the dotcom death and an international crisis in the telecommunications industry. The Estonian telecommunications system of innovation can here be seen to develop in somewhat ambiguous and often unexpected ways.

The extensive historical study is followed by two analytical and concluding chapters. Chapter 8 provides a summary and final analysis of the Estonian case in its entirety, using the theoretical and methodological tools developed in Chapters 2 and 3. Chapter 9, finally, is devoted to elaborating the implications of the research results for the further development of theory and policy.

NOTES

1. Thus in 2001, the OECD average in firms' investment into R&D amounted to 1.58 per cent of GDP, whereas the corresponding figures for Central and East European OECD members were 0.24 per cent for Poland, 0.38 per cent for Hungary, 0.44 per cent for Slovakia and 0.78 per cent for the Czech Republic.
2. As an alternative to 'creative' and 'imitative' innovation, one could use the notions of 'generating' and 'diffusing' innovations. In addition, as suggested by Carlsson et al. (2002: 243), one could study the 'use' of innovations. The use of innovations will in this study only be taken into account to the extent that it influences the activities in the system of innovation.

2. Conceptualizing the Transformation and Reorientation of Systems of Innovation

1. INTRODUCTION

The purpose of this chapter is to provide a conceptual and theoretical background to the dynamics of innovation in the context of radical transformation and reorientation. The chapter builds on recent theorizing on innovation and, in particular, systems of innovation, while also focusing on important theoretical issues that have so far hardly been dealt with in depth in previous research. To do this, the chapter combines arguments and findings from different disciplines, such as economics and management of innovation, history and sociology of technology, social systems theory and political science.

The proposed approach aims to supplement earlier perspectives developed in research on East European transformation, such as 'transition economics' (e.g., Roland, 2000) or the analysis of 'national innovation capacities' (as proposed by Radosevic, 2003). The main unit of analysis in the suggested approach is the system of innovation seen as a whole, whereby one argument, to which there will be reason to return, is precisely that such systems have to be understood *holistically*, i.e., in their entirety. This means that one should take into account, for example, not only business firms, but also a variety of other organizations; not only transactional relations between firms, but above all collaborative and competitive links; not only R&D and production, but a much broader array of activities; etc. How this can be done is discussed at length. The approach is neither firm- nor policy-centered; rather, the activities of firms, governments and other actors are put in relation to the system as a whole. In particular, the approach goes far beyond the traditional interest in East European 'reform policies' and their impacts.

The chapter is organized as follows. Section 2 discusses 'systems of innovation' as a theoretical tool along its most important dimensions. Building on this, the ensuing sections turn to the issue of transformation and reorientation of systems. Section 3 discusses transformation from the perspective of the system's *internal logic*, then section 4 develops a theoretical perspective of the ways in which systems change can be understood in relation to the system's *environment*. The latter aspect relates to the issue of systems reorientation. The last and concluding section summarizes the conceptual and theoretical ideas developed.

2. WHAT IS A SYSTEM OF INNOVATION?

There is today considerable evidence that an understanding of the dynamics of innovation can benefit substantially from systemic perspectives on the innovative process (e.g., Freeman, 1987; Lundvall, 1992; Nelson, 1993; Edquist, 1997a, 2004). In the context of Eastern Europe, such a systemic understanding can be argued to be of even greater importance. The reason is that processes of technological change and innovation as they occurred in the former socialist countries tended to take place with a fundamentally different logic as compared to innovation in capitalist countries (see Box 2.1 below), and that the collapse of socialism in the years around 1990 therefore was precisely a process of *systems* change.

Therefore, the aim of this section is to conceptualize, in a general way, systems of innovation in their different dimensions. In the following sections, the study turns more specifically to the issue of system evolution, and how it can be understood in relation to the interplay between domestic transformation and international reorientation of East European systems of innovation.

2.1 Components and Relations in Systems of Innovation

Like all systems, systems of innovation are constituted by various *components* and the *relations* between them. Together, they define the *structure* of the system. The most important components in systems of innovation are usually seen to be *organizations* and *institutions* (Edquist, 2004: 188).

However, there is some confusion with respect to the notions of institutions and organizations. If one takes North's distinction between institutions as 'the rules of the game' and organizations as the 'players' or 'actors' (North, 1990) – as this study does – it appears that some authors (notably Freeman, 1987: 1; Lundvall, 1992: 2) with their respective use of the term 'institution' actually include certain types of organizations under this label. Edquist and Johnson (1997) have argued convincingly that innovation research should follow North's distinction, not least because the influence of institutions and organizations on each other is crucial for our understanding of systems of innovation. If these two groups of components are blurred, it is obviously difficult to speak about their potential influence on each other.

On the other hand, despite the 'conceptual diarrhea' in this respect (Edquist, 1999), there is a striking agreement in the literature that both organizations and institutions *of very different kinds* need to be taken into account in the analysis of systems of innovation. With respect to organizations, it has been increasingly emphasized that not only firms, but also a variety of *non-firm organizations* are important actors (Freeman, 1987: 1; Edquist and Johnson, 1997: 58; Carlsson et al., 2002: 234; Malerba, 2002: 250). In systems of innovation research, firms are still recognized as belonging to the most important organizations in the system. However, in the absence of non-firm organizations, it is difficult to arrive at a clear understanding of the dynamics of innovation and the evolution of systems. To non-firm organizations belong a wide array of actors stretching from educational and research organizations such as universities, academies and research institutes to government agencies, industry associations, labor unions, lobby groups, standard committees, etc.

It should perhaps be pointed out that the strong emphasis on the need to include non-firm organizations should not be confused with seeing systems of innovation as including *only* non-firm organizations, i.e., excluding business firms. The system of innovation should thus not be confused with, for example, the 'public support system'. Similar to traditional approaches in economics, such a perspective would miss a key argument of the very idea of systems of innovation, namely the interrelations between often radically heterogeneous actors in both the private *and* the public sector.

The components referred to above as 'institutions' – i.e., the 'rules of the game' – can, like organizations, be of very different kinds. Above all, they can be both 'formal' and 'informal'. *Formal institutions* should not be confused with organizations such as universities or government agencies, although this happens frequently in the literature as well as in everyday language. Instead, formal institutions are codified rules such as laws, official standards, regulations and recommendations. These are consciously created and upheld by organizations; for example, national patent laws are in most countries approved by national parliaments and upheld by patent offices and courts. Standards may be created by public agencies but also agreed upon by the members of industry organizations, etc. Organizations thus have a strong and direct impact on the dynamics of institutions in systems of innovation. On the other hand, the purpose of formal institutions is usually that they should influence the behavior of organizations in certain ways. Hence, the relations between organizations and institutions may matter considerably for the dynamics and evolution of the system (see section 3.1).

Informal institutions include uncodified traditions, norms, routines, taboos, etc. A typical informal institution is, for example, the norm of honesty in and among organizations or, in contrast, the acceptance of corruption. In contrast to formal institutions, informal institutions can seldom be created suddenly or rapidly, and they therefore almost always show strong continuities. This is because they are necessarily – though to a varying extent – deeply rooted in human consciousness, business history, legal traditions, etc. In times of radical political and economic change, such as in Central and Eastern Europe in the years around 1990, this may give rise to substantial tensions and contradictions between the formal and the informal sets of institutions, since old informal institutions may continue to exist long after the accomplishment of formal changes.

The *relations* in systems of innovation are crucial, since it is mainly the relations that shape the dynamics of the system (Carlsson et al., 2002: 234). The most important relations in a system of innovation include those between organizations and institutions, as well as inter-organizational links such as cooperative, transactional and competitive relations among firms and other organizations (Edquist, 2004: 195ff.). Nowadays it is common that firms both compete and cooperate with each other at the same time; one may then speak of 'coopetition'. The fact that both cooperative and competitive relationships between organizations have grown stronger and more important during the past decades is one of the strongest motivations for using a systemic approach to the analysis of innovation. This in turn emphasizes the need for understanding the relations between organizations and institutions, since institutions may strongly influence both cooperative and competitive relationships.

An important characteristic of systems of innovation is that inter-organizational links are seen to consist of both *market* and *non-market* relations (Malerba, 2002: 251; Carlsson et al., 2002: 234). Inter-organizational relations can therefore not be traced fully by tracing economic transactions. In addition, inter-organizational links may be both *formal* and *informal* (Freeman, 1992: 99). Due to the often large number of external links that organizations build with other organizations, it is often purposeful to study inter-organizational relations in terms of *networks* rather than of simple bilateral links. Networks contribute in important ways to the structure of systems of innovation, and are crucial for their dynamics (Carlsson et al., 2002). They are made up of a variety of inter-organizational relations such as strategic alliances, research consortia, joint ventures, licensing agreements, production sharing, direct investments, university-industry-government relations, organizational arrangements for financing innovation, subcontracting and other user-producer links (Tidd et al., 2001: ch. 8; Freeman, 1992: 99). Since networks are usually considered to be more than the sum of these individual relations, it is important to see the relations as parts of the network. The relations within a network are influenced by each other and are therefore unlikely to be properly understood without the broader picture of the network.

2.2 Activities in Systems and Styles of Innovation

As mentioned above, components and relations, taken together, define the *structure* of systems of innovation. However, as this book is primarily concerned with the dynamics of innovation and the evolution of systems – rather than with analyzing the system as it appears at one point in time – it is above all necessary to conceptualize the actual *processes* that take place in the system. To deal with this issue, the present study builds upon the concept of 'activities' in systems and develops the notion of 'styles of innovation', whereby the latter is defined as the aggregate pattern of the activities and the relations between them as it appears in a specific system and at a specific point in time.

The point of departure is here the overall *function* of systems of innovation, which is the creation, diffusion and utilization of innovations (Edquist, 2004: 190). For a system of innovation to perform this overall task, a variety of activities take place in the system.[1] The activities describe what is 'happening' in the system and they can be interpreted as the *determinants* of the overall function.

There is no consensus in earlier research as to what the relevant activities in a system of innovation actually are or how important certain activities are in relation to each other, although several authors (notably Rickne, 2000; Liu and White, 2001; Bergek, 2002; Jacobsson, 2002; Edquist, 2004) have tried to provide listings of what, in their view, constitute the most important activities. The argument in the present study is that activities of very *different* kinds have to be taken into account. In particular, the activities that take place in a system of innovation are much broader than the notion of 'innovative activities' in a traditional sense – in particular if the latter are defined in terms of mere 'R&D activities'.

In the present study, I have chosen to address eleven different activities when analyzing systems of innovation. The choice of these is based on my own readings and interpretations of the above authors, as well as related insights from research on innovation management, organization theory, policy studies, history and sociology of technology and other fields. I have also complemented these by drawing on various case studies of innovation processes in specific organizations (including firms), industries, countries, etc. (e.g., Nelson, 1993; Utterback, 1994; Tidd et al., 2001; Jacobsson et al., 2004; Hughes, 1983; Edquist et al., 2000; Edquist, 2003; Carlsson, 1995a;

Malerba, 2004; McKelvey, 1996; Malerba and Orsenigo, 1996; Hobday, 1995; Freeman, 1987, 1992; Freeman and Soete, 1997; Bell, 1997; David, 1985; Dahmén, 1988; Berner, 1999; Bijker et al., 1987; Kline, 1991; Pavitt, 1984; Hippel, 1988).

The eleven activities are the following:

- **Formulation of visions**. This includes activities pursued by firms and other organizations to identify market and technological opportunities – as well as to recognize limitations in these opportunities, and actors' critical reflections on their own suitability to take part, in some way or the other, in the risky creation and exploitation of opportunities. The formulation of visions also includes the often very difficult recognition of shortcomings in old behaviors. The formulation of visions should not be interpreted in purely technological terms; visions, as understood here, also include the scope for creating new partnerships, alliances, collective identities, etc.
- **Articulation of demand for new, improved and cheaper products**. This includes, on the one hand, demand for already existing products that are in a fluid stage of development and for which market growth may be a crucial prerequisite for the process of further improvements. On the other hand, it also includes the demand for products that do not yet exist. The articulation of demand is an integral part of what in evolutionary economics is referred to as the selection process, since the differing demands for alternative products typically have a considerable impact on their chances for survival.
- **Creation of new knowledge**. This refers in particular to the creation of new scientific and technological knowledge underlying new products and processes. It includes R&D activities in a traditional sense, but also the combination of old pieces of knowledge in new ways, with the purpose of creating new products and processes. Knowledge typically accumulates in incremental and path-dependent ways, but radical shifts in knowledge-creating activities may take place by abandoning certain knowledge areas and deliberately choosing to focus on alternative knowledge paths.
- **Competence-building**. Building competencies is here associated with education and training. It is a matter of transmitting already existing knowledge, whereby the purpose is not primarily to create new products and processes, but rather to build up competencies that can later be used in the context of knowledge-creating activities. It may take place in a multitude of different settings, in particular schools and universities, and through internal training in firms and other organizations.

- **Formation of new firms and other organizations and market entry and exit**. Organizations may be formed in a variety of ways, including more traditional forms of entrepreneurship as well as through the creation of spin-offs from existing firms, universities, research institutes and other organizations, and through the establishment of subsidiaries of foreign firms, etc. With respect to market entry, new firms are typically accompanied by already existing firms that choose to diversify into new markets. Conversely, firms may exit from markets both through total bankruptcy, through mergers and acquisitions or by selling off non-core product areas.
- **Adaptation of organizations to accommodate for innovation**. Innovation is a process of change that is often very turbulent. Firms and other organizations therefore have to adapt themselves internally in a multitude of ways in order to respond to new opportunities and threats. It can be a matter of, for example, hiring new staff, forming new teams, reformulating strategies, changing organizational culture, responding to institutional change, etc.
- **Networking**. This includes creating, handling and exploiting collaborative links and networks, whereby the partners can be not only firms, but also universities, government agencies, regulatory boards, trade unions, etc. It also involves responding to the activities of competitors, handling external conflicts, breaking old relationships that have become dysfunctional or obsolete, etc. Networks are also channels of communication, which help organizations scan their environment for important signals in the system. This activity is closely related to the internal adaptations referred to above, since internal processes of change typically require mobilization of resources from outside or even acquisition of entire firms to support the internal innovation process.
- **Provision of finance**. In principle, all other activities discussed here are likely to be shaped by the extent to and ways in which funding is available. This goes far beyond R&D expenses in the traditional sense of this concept. Competence-building may be heavily dependent upon the financial situation of universities; new firm formation on the availability of venture capital; acquisition of external resources may involve licensing fees, investments into joint ventures and alliances; public policy might be meaningless in the absence of financial instruments to support and direct various activities in the system of innovation; etc.
- **Consultancy, advice and lobbying activities**. These activities play a crucial part in the information flows in most systems of innovation, although such flows may look very different in different cases. Consultancy, advice and lobbying activities may take place according

to both formal and informal arrangements, they may involve foreign and domestic actors, public and private organizations, etc. Lobbying may be highly controversial, but may also be a key driving force in system evolution.

- **Creating, changing and abolishing institutions**. The rules of the game do not change automatically, but as a result of the activities of various actors, involving government agencies and business firms as well as various other organizations. Some organizations have among their main tasks explicitly the responsibility to engage in processes of institutional change, but these actors are seldom able to handle such changes on their own. Complex networks of actors may therefore emerge in the process of institutional change, whereby this activity may also relate closely to consultancy, advice and lobbying activities.
- **Formulation of public policy**. Public policy in the field of innovation can have substantial impact on the evolution of systems of innovation and should therefore be seen as an important activity taking place in systems. Public policy may be closely related to the creation, change and abolition of institutions and to the provision of public financial support. Since the activities in systems of innovation can be interpreted as the determinants of innovation (see above), public policy may seek to influence all other activities listed here – including the ways in which policy itself is being worked out.

A main argument in the present study is that systems of innovation are unlikely to be satisfactorily understood unless – in the above sense – a wide array of activities are taken into account when analyzing real-world systems, and in particular their dynamics and evolution over time (as will be discussed in further detail in sections 3 and 4 below).

Furthermore, the list of activities above does not necessarily indicate *how* or *by whom* they are carried out. The activities may have radically varying characteristics, be carried out in very different ways and by very different types of actors, depending on factors such as national, sectoral and historical settings, as well as on the evolutionary paths of systems (as discussed in section 3.1 below). This relates also to what in systems of innovation research is usually referred to as a lack of 'optimality' (e.g., Edquist, 1997b: 19f.); there is, as far as we know, no 'ideal' way to carry out the above activities. For example, R&D activities are typically carried out by vastly differing organizational arrangements in different – but perhaps equally advanced – countries. Similarly, the character of networking activities can be seen to differ vastly between different sectors and technological fields (typically as a result of different types of knowledge bases) (e.g., Breschi and Malerba, 1997).

From this perspective, it is hardly possible – nor desirable – to seek to identify any 'most suitable' character of the above activities and the ways in which they 'should' be carried out. A more important and compelling task is to analyze the ways in which the activities contribute to the overall function of the system, and how they develop matches or mismatches with each other. For example, the creation of radically new knowledge through aggressive R&D risks being useless for the system as a whole, unless new matching competencies are built up, relevant new firms are created to exploit the new knowledge, formal institutions are altered to remove legal obstacles, complementary assets are secured through networking activities, etc. In other words, the activities cannot be understood in isolation from each other, and the focus of analysis should therefore be on how the activities in systems of innovation contribute to the functioning of the system as a whole.

As a way to conceptualize the multitude of possible characteristics that activities can take in different systems of innovation, I will here introduce the notion of 'style of innovation'. The concept of style has proved a useful analytical tool within a number of research fields within the social, historical and economic study of science and technology. For example, in the history of science, it has proved useful to distinguish 'styles of scientific thinking and reasoning', which, similar to activities in systems of innovation, have been seen to differ widely between different disciplines, national contexts, historical periods, etc. (e.g., Hacking, 1994; Gavroglu, 1994). In the history of technology, Thomas P. Hughes has similarly discussed what he calls 'technological style' or the 'style of evolving systems' as a way to capture the fact that technological change takes place in different ways across national settings in a certain sector (Hughes, 1983: 404ff., 1987: 68ff.).

In the field of innovation studies, too, the concept of style has been used occasionally, though it has seldom been operationalized in any clear way. Most commonly, it seems to have been understood in terms of one or a few of the activities referred to above, although different authors focus on different activities. For example, Tylecote et al. (1998) and Estades and Ramani (1998) focused on differences in financing; Hond (1998) on how organizations adapt to institutional change; and Irwin et al. (1998) on styles of networking. At the firm level, styles of innovation have also been related to issues such as differing management traditions in different countries and to varieties of innovation strategies.[2] At the level of national systems, styles of innovation have often been understood as issues of technological specialization (Vertova, 1998) or the ways in which institutional configurations evolve and influence 'styles of innovativeness' (i.e., performance characteristics) (Hollingsworth, 2002).

In this book, the concept of style is used in relation to the system as a whole and at the level of activities. It is defined as the aggregate pattern of activities and the relations between them, as it appears in a specific system and at a specific point in time. It follows that the style of a specific system reflects not merely the strength and nature of R&D and other knowledge creating activities, or the 'style of management' in a certain country or sector. Instead, styles of innovation, as interpreted here, also reflect factors such as the way in which institutional change is handled in the system, what types of visions underly systems evolution, how new firm formation is enabled, how competence-building is organized, what role lobbying plays for the functioning and evolution of the system, from what types of sources innovation is financed, etc.[3]

To illustrate this roughly, one may, for example, look at the different styles of innovation in socialist as opposed to capitalist systems of innovation in the Cold War era, i.e., two ideal types of systems. This is done in Box 2.1 below.

BOX 2.1. The peculiarities of socialist as opposed to capitalist styles of innovation

Drawing on the eleven activities enumerated above that need to be attended to when analyzing systems of innovation, it appears the peculiarities of socialist styles of innovation are particularly obvious. For example, the *formulation of visions* under East European socialism was an activity associated mainly with central planning, which stands in sharp contrast to capitalist systems, where the formulation of visions is typically a decentralized process in which nearly all actors in the system are engaged in more or less important ways.

The *articulation of demand* was a notoriously weak activity under socialism, in the sense that actual demand – at least consumer demand – hardly affected the other activities and did not contribute in any essential way to the evolution of the system.

The *creation of new knowledge*, in contrast, was a highly developed activity, which was mainly carried out by the well-funded academies of sciences and a variety of industry-specific R&D organizations. A ubiquitous problem with the new knowledge being generated, however, was the severe difficulties in exploiting it for actual new processes and products, thus pointing at substantial mismatches in the relations between new knowledge creation and other activities in the system.

At the same time, the difficulties in managing *networking* activities meant in practice that the creation of new knowledge hardly involved any interactive learning. New knowledge creation sought also to a great extent to build upon the exploitation of Western technological advances; the general strategy was to first imitate Western technologies and then develop them further domestically. However, this turned out to be highly troublesome.

Although new knowledge creation did involve universities, albeit to a much lesser extent, clearly their most important task was to engage in *competence-building*. Arguably, this was the most successful and impressive activity in socialist systems, and the strength of former socialist educational sectors are often referred to as the main inherited strength in post-socialist countries.

Furthermore, *the formation of new firms and other organizations* had little in common with the corresponding activity in capitalist systems. It was largely determined by planning bodies, and there was therefore little scope for entrepreneurship in the Western understanding of the term. This meant that innovation in socialist systems typically took place without any notable participation of small firms – even in sectors where one would 'normally' expect a variety of specialized suppliers to play a natural and important role.

There did exist variety of *inter-organizational links and networks* in socialist systems, but the difficulties in handling formal innovation-related relationships tended to put the emphasis on managing networking activities through informal networks and personal channels of communication. Vast resources were often spent on exploiting these informal connections in order to mobilize resources for the purpose of innovation. Naturally, responding to *competition* was hardly an activity at all in socialist systems.

The *formulation of public policy* was an activity that was strongly influenced by linear science-push thinking, and therefore the Academy of Sciences typically played an extremely powerful role in policymaking activities. Due to central planning, linear thinking had an enormous impact on all other activities in socialist systems, including the tendency to neglect the need for interactive learning between basic research and more market-oriented developmental activities.

Source: Amann and Cooper (1982); Hanson and Pavitt (1987); Sandberg (1989).

As will be shown in sections 3 and 4, the concept of style is a helpful tool when it comes to analyzing the evolution and dynamics of systems and the relationships between different systems, or between a system and its environment. This is particularly important for the present study, since it focuses explicitly on the interplay between domestic transformation and international reorientation.

2.3 Systems Boundaries and Typologies of Systems

In the preceding sections it was argued that a wide range of organizations, institutions, relations, and activities should be taken into account when analyzing the nature and dynamics of systems of innovation. On the other hand, however, a system of innovation does certainly not include *everything*. A key characteristic of any system (not only systems of innovation) is that it has a *boundary*, implying that some things belong to the system while other things do not.

Boundaries of systems of innovation can be defined in terms of geography and/or in terms of product areas or technological fields.[4] If the boundary of a system is defined in terms of *geography*, one may speak of, for example, 'national systems of innovation'. National borders are then seen to coincide with the system boundary. Organizations, institutions and activities in foreign countries are in this interpretation not seen as being part of the system, but belong, instead, to the *environment* of the system, a concept that will be of great importance for this study. Based on systems theories in political science and sociology, one may discern two somewhat different but related ways of determining whether national boundaries can be said to define system boundaries:

- First, following the political scientist Karl Deutsch, one might suggest that national borders define the boundary of a system if the intensity and number of domestic relationships (i.e., among domestic organizations and institutions) is much greater than the intensity and number of transnational relationships (i.e., between domestic and foreign organizations and institutions) (Deutsch, 1963: 205). I will refer to this as a *Deutsch-type boundary*.
- Alternatively, building on the reasoning of the sociologist Niklas Luhmann, one may argue that national borders define the boundary of a system if interactions with the foreign environment have a different impact on that system as compared to the impact on other national systems, in terms of structure and style (Luhmann, 1984: 17). I will refer to this as a *Luhmann-type boundary*.

An interesting issue is to what extent and in which ways these two types of boundaries are related to each other. For example, does the existence of a Deutsch-type boundary automatically imply that there also exists a Luhmann-type boundary, i.e., do strong domestic relations lead a country to develop a system with its own, nationally specific structure and style? Or conversely, does the existence of a nationally specific structure and style automatically imply that domestic relations are much more intense than relations across national borders? Clearly, empirical research on specific countries is needed to shed light on this issue. However, in a preliminary way, I would here like to suggest that a national system, strictly defined, has to fulfill *both* the above criteria.

The actual existence of national systems in that sense is indicated by the often radically differing organizational and institutional structures and styles of innovation in different national contexts, and of the continuing importance of strong domestic relations – even in an era of globalization. Even countries which show strong relationships to each other, such as Sweden and Denmark, have been found to differ greatly in their structures and styles (Edquist and Lundvall, 1993: 5f.). I will return to the discussion of national boundaries in section 4.4 below, in relation to the issue of inter-system integration.

In analogy to national systems, one may also speak of 'regional' and 'local' systems of innovation, whose boundaries are also geographical, and in which organizations, institutions and activities outside a certain *region* are not seen to belong to the system, but to its environment. In contrast, the rationale for delimiting systems of innovation in terms of *product areas* or *technological fields* stems from the empirical observation that innovation often takes place in very different ways depending on what sector of the economy or what technological or knowledge field is dealt with. 'Sectoral systems of innovation' and 'technological systems' can thereby be defined using the same criteria as suggested above for national systems.

As argued by Edquist, systems of innovation may be regional, national or even supranational, while at the same time they may be sectoral within any of these geographical demarcations. In practice, the choice of framework – national, regional, sectoral, technological, etc., or a combination of these, depends very much on the research interest (Edquist, 1997b: 12). Since the present study focuses explicitly on the telecommunications sector, while at the same time concentrating specifically on the national case of Estonia, a combination of sectorally and nationally delimited systems of innovation is applied. In the rest of this chapter, the theoretical discussion therefore addresses *national variants of sectoral systems*. This helps us analyze both the internal dynamics of the Estonian telecommunications sectoral system of innovation, and the relation between the Estonian and other systems (in particular the influence of the foreign environment on the Estonian system). In other words, the focus on national variants of sectoral systems seems to offer a useful approach when the interest is in both domestic transformation and international reorientation.

3. THE TRANSFORMATION OF SYSTEMS OF INNOVATION

Systems of innovation analyses have often been criticized for being too static, i.e., for not taking into account how and why systems evolve and change over time (see, e.g., Balzat and Hanusch, 2004: 205). Moreover, in cases where the interest has been explicitly in system evolution, it has mostly been a matter of the emergence of 'new' systems – either around a certain new technology (e.g., Carlsson, 1995a, 1997; Rickne, 2000; Jacobsson et al., 2004), or in developing countries that for the first time ever see the emergence of systems of innovation (e.g., Nelson, 1993: chs. 11-15; Muchie et al., 2003). Much less effort has been spent on how already existing systems of innovation undergo transformation under (often radically) changing external circumstances. It is the latter type of transformation that stands at the center in the case of Central and Eastern Europe, where systems of innovation do not merely emerge out of nothing, but rather undergo a radical transformation.

In the following section, the interest is in how systems of innovation are transformed in response to a radical political change, such as the Soviet collapse. In analogy with the concept of 'technological discontinuities', which refers to the emergence of radically new technologies, I will interpret this as an 'institutional discontinuity', which can then be defined in terms of the emergence of radically new institutions. The analogy is relevant, since both technological and institutional discontinuities represent strong disturbances from the perspective of the system, with the power to radically transform it, in terms of both its structure and its style.[5]

3.1 Structural Transformation

The structure of a system of innovation in terms of its components and relations is transformed through reconfigurations in the organizational and institutional landscape. This involves changes in and destruction of existing components and relations, but also the emergence of new components and new relations. Structural transformation of systems of innovation is thus reflected in the emergence of new types of innovating firms (and other organizations) and the disappearance and transformation of old ones. It also includes the transformation of inter-organizational networks, through the creation and destruction of various types of links between organizations. In the same way, the institutional landscape undergoes transformation through the creation, destruction and adaptation of institutions, and through the changing relations between different institutions. In addition, the links between organizations and institutions undergo changes.

Since components and relations may be of very different kinds (see above), structural transformation may take very different forms. It includes not only changes in the population of firms, but also the creation, adaptation and destruction of government agencies, research institutes, universities, industry organizations, lobby groups, etc. Similarly, the structure of relations may be transformed in a variety of ways, including changes in collaborative, transactional and competitive relations between firms as well as between firms and non-firm organizations, between formal and informal institutions, and between organizations and institutions.

These changes in the structure of the system are not independent of each other. The change in a component or a relation has often far-reaching consequences for other components and relations. This is, of course, one of the most fundamental aspects of any systems approach to innovation. Through the activities occurring in the system (see section 3.2), components respond to changes in other components and in their interrelations, in order to 'fit' in the system as it undergoes transformation and develops a new structure. Components are therefore not free to change in arbitrary directions, but are for their survival forced to take into account the state of the system as a whole.

When a radical change occurs, such as the Soviet collapse, structural tensions result in the system. For example, if a set of formal institutions are suddenly replaced by a new set of formal institutions, this results in an initial mismatch between these and the rest of the system. Components and relations, which have so far been adapted to the old situation, will sense a need to respond to the radical change. However, the extent to which they can actually do so is limited by a number of factors. For example:

- *Informal institutions*, in contrast to formal institutions, seldom change suddenly or rapidly, but rather evolve in incremental ways (North, 1990: 6). This is because informal institutions are closely related to behaviors and habits that take time to learn and unlearn. For example, in a system where market forces for decades have not been allowed to play any important role in the system of innovation (such as in socialist systems – see Box 2.1), it may require much effort and time to develop a new habit of thinking about the role of the market. Unfortunately, human cognitive capacity is severely limited, and cannot be 'reprogrammed' overnight. This is not to say that change is not possible at all; however, it may require years, decades or even generations for tensions between formal and informal institutions to be resolved.
- *Formal institutions* may in principle be altered instantly, for example, through a change in legislation. In practice, however, institution-building is a difficult activity which requires advanced and relevant

competencies, and formal institutional change is therefore dependent upon the availability of such competencies. For example, the creation of a legal act regulating the Internet-related use of databases, is likely to require far-reaching knowledge about the Internet and of database technologies. There is also a need for knowledge about how a new formal institution fits into the overall structure of formal institutions, in order to minimize the emergence of new contradictions among various legal acts and other formal rules, norms, standards, etc. If institution-building competencies are weak, it may take a long time before formal institutions can adapt to changes in other parts of the system. This is particularly important in times of rapid technological change, which often make existing institutional arrangements obsolete. If a system of innovation then lacks the necessary institution-building competencies, the result may be difficult bottlenecks that prevent a more far-reaching innovative dynamism.[6]

- Similar to formal institutions, *organizations* are also consciously and directly created by people, involving both individuals and other, already existing organizations (e.g., North, 1990: 5; Edquist and Johnson, 1997: 47). Organizations are therefore dependent upon the availability of individuals and organizations that can create and develop them, as well as upon the availability of other resources in the system (such as finance and consultancy services). When such factors are lacking, it may be extremely difficult for organizations and inter-organizational networks to respond to important changes in other parts of the system. For example, a new formal institution designed to stimulate the creation of small technology-based firms may not lead to anything, unless there exist people with relevant knowledge, complementary organizations that provide finance and advice, etc. In addition, some types of organizations – for example, universities – are so complex that it may take a very long time to establish and transform them in response to institutional discontinuities.
- *Relations* in systems of innovation tend to evolve in incremental and path-dependent ways. The source of innovation networks is typically found in inter-organizational relations that have evolved over a long period of time (Tidd et al., 2001: 215). The high risks – for example, in terms of information leakage – associated with acting in inter-organizational relations with the purpose of generating innovation, means that such relations often need to build on trust and effective routines, which take time to build. Conversely, links that have been successful in one context may be difficult to break, even if a changed context makes the existing links obsolete. The activities of organizations for innovation may have been built up in such close

relation to external partners that the resulting systemic interdependencies put serious constraints to change (ibid.: 216). The adaptation of networks to an institutionally new situation is therefore likely to require considerable time and effort.

Hence, the structure of a system of innovation cannot be expected to respond in any automatic, let alone predictable, way to an institutional discontinuity. Instead, the pace at and extent to which structural tensions, following an institutional discontinuity, can be smoothly resolved, depends on factors such as the complexity and strengths of existing components and relations, and the availability of knowledge, competencies and experiences that are relevant in handling structural transformation. An important implication of this perspective is that empirical research has to address the *history* of the system if one wants to understand the dynamics of its transformation, i.e., it is important to ask 'from where the system comes'.

In the same way as a technological discontinuity typically results in a more fluid and turbulent stage, an institutional discontinuity is likely to give rise to a more loosely defined system structure. However, as components and relations continuously adapt to the new situation in a circular and cumulative way (Myrdal, 1957: 16ff.), the system is likely to gradually take a more stable shape. Components and relations develop interlockings by adapting to each other, and once a more stable new structure begins to crystallize it becomes more and more difficult to change the structure in more far-reaching ways. This can be seen in analogy to the gradual maturation of a (sectorally or technologically defined) system of innovation in response to a technological discontinuity.[7] Therefore, the key to understanding the emergence of a new structure should be searched for in the early years following the institutional discontinuity.

However, if the inertia of the earlier system is strong, institutional discontinuities may – paradoxically – have little impact on the system. The existing structure may simply be so strong that further changes are made difficult in the whole system, which therefore remains very much as it was. For example, a formal institutional change demanding the split-up of large conglomerates in Eastern Europe after the collapse of socialism may be followed by a reintegration of actors into networks that in practice are almost identical to the old conglomerates (Dyker and Radosevic, 1999). In the extreme case, the power of the old structure may lead to an abandonment of the new formal institutions (in our case this corresponds to a revival of more Soviet-type institutions). This may be seen in analogy to technological discontinuities that fail to give rise to far-reaching systems transformation.

3.2 The Emergence of New Styles of Innovation

Just as an institutional discontinuity has the power to open up far-reaching opportunities for structural transformation of systems of innovation, it also opens extensive opportunities for changes in styles of innovation. The institutional discontinuity is likely to stimulate wide-ranging changes in the ways activities are carried out, and thus lead to a more fluid style of innovation. In such a state, where both the activities and their interrelations are undergoing changes, they will be easier to influence, as activities that have earlier been interlocked in a more mature style are likely to lose some of their momentum.

Sooner or later, however, the activities start to adapt to each other in interlocking ways, through processes of positive feedback that may lead to both virtuous and vicious circles. This is because a change in one activity is likely to demand adaptations of other activities, in order for the activities to fit with each other. For example, a new articulation of demand may stimulate the creation of new knowledge, and if this is associated with the formulation of new visions, there may be scope for public policy-making activities that, for example, may support the process through financial means, investment in competence-building or other activities. In the ideal case, the activities influence each other in such a way that mismatches between them are resolved. This then results in an increased innovative dynamism (Myrdal, 1957: 19). It may take a long time – years or decades – for activities to develop matches between each other in such a way that far-reaching creative innovation and system dynamism are enabled.

But the activities may also develop in relation to each other in such a way that the system is locked into more problematic styles of innovation. For example, lack of demand may lead to shut-down of knowledge-creating activities, so that the style moves in more imitative directions, for example, through non-innovative subcontracting agreements as the main way of staying alive after the institutional discontinuity.

The way in which a style responds to a discontinuity is also dependent upon the already existing system. To understand this, one may connect to the argument in social systems theory that a system is capable of reacting to change only in accordance with its own logic, rationality and history (Luhmann, 1986: 15ff.). In other words, depending on the structural configurations and the often long-established traditions of carrying out certain activities in a system, different systems are likely to deal with new problems in very different ways.

For example, although the collapse of socialism was a radical political change that occurred in a large number of countries at roughly the same time, systems of innovation in different post-socialist countries did not respond to the radical political change in the same ways. In other words, different systems develop different styles of innovation in dealing with the problems and opportunities associated with the new institutional situation. One country may have a particularly strong R&D base that is deemed relevant for the post-socialist context, and may therefore seek to build a style that is based on strong knowledge creation activities. Other countries may have a much weaker R&D base, but may have a tradition of flexibly adapting their institutional environment to new technologies, which may give them a somewhat different advantage. Yet other systems may see that the new technology is associated with opportunities that require inter-organizational collaboration of a kind in which it has earlier experience. While it is hardly possible to identify any optimal way of responding to new opportunities, it therefore seems natural that different systems interpret the new political situation and its opportunities in different ways with respect to potential innovation. This is in line with the argument above that it is not possible to define any 'ideal' or 'optimal' system of innovation (see, e.g., McKelvey, 1997: 202; Edquist, 2004: 185).

In the same way as for structural transformation, the immediate years following the institutional discontinuity may be expected to be of crucial formative importance for the transformation of styles. Steps taken in one way or the other with respect to the different activities during these years will undoubtedly be followed by adaptations of the activities to each other in such a way as to fit with the new formal institutions. As a consequence, major future adjustments in any activity will be much more difficult to carry out than in the very first few years of capitalism. Hence, the opportunities related to the 'new freedom' will quickly be followed by the risk of becoming locked into new – advantageous or disadvantageous – styles of innovation. Therefore, in order to understand the emergence of a new style, empirical research will have to study in detail the transformation of activities during the immediate period following the institutional discontinuity.

4. THE INFLUENCE OF FOREIGN ENVIRONMENTS ON SYSTEM EVOLUTION

Systems of innovation are seldom *closed* systems. In other words, systems of innovation communicate and interact with their environment. Thus a system of innovation has an influence on its environment, and the environment has also an influence on the system (Hughes, 1987: 53). In this section I deal specifically with the latter aspect, whereby the focus is on environments that are defined by spatial boundaries of systems. The discussion here concentrates on *the influence of the foreign environment upon a nationally and sectorally defined system of innovation*. In the context of East European transformation and reorientation, this is of crucial importance, since it is to a decisive extent a matter of developing new and destroying old relations with foreign systems of innovation.

4.1 Foreign Environments and Spatial Boundaries

In earlier research on systems of innovation, spatial boundaries have been thought of in two somewhat different ways. The *first* is inspired by the long research tradition dealing with sectoral patterns of innovation, economic geography and industrial organization. The focus is on the geographical dimension of industrial organization and how this dimension varies from sector to sector. For example, in sectors where the knowledge base is largely tacit and complex, it may be argued that innovating organizations are likely to show a high geographical concentration, whereas in the case of a more codifiable and simple knowledge base, innovators may be geographically more dispersed (see, e.g., Breschi, 1995).

This understanding of spatial boundaries can be interpreted as a critique against the national and regional systems of innovation literature, where the geographical area to be studied is defined *ex ante* by focusing on specific national or sub-national territories. Carlsson and Stankiewicz (1991) emphasized this aspect in their seminal work on 'technological systems', stressing that such systems differ from geographically defined systems of innovation in that 'the boundaries of technological systems do not necessarily coincide with national boundaries' (Carlsson and Stankiewicz, 1991: 112). Breschi and Malerba (1997), in their original delineation of sectoral systems of innovation, similarly criticized the national and regional systems of innovation approaches for taking 'the geographical boundaries of the innovation system as given' (ibid.: 130). Following the argument above on sectoral patterns with respect to the organization of innovation in different industries, Breschi and Malerba argued that 'the boundaries of SIS [sectoral innovation systems] are endogenous: they emerge from the specific conditions of each sector. As a matter of fact, different industries may have different competitive, interactive and organizational boundaries which are not necessarily national' (ibid.: 131). Carlsson and Stankiewicz similarly suggested that the spatial boundaries of technological systems 'may well be different from one techno-industrial area to the next' (Carlsson and Stankiewicz, 1991: 112).

The *second* interpretation of spatial boundaries refers to the potential co-existence of two or more – isolated or communicating – systems of innovation in the 'same' sector, but in different parts of the world. This possibility was observed by Breschi and Malerba (1997: 132) and relates also to Edquist's (1997b: 12) argument that sectoral and technological systems may be found at various geographical levels, i.e., a sectoral system may itself be global, national, regional, etc. For example, nuclear power technologies developed rapidly after the war in both capitalist and socialist worlds, but the different and separated nuclear power systems of innovation favored radically different technological solutions in Eastern and Western countries, and the innovative process was organized in completely different ways and was subject to totally different institutional arrangements. The activities in the respective systems, too, were carried out in totally different ways. In other words, Eastern and Western systems of innovation in the nuclear power sector differed considerably from each other.

This argument may be generalized to include not only the division between East and West, but also between individual countries. It is for this reason that one may speak about 'national variants' of a sectorally or technologically delineated system. In practice, it can be noted that this is exactly what has been the preferred research approach both in the technological and sectoral systems literature. Carlsson and his colleagues, for example, in their important book on factory automation as a technological system, focused specifically on the *Swedish* technological system (Carlsson, 1995a). In a concluding chapter, the Swedish system was compared to the 'same' technological systems in the United States, Japan and Germany, thereby strengthening the sense of technological systems as something embedded in national settings. Despite the explicit argument put forward by Carlsson and Stankiewicz (1991) that technological systems should *not* be seen as limited to specific countries, this is in reality exactly what is done when it comes to concrete empirical research. In the same way, Breschi and Malerba (1997) discuss and compare sectoral innovation systems in the United States, Japan, Germany, United Kingdom, France and Italy. Despite starting from an explicit critique of the national systems literature that 'takes the geographical boundaries of the innovation system as given', both the technological and the sectoral systems literature in practice argue for an emphasis on national environments and national systems – though at the sectoral and technological levels.

4.2 Introducing a Transnational Dimension

On the one hand, it may be regarded as an interesting result in itself that research on sectoral and technological systems reveal a number of differences not only across sectors, but also across countries in the 'same' sector. On the other hand, however, the focus on comparisons among countries reflects one of the major deficiencies in the whole literature on systems of innovation, namely the neglect of the *transnational* dimension. This is to say that while it is well-known that systems of innovation differ between countries, much less is known about the mechanisms through which a system in one national setting evolves in relation to the 'same' system abroad, i.e., about how different systems influence each other.

From a formal-theoretical point of view, a system cannot be influenced directly by another system, but only by the environment, in which the system, however, may discern other systems. Building on the analysis of how styles of innovation emerge (section 3.1), where it was emphasized that a system is capable of adapting to changes only in accordance with its own structure, rationality and history, it can be seen that a system is unlikely to be influenced by the environment in any random way; instead, the way in which the system responds depends very much on its own internal characteristics. Two systems with different structures and styles are therefore likely to react in different ways to the same outside pushes and pulls (see section 3.1).

More interesting, however, is the question as to what the forces actually are that determine the impact of the environment upon a system. In social systems theory, Luhmann argued that the environment can only have an impact upon a system if there is 'resonance' at the system boundary (Luhmann, 1986: 15ff.). The metaphor comes, of course, from physics, where the oscillations with a certain frequency in a medium (a system's environment) can make another medium (the system) with the corresponding eigenfrequency oscillate. Socially, resonance can also be said to occur when people react to messages from each other (through language or other media). Naturally, it is easier to communicate if there exists a common or related language.

One plausible interpretation of this argument could in our case be that a system of innovation reacts to signals especially from those systems in the environment with which it has strong economic, political, cultural or other similarities and ties. This is because such external systems are usually the ones with which it is easiest to communicate in a meaningful way. For example, a system of innovation in a European country is likely to react more strongly to signals from other European systems, but more weakly to signals emanating from Japanese or Chinese systems, with which it has less strong economic, political and cultural links. This is in line with research on, for example, management of innovation, where it has been found that European firms find it considerably more difficult to learn from Japanese firms than from European and North American firms (Tidd et al., 2001: 94).

This may also relate to the discussion in section 2.2, where Soviet-type systems of innovation were discussed. It was noted that the activity of new knowledge creation to a great extent sought to build upon the further exploitation of Western technological advances; the general strategy was to first imitate Western technologies and then develop them further domestically. In reality, however, Soviet-type systems showed great difficulties in 'learning from capitalists'. Imported technologies hardly ever acquired a dynamics of their own, but worked as static enclaves in the centrally planned economy (Sandberg, 1989). For example, modernizations were made only with new support-package deals, and diffusion to other sectors than the military or space industries were minimal (ibid.). One possible interpretation of this would be that the great differences in the political and economic environments in East and West during the Cold War prevented a more efficient communication between socialist and capitalist systems of innovation.

Freeman, though not referring to social systems theory, has connected this line of reasoning to theories of catch-up in innovation, arguing that catch-up countries that are located in geographical proximity to leading systems of innovation or that have strong cultural, economic, political and other links to them, should have a clear advantage as compared to catch-up countries located further away from and having less affinities with advanced systems (Freeman, 1994; 2002). This idea is supported by historical evidence showing that the most successful catch-up countries in different historical periods have nearly always been located in cultural or geographical proximity to the most advanced national systems of innovation.

Thus the British leadership in the 18th and early 19th centuries was followed by a rapid catch-up process taking place in the culturally close United States and the geographically close Belgium, Germany and other continental European countries. Scandinavia and Eastern Europe, in turn, both of which lie culturally as well as geographically close to Germany, were then the next to follow. In the 20th century, a similar pattern can be observed in East Asia, where the 'Tiger' economies have been located in proximity to the leading country of that region, Japan. Empirical evidence shows that Japan and later South Korea have played key roles in the catch-up process of other countries in the region (Freeman, 1994: 214f.). In particular, China profited substantially in the 1980s and 1990s from linguistic, cultural and close family ties to Hongkong and Taiwan (Hobday, 1995: 23).

From a somewhat different perspective, Bresnahan and his colleagues came to the same conclusion. They studied the emergence of ICT-related clusters in a number of countries, whereby they stress the importance of connections to advanced ICT markets – in particular the US market. The highly successful clusters in Ireland, India, Israel and Taiwan are seen to 'come from regions that, for one reason or another, have easier potential interactions with the US market (language, cultural connections, diaspora, etc.)' (Bresnahan et al., 2001: 843). This also relates to the issue of innovation networks, which play such important roles in most modern systems of innovation. For such networks to work efficiently, personal relationships of trust and confidence (and sometimes of fear and obligation) play important roles, and therefore 'cultural factors such as language, educational background, national loyalties, shared ideologies and experiences, and even common leisure interests continue to play an important role in networking' (Freeman, 1992: 100).

As a contrast, one may also note the *absence* of strong catch-up in South America and Africa, which are geographically and culturally farther away from the world-leading systems of innovation (Freeman, 2002).

From the above discussion it may be concluded that strong links to an environment of leading systems of innovation seems to be a clear advantage. On the other hand, there are many cases of systems of innovation that have not been able to exploit this vast potential. For example, not all East Asian countries have managed to develop successful systems of innovation. Similarly, a country such as Mexico has found it extremely difficult to profit in terms of innovative performance from its proximity to the United States. In those catch-up countries that *have* been successful in the 20th century, however, it seems to be a common characteristic that the *government* has played an important and active role. This has included both important political choices with respect to issues such as technological and sectoral specialization, but also in a crucial way heavy investments in education. The success in exploiting geographical and culture proximity to leading systems of innovation systems thus appears to depend on these activities (see, e.g., Hobday, 1995; Freeman, 1994, 2002).

4.3 Inter-System Integration

A further step in the conceptualization of system-environment interaction is the question of whether and in what sense this process may lead to the *integration* of a system of innovation with foreign systems. To investigate this issue, one may build on the discussion of system boundaries in section 2.3. As a parallel to the two different types of boundaries that were defined there, I would here like to introduce two different types of inter-system integration:

- Following Deutsch's argument, a national system exists only to the extent that the intensity and number of domestic relationships (i.e., between domestic components) is much greater than the intensity and number of transnational relationships (i.e., between domestic and foreign components) (Deutsch, 1963: 205). Linking this argument to inter-system integration, one may state that a system undergoes a process of integration with another system if the relative number and intensity of transnational relationships is increasing in relation to the intensity and number of domestic relationships. The final stage of such a process is a situation where domestic relations are no longer more important than system-environment relations (i.e., there no longer exists any Deutsch-type boundary). I will in this study refer to this as *Deutsch-type integration*.

- Following Luhmann's argument, two or more national systems exist if interactions with the foreign environment have different impact on the two systems (Luhmann, 1984: 17). Relating this to the issue of intersystem integration, one may then say that two systems are becoming integrated if they respond in increasingly similar ways to interactions with the environment, in terms of adaptations in the structure and style of the system. For example, a Swedish and a Danish system of innovation are in this sense only integrated to the extent that the structure and the style of these systems show strong similarities (which is true only to a minor extent). The final stage of this type of intersystem integration is a situation where differences in structure and style between two or more systems completely vanish (i.e., there no longer exists any Luhmann-type boundary). I will in this study refer to this process as *Luhmann-type integration*.[8]

Conversely, one may use the same arguments to define two different types of intersystem *disintegration*. Luhmann-type disintegration thus refers to a situation where two systems develop increasing dissimilarities over time, i.e., their structures and styles diverge, whereas Deutsch-type disintegration means that cross-border relations are becoming increasingly unimportant over time in relation to domestic relations in the system.

An interesting issue is here to what extent the two types of (dis)integration processes are interrelated and dependent upon each other. For example, does globalization in terms of an increasing importance of international linkages (i.e., Deutsch-type integration) automatically also lead to a process of convergence of structures and styles of different national systems (i.e., Luhmann-type integration)? Or, conversely, can harmonization within the European Union (i.e., a case of Luhmann-type integration) be used as a way to stimulate the build-up of new international linkages among various national systems in Europe (i.e., a case of Deutsch-type integration)? More empirical research will be needed to answer these questions. The present study seeks to make a small contribution to this issue.

5. SUMMARY AND CONCLUSION

Looking at the challenge that East European countries have come to face after the events in the years around 1990, and seeing innovation as a result of far-reaching systems change, it is clear that the transformation and reorientation of East European systems of innovation is an extremely difficult process. However, the conceptual framework elaborated in this chapter offers interesting opportunities to better understand this process.

The chapter has pointed to the need to study both the internal dynamics of the system and the influence of the foreign environment when seeking to grasp the transformation and reorientation of systems of innovation. This is true for all systems, but it is even more strongly pronounced in the case of Eastern Europe, whose most recent history is so different from the West and where the international reorientation is so radical. The study has pointed to the need to see systems not only as a structure, but rather to analyze their evolution in terms of what is 'happening' in the system, i.e., to study the activities and styles to which they give rise. It is necessary to study a variety of actors as well as institutions, and a variety of activities – for example, not only R&D, and not only business-related activities – and thereby also take explicitly into account activities such as competence-building (not least university education), policymaking, the formulation of visions in various organizations, the creation and abolition of institutions, etc.

Moreover, the systemic approach followed in this chapter has showed that firms and other organizations cannot be expected to suddenly begin to innovate in radically new ways and along totally new trajectories. Rather, actors – with inherited competencies and behaviors that may be embedded in both organizations and individuals – are likely to create their future by building largely on past experiences. These may be combined with new Western inspirations, but the systemic view suggests that their own heritage is unlikely to be suddenly unlearnt and abandoned. Conversely, new knowledge, as well as new kinds of components and activities, take time to build and develop. While it might be tempting to associate the massive invasion of Western products and capital in Eastern Europe after 1989 with an obsoleteness of historical systems of innovation, new patterns of trade and investment do not necessarily mean that local cultures, behaviors and competencies are replaced by imported ones. The conceptual and theoretical discussion in this chapter rather points to the necessity of studying the underlying history of systems, if we are to understand the nature and logic of their transformation and reorientation.

At the same time, the radical institutional changes in the years around 1990 undoubtedly mean that radically new opportunities for systems change and transformation are opening up. The chapter emphasized in this connection that it is important to address the immediate years following the collapse of socialism, as it is this turbulent period that provides the formative stage for the emergence of post-socialist systems of innovation.

Furthermore, linkages to systems of innovation in other countries are likely to strongly shape East European systems. It was noted that successful catch-up systems of innovation have typically been located in geographical and cultural proximity to leading systems of innovation. This was explained theoretically by the greater possibilities and the greater ease for such catch-up systems to communicate in a meaningful way with the more advanced systems. Such communication, in turn, is not only necessary, but entirely crucial in the context of innovation, since innovation is a social process of interaction and learning in which factors such as trust and common frames of reference play important roles. While the potential for learning from abroad is enormous, the ability to actually exploit this potential is therefore dependent upon a variety of 'soft' factors related to culture and history. Therefore the study of East European transformation and reorientation should focus not only on domestic development, but also on the relation between domestic transformation and international reorientation.

Related to the last argument, the introduction of the concept of 'style of innovation' means that the process of East European catch-up with the West is about much more than 'how fast' Central and East European countries can converge with the West. A much more interesting question seems to be what kind of styles of innovation are emerging in the Eastern countries. This is an empirical and explorative issue, whereby one may, for example, look in particular for more and less 'creative' as opposed to 'imitative' styles of innovation in the post-socialist era.

In conclusion, by elaborating on the dynamics of innovation in a systemic and historical perspective, and relating this to the difficult context of East European transformation and reorientation, this chapter has offered a theoretical foundation and conceptual basis for analyzing systems of innovation in Central and Eastern Europe. Based on this conceptual framework, I will in the next chapter propose a methodology for empirical research, through which the theoretical elaborations here can be applied to a real-world context, and further adjusted and improved.

NOTES

1. Some authors prefer to speak also of the activities in terms of 'functions', which are then understood as subtasks relating to the overall function of the system of innovation. To avoid confusion, I here follow Edquist's (2004) suggestion to speak of the overall function of the system and the activities going on to serve this overall function.

2. Thus Kline (1991) discussed the cultural basis underlying differences in the ways US and Japanese firms innovate, and the prospects for US firms to adopt some of the more successful elements of the Japanese style. Similarly, Lundvall and his colleagues (Lundvall et al., 2002) recognized that one source of inspiration in their seminal work on national systems of innovation was the international comparison of national styles of management of innovation pursued at SPRU.
3. Hence, the concept of style is also broader than the notion of 'technological paradigms' in previous research on innovation (e.g., Dosi, 1988) and, in the sociology of technology, 'technology traditions' (e.g., Berner, 1999).
4. A somewhat different way of thinking about boundaries is to ask which organizations, institutions, relations and activities within a specific geographical or sectoral context are actually to be interpreted as belonging to the system of innovation, as opposed to activity fields within that context which are not related to processes of innovation. For example, not all business firms are relevant in the study of innovation, but only those that actually have a notable direct or indirect influence on innovative outcomes. See Edquist (1997b) for further discussion on this issue.
5. For an overview of the concept of technological discontinuities, see, e.g., Anderson and Tushman (1990) and Utterback (1994).
6. This argument relates to classical ideas on the interrelations between technological and institutional change, perhaps most brilliantly expressed in the Marxian dialectics (e.g. Marx and Engels, 1848; see also Rosenberg, 1982). It also relates to the interest expressed by North (1990), Rosenberg and Birdzell (1986), Murmann (2003) and others, in how differences in the institutional histories of countries can be seen as a major determinant of their economic performance.
7. See, e.g., Utterback (1994), who analyzes this development in terms of a 'fluid phase', in which the new industry is still open to far-reaching transformation and social shaping, before a 'dominant design' emerges and the different components in the industry become more stable and much more difficult to change in any radical way ('specific phase').
8. Another example of this is the development towards a harmonized, common European system of innovation, as discussed by, for example, Caracostas and Soete (1997).

3. Historical Case Studies: A Methodological Discussion

1. OPERATIONALIZING TELECOMMUNICATIONS AS A SYSTEM OF INNOVATION

I have in this book chosen to define the telecommunications sector from the *product* side rather than from the perspective of underlying telecommunications *technologies*. Therefore, what is meant by telecommunications products must be specified.

I will here stress the need to take into account both telecommunications *equipment* and telecommunications *services*. Edquist (2003) and others have argued convincingly that these products should be studied together as being part of the same sectoral system, as they are closely related to and dependent upon each other. Together, equipment and services are therefore seen to constitute the products of the telecommunications system of innovation.

A limitation of the study is that it only takes into account two-way telecommunications. In other words, broadcasting in the sense of provision of radio and television services has as a rule not been included in this study.

Telecommunications equipment refers first and foremost to hardware installed and used in telecommunications networks. It includes telephone terminals, cables, exchanges, routers, radio base stations, modems, antennas, electronic components, etc. It can be noted, however, that such hardware has come to incorporate more and more software. Software that is embedded in telecommunications equipment is therefore defined as belonging to the equipment side of telecommunications products.

Telecommunications services were for a long time limited to a very narrow range of services, mainly telegraphy and fixed voice telephony. Over the past few decades, however, with the development of digitalization, the rise of mobile telephony and the Internet revolution, telecommunications services can now be divided into at least four separate 'layers' (which are not necessarily provided by the same actors). The four layers are (Casper and Glimstedt, 2001: 267; Corrocher, 2003):

- The *network layer* (provision of physical network capacity);

- The *connectivity layer* (provision of access to telecommunications networks, including Internet access provision);
- The *middleware layer* (provision of solutions that *enable* various Internet applications, i.e., web browsers and other navigation services, search engines, electronic payment systems, transformation of web contents to WAP format, information security products, etc.); and
- The *applications layer* (provision of web design services, web-based e-mail, solutions for Internet banking, e-government, e-commerce, etc.).

All these layers are taken into account in this book. In addition, one might mention the provision of 'Internet content', which then includes areas such as online newspapers and informative homepages. The present study takes into account Internet content only to a limited extent, mostly in cases where innovation in Internet content is strongly and interactively related to the rest of the telecommunications system of innovation.

In terms of organizations, all actors engaged in the (renewal of the) production of equipment and the (renewal of the) provision of services in the above sense should be seen as belonging to the sectoral system of innovation in telecommunications. This includes not only the specific firms that supply the sectoral products, but also organizations that are engaged in carrying out the activities in the system which are necessary for the supply of these products, such as research institutes, government agencies, educational organizations, standard-setting bodies, providers of finance and industry organizations (cf. Chapter 2). Similarly, consumers of telecommunications services can be thought of as an important group of actors, to the extent that they may play important roles in the articulation of demand for new services and thus influence the evolution of the system. However, as noted in the theoretical chapter, there is no reason to expect *ex ante* that a certain type of organization or institution will play a role for the creation of new telecommunications products.

2. HISTORICAL CASE STUDIES AS A RESEARCH STRATEGY

As stated in the introductory chapter, this book seeks to contribute to a better understanding of the forces and processes that account for the emergence of new – and the transformation of old – systems of innovation in Central and Eastern Europe. The purpose of the study is both to investigate the dynamics and evolution of the specific system of innovation for telecommunications in Estonia from the late Soviet period to Estonia's EU accession, and, above all, to provide an in-depth explanation of *how* innovation has been enabled to occur in the system. Under such circumstances, i.e., when the questions to be researched are of a 'how?' and 'why?' type, the most suitable overall methodology is generally recognized to be *case studies*. The reason is that such questions ('how?' and 'why'?) deal with 'operational links needed to be traced over time, rather than mere frequencies or incidence' (Yin, 1989: 17f.). This should be understood in contrast to alternative methodologies such as experiments, which are seldom possible in social science and the humanities, or surveys, which are more suitable for studies focusing on questions such as 'who?', 'what?', 'how many?', etc. (ibid.). For this reason, the research strategy outlined here follows a case-study approach.

The main characteristic of case-study methodologies is that they seek to develop as full an understanding of the case(s) as possible. Case studies aim to understand the case in depth, and in its natural setting, recognizing its complexity and its context. They also have a holistic focus, aiming to preserve and understand the wholeness and the unity of the case (Punch, 1998: 150). Similar to many other case studies, the present investigation understands case study research more as a *research strategy* than as a *method*, in the sense that it can itself draw on a variety of methods of data collection and analysis – both quantitative and qualitative. For example, a case study can involve elements such as surveys and statistical analyses, which are then used to illuminate the overall case. In case study research, there is also a useful variety to be exploited in terms of data and data sources (as discussed further in section 3).

In general, case studies usually focus on one case ('single-case study') or a small number of cases ('multiple-case study'). The present study applies a single-case study methodology. The case consists of the telecommunications systems of innovation in Estonia and is treated as a case of Central and East European systems of innovation. Central and East European systems of innovation thus form the overall population from which the case has been selected.

An alternative to the single-case approach would have been to focus not only on the case of the telecommunications system of innovation in Estonia, but to take into account further Central and East European systems of innovation – either in different sectors or different countries. However, although such studies might give interesting results, they do not necessarily give an answer to the questions of *how* different systems have evolved and been transformed in their specific ways, and *how* creative innovation has actually been enabled to occur. Since it is these latter aspects which stand at the center of the present study, the empirical study builds on a deeper, historical investigation of a single case (the Estonian telecommunications system of innovation), rather than on a comparison between several cases. This allows us to gain an understanding of the emergence of one specific system *as a process in time* and to investigate the transformation and evolution of the structure and the style of the system. What is at stake here is a broad explanation of 'how things got where they are', a characteristic question of historical and evolutionary approaches to innovation (Nelson, 1995).

Periodization

The empirical part of the study applies a periodization based on political and economic developments in Estonia from the Soviet period to EU accession. Four different periods are distinguished in this way, each characterized by specific political and economic circumstances in Estonia:

1) The Soviet era up to 1985. This period was characterized by Estonia being a Soviet republic, which means that the Estonian sectoral system of innovation in telecommunications was subject to central planning and to a large extent to decision-making that took place centrally in Moscow rather than domestically;
2) The new situation following Gorbachev's coming to power in 1985. The general institutional and organizational conditions in Estonia thereby underwent a series of important changes. The period ended with the establishment of national re-independence and the creation of a fundamentally new institutional structure based on the idea of a market economy system and Western-type democracy;
3) The new period of national independence, which I will here count from about 1993, when the new institutional and organizational basis had crystallized in the form of a new constitution, a new independent currency, free elections, a total freedom to build linkages with the West, etc.;
4) The period following the confirmation of Estonia as an EU accession country in 1997, the steep economic downturn following the stock exchange crisis in the fall of that year, and the Russian financial crisis a few months later, which radically changed the basic political and economic conditions in Estonia. The period ends with the enlargement of the European Union in May 2004.

Dynamics and Evolution

Given this periodization, the study investigates – in four detailed empirical chapters, with each chapter corresponding to one period – the dynamics and evolution of the Estonian telecommunications system of innovation. In accordance with the arguments developed in the theoretical chapter, the dynamics and evolution of the system is operationalized through the structure of the system (in terms of organizations and institutions) and the activities occurring in it, i.e.,

- Formulation of visions;
- Articulation of demand for new, improved and cheaper products;
- Creation of new knowledge;
- Competence-building;
- Formation of new firms and other organizations and market entry and exit;
- Adaptation of organizations to accommodate for innovation;
- Networking;
- Provision of finance.
- Consultancy, advice and lobbying activities;
- Creating, changing and abolishing institutions; and
- Formulation of public policy.

As stated in the introductory chapter, a specific interest of the study is the emerging features of post-socialist styles of innovation. This interest is pursued by studying how the style of innovation in terms of the above activities undergo transformation over time.

Furthermore, the empirical study also investigates at what points in time and in what ways the role of inherited competencies and East-West linkages influence the activities. This corresponds to the first two major questions raised in the introductory chapter. They are approached in an exploratory manner, where the goal is to contribute to an improved theoretical understanding of the mechanisms and processes that account for the role of Soviet-era history in post-Soviet developments, and for the role of East-West interaction in the build-up of East European systems of innovation, respectively.

System Performance: Creative vs. Imitative Innovation

The performance of the Estonian telecommunication system of innovation is in this book evaluated along the distinction between creative and imitative innovation. Imitative innovation is evaluated along dimensions such as the speed of introduction of mobile and Internet services, the speed of digitalization of networks, the adoption of improved production processes in the electronics industry, etc. Creative innovation is evaluated through the introduction of globally unique goods and services on the market, and the renewal of processes in unique ways.

3. DATA AND DATA SOURCES

In order to strengthen the internal validity of the study, a variety of empirical material has been collected. The empirical core is formed by *documentary data* and *interviews*. In addition, while the historical case-study is of a qualitative character, it also draws, though to a lesser extent, on *statistical data*. Data and sources of data are discussed in the following sections.

Interviews

Interviews have been carried out with representatives of organizations, who in the case of business firms have been primarily general managers, R&D managers and project managers; in the case of non-firm organizations the corresponding representatives have been directors of research institutes and competence centers; professors, lecturers and rectors of higher education organizations; ministers of economic affairs and of communications; and heads of other government agencies and non-firm organizations, such as industry associations. The selection of interviewees has been based on the desire to meet with persons who have both a deep and broad overview of developments of different types of organizations.

The main principle for selecting organizations has been by subsector in the Estonian telecommunications system of innovation, in such a way that representatives from all relevant subsectors were covered. Based on the portals http://www.ee and http://www.neti.ee, which provide comprehensive listings of firms as well as non-firm organizations, an initial mapping was done of all telecom-related organizations active in Estonia. This has then been continuously complemented and updated with the help of press material and other documentary sources, in particular with the goal to also include organizations that are no longer active. In cases where a subsector is represented by only a few organizations (such as the operation of mobile networks), all of these were contacted for interviews, whereas in subsectors with many actors a sample of a few organizations were selected in a random way. A complete list of interviewed organizations (by subsector) is provided in Appendix I.

A total of 56 interviews were carried out, each of which as a rule lasted between one and three hours. The interviews were carried out in Estonian, Swedish and English, and have in most cases been recorded on tape and later transcribed. A semi-structured interview strategy has been followed, using an interview guide based on broad areas of questions relating to, for example, the history of organizations, the structure and organization of innovative activities, the role of external (cooperative and competitive) relationships, linkages to foreign systems of innovation, etc.

Where interview data have been used in the present text, interviewees have been asked to confirm (or reject) their statements. A number of interviewees have also been asked to comment on certain parts of the historical text, in order to avoid misunderstandings and thus increase the internal validity.

Documentary Data

Documentary data used in the book consist of material gathered in connection with interviews and other organization-specific material such as annual reports and press-releases; industry-specific journals, in particular with direct contributions from representatives of firms and other organizations; conference proceedings; yearbooks and other publications by universities, research institutes and government agencies; legal documents; contracts and other agreements; etc. Secondary material from earlier research on Estonia has also been used when appropriate for rough background information.

Statistical Material

Statistical material includes data collected mainly from publications by the Statistical Office of Estonia, but also data compiled by international statistical bodies such as Eurostat, OECD, the US Patent Office, the International Telecommunications Union (ITU), etc., as well as reports carried out by the EU and other organizations on the new EU member states, surveys presented in the Baltic IT&T Review, the IPTS Report, Transition Report, the Enlargement Future Report Series, the Global Information Technology Report, other journals and yearbooks and statistical data and analyses presented in previous research on Estonia.

4. METHOD OF DATA ANALYSIS

Units of Analysis

The *main (holistic) unit of analysis* in this study is the system of innovation. The ultimate focus is thus on the system as a whole, rather than on any of its components or activities. The interest in the system includes the relations between the system and its environment. As already indicated, the Estonian telecommunications system of innovation is treated as a case of post-socialist systems of innovation in Central and Eastern Europe.

Embedded units of analysis are the structure of the system, the style of innovation and the system's performance. The structure of the system is analyzed in terms of further embedded units of analysis in accordance with the theoretical elaboration on systems of innovation, where the structure was defined in terms of components (organizations and institutions) and relations. The style of innovation, too, was discussed extensively in the theoretical chapter, where it was defined in terms of a wide array of activities. System performance, as discussed in section 2, is analyzed along the distinction between creative and imitative innovation.

Historical Analysis

The analysis is based on developments and changes over time. In particular, the study focuses on how the structure, the style and the performance of the Estonian telecommunications sectoral system evolve over time. Following the model of periodization discussed in section 2, the Estonian system is analyzed in four different time periods. The system is then compared in the different periods.

An alternative approach would have been to compare the Estonian system at one point in time with other systems at the same point in time. This approach has been avoided, since the primary interest of this book is in understanding system dynamics and evolution over time, rather than in describing and comparing different systems at one point in time. While the study refers to parallel developments in other countries in order to emphasize the emergence of features that are unique to the Estonian context or, conversely, that resemble Western developments, the core of the historical analysis remains firmly with the Estonian developments themselves.

The empirical material has in the historical part of the study been structured along a number of segments of the telecommunications sectoral system of innovation. These sectoral segments have been selected to cover the entire telecommunications sector as it was operationalized in section 1. The segments are:

- Institutional and policy developments;
- Fixed voice telephony;
- Mobile telecommunications;
- Data communications;
- Equipment production; and
- Education and research.

These six segments are discussed at length in each of the historical chapters, where they form the main sections of Chapters 5, 6 and 7 (though sometimes in a somewhat different order than indicated above). Chapter 4, which has the character of an introduction to the ensuing three chapters, has a somewhat freer structure and also includes the 'pre-history' of Estonian telecommunications, from tsarist times to World War II.

For each of the four historical periods, the corresponding empirical chapter concludes with a summary and an analytical discussion, which addresses both the evolving structure of the system in that period and the character of the Estonian style of innovation in the period.

In the two final chapters, major summaries are made and conclusions drawn. The first of these chapters links together the four time periods constituting the Estonian case. The analysis is carried out in such a way that, first, the evolution of the system's performance is evaluated. Secondly, the structure of the system and its evolution over the four time periods is analyzed and explained. Thirdly, the evolution of the Estonian style of innovation is dealt with. Finally, the chapter returns to the first two major questions that were posed in the introductory chapter, discussing in detail what the Soviet heritage has meant for the post-socialist system of innovation, and how Estonia's reorientation towards the West has influenced the system.

The material is also analyzed with the intention to develop theory further. This is done in the final chapter.

5. METHODOLOGICAL PROBLEMS ENCOUNTERED

As noted above, the empirical emphasis of the study is on the in-depth historical analysis of the telecommunications system of innovation in Estonia and its reorientation towards the West. The main strength of the approach elaborated here is clearly the depth of this single case. Conversely, it may be seen as a weakness that the book does not take into account other countries or sectors to the same extent, and a methodological issue is thus to what extent the study can be expected to have any *external validity*. In other words, to what extent does the in-depth study of Estonian telecommunications have the potential to provide us with new knowledge about the overall problem of the dynamics of innovation in Eastern Europe?

This issue should be viewed from the perspective of the in-depth focus on a single case as being related to the emphasis in this study not primarily on 'theory-testing', but rather on 'theory-building', i.e., on developing new knowledge from the single case which may be used to construct further theoretical questions and propositions. This is particularly desirable in the present area of research, as very few extensive studies have been previously conducted of the actual dynamics of innovation and systems of innovation in Central and Eastern Europe. As noted in the introductory chapter, earlier studies tend to draw more on theories developed from empirical research on already advanced Western countries, whereas theories addressing the particular problems of innovation in the transition context are much more weakly developed.

This is not to say that 'Western' theories of innovation are useless in the Eastern context – on the contrary, as argued extensively in Chapter 2, they may offer valuable conceptualizations of systems of innovation that are useful also in the context of transformation and reorientation. But at the same time, there is a clear need to take into account the actual empirical phenomena of transformation and reorientation when seeking to conceptually and theoretically develop a realistic understanding of the dynamics of innovation in Eastern Europe. A single, but carefully researched case has the power to shed light on processes and mechanisms to which attention has hardly been paid before, but which may still play key roles in the transformation process.

With respect to the *internal validity* of the study – that is, the extent to which the material presented actually reflects the real-world transformation and reorientation of the Estonian telecommunications system of innovation – the overall strategy has been to proceed as deeply as possible in the collection of empirical material while at the same time retaining the breadth of the case (covering, for example, not only fixed and mobile telephony, but also the Internet; not only firms, but a variety of non-firm organizations; not only R&D, but all relevant activities; etc.). The consistency of the vast historical material has as far as possible been sharpened through extensive triangulation, i.e., by using different, mutually independent sources that address the same phenomenon. In particular, I have sought to relate interviews, documentary data and statistics to each other, in order to carefully assess their coherence. Sometimes contradictions between interview data and documentary data have been found, in which cases the collection of data has been followed up by further data collection in terms of both interviews and documentary data. Remaining contradictions are pointed out in endnotes in the empirical chapters, thus leaving the floor open for further, complementary research.

An important question is the extent to which the available data really reflect the issues of interest in this study. One problem has thereby been the *terminology* used in the study of innovation. The term 'innovation' is itself highly problematic when it comes to empirical research, as it has become a common word not only in academic studies of innovation, but also in everyday language in politics and business. The notion of 'innovation' is therefore not necessarily assigned the same meaning as in innovation research. In business as well as in politics, it is, in addition, a term loaded with value. For these reasons, I have tried to avoid the term 'innovation' when carrying out interviews, 'translating' it into alternative terms that better fit the specific organizations interviewed, such as the 'creation of new products', 'the improvement of technologies', 'the further development of services', etc. Conversely, I have from interviews and documentary data sometimes 'translated' everyday language and specific jargons of different communities into the terminology of innovation research that was elaborated in Chapter 2.

A surprisingly minor problem in the empirical research was the availability of interviewees. Having decided to establish contacts with persons presently active in the Estonian telecommunications sector, it was easy to actually get in contact, and only exceptionally have potential interviewees declined to participate in the study. Somewhat more difficult was the establishment of contacts with persons no longer active but who have earlier played important roles. However, these could often luckily be reached by way of other interviewees. It also turned out to be a problem in a few cases to get in contact with interviewees for post-interview comments and data checks. Due to the rapid and often turbulent development of the sector in recent years, some persons turned out to have changed their working places or retired, thus contact could not always be successfully re-established.

It turned out to be an advantage to come as a foreign, Swedish researcher to Estonia, as people in general appreciated very much a foreign, genuine interest in their activities. The relaxed atmosphere made it easier to concentrate on the core issues. Although my knowledge of the Estonian and Russian languages are far from perfect, I did not experience any great difficulties in using these languages in interviews and in connection to the evaluation of documentary data. Although many people in Estonia have very good knowledge of the English language, it turned out to be very important to be able to read and speak Estonian and Russian, as many documents – such as the journals *Arvutustehnika & Andmetöötlus*, *Arvutimaailm*, *Vestnik Sviazi*, *Tallinna Tehnikaülikooli Aastaraamat*, as well as local newspapers, firm-specific sources, Soviet-era statistical material, etc. – were available only in local languages.

4. The Origins of Telecommunications in Estonia (1855-1985)

This chapter introduces the history of the Estonian telecommunications sector, from the earliest experiments with telegraphy and telephony in the 19th century, through the period of national independence in the interwar years, and further through the Soviet era up to the early 1980s. This forms the point of departure for the main part of the historical study, which analyzes the transformation and reorientation from the later 1980s to the present and which is dealt with in Chapters 5 to 7.

1. FROM TSARIST PROVINCE TO INDEPENDENT REPUBLIC

It was in connection with the Crimean War in the 1850s that Estonia made its first acquaintance with telecommunications in a modern sense, i.e., communications over a distance by means of electrical signals. In Western Europe, the first long-distance telegraph lines had been opened in the late 1840s, and the war that broke out a few years later and which had front lines in both the Black Sea and the Baltic Sea region, served to make military leaders in tsarist Russia aware of the administrative and military advantages of building telegraph lines between the most important cities and other locations in the centrally governed empire. In Estonia, this led to the creation of connections from Tallinn to St. Petersburg in 1855 and to Riga in 1857 (Sirkel, 1999: 14).

During the following decades, the new technology also came to play an important role for the *economic* development of Estonia. Telegraph lines were built between the larger Estonian cities, and although direct connections to most West European countries were opened only after World War I, Estonia had then, via its connections to St. Petersburg, Riga and Helsinki, in practice already been part of a pan-European telecommunications network for more than half a century.[1]

The organizations that drove this initiative and took a leading role in its development were above all the imperial authorities in St. Petersburg. Local

Estonian actors were as a rule not allowed to play any critical roles, as the creation of a long-distance telegraph network was much too closely related to the strategic and military functions of the new technology. Estonian firms were thus limited in their scope of action to *using* and *operating* the new technology, whereas they were not involved in any technological development activities and not even in network construction or installation of equipment. The central imperial authorities, for their part, initially procured most of the equipment from the rapidly expanding Siemens works in Berlin (where Russia was probably a key demand-side driver of the success of this company's telecom branch). In 1882, however, procurement arrangements shifted to St. Petersburg, where Siemens itself built Russia's first large telecommunications equipment factory. It was also Siemens that constructed the already mentioned telegraph line between the Russian capital and Tallinn (Kaasik-Aaslav and Maripuu, 1999: 41-43).

In contrast to the telegraph, the *telephone* had, from a technical point of view, in its infancy a much more *local* character, and it served initially rather as a complement to the telegraph than as a competing technology. The first telephone call in Estonia is said to have occurred already in 1877 in Tartu (*Postimees*, 14 December 1877), but the systemic introduction of telephony was an act of the 'Börsenkomitee' (the successor of the Hanseatic-style city council) in Tallinn, which in 1884 applied for permission to build a telephone network in the city (Kaasik-Aaslav and Maripuu, 1999: 49). The realization of this network, which was opened in 1886, was a complex process that involved bureaucratic negotiations with St. Petersburg and Riga, as well as the installation of apparatus from Western companies (ibid.: 49-57). Though local from a technical point of view and driven by local interests, the development of telephony in Estonia was thus dependent upon a transnational network of relationships, tying it to the bureaucracy of Russia as well as to the emerging telephone industry in the West.

The fact that the driving forces in the development of the first telephone networks were of local character, together with the economic rather than military function of telephone services, meant that the prospects for *local entrepreneurs* to be involved in the build-up of an Estonian telephony sector were considerably better than in the case of telegraphy. First and foremost, this shift took place with respect to the provision of telephone *services* in the network and connectivity layers (which at this time were still the only layers of services). Somewhat later, a similar shift took place in the manufacturing of equipment.

With regard to network operation services, the local character of the Estonian networks manifested itself in the initially self-evident use of German as the only working language when a subscriber demanded a connection from the switchboard operator – a peculiar feature that was related to the dominance of Baltic Germans in the local political and economic elite (Kaasik-Aaslav and Maripuu, 1999: 59).

More surprising, however, was the fact that the Börsenkomitee procured the construction of the telephone network from a *local* company – the Kurt Siegel engineering firm in Tallinn – and not directly from the equipment manufacturer, as was the typical case in cities such as Riga, Stockholm, Helsinki and even St. Petersburg. This seems to have had the positive effect of maintaining a freedom in the choice of technology, allowing the Siegel firm to abandon the original idea of using Bell technologies and, instead, turn to the newly established Swedish firm L.M. Ericsson, when it turned out that the latter's equipment – in Siegel's own formulation – worked better in the troublesome Baltic Sea climate (ibid.: 53-55).

The Estonian telephone networks themselves remained very small throughout the tsarist period. On the other side of the Baltic Sea, there were 5,000 telephones in use in Stockholm already in 1885 (Kaijser, 1994: 19), i.e., before the very launching of the Tallinn network, while at the time of the fall of the tsarist empire in 1917, there were still not more than about 2,500 subscribers to the public networks in Estonia, whose population was about the same as that of Stockholm (Merilaid, 1938: 64).

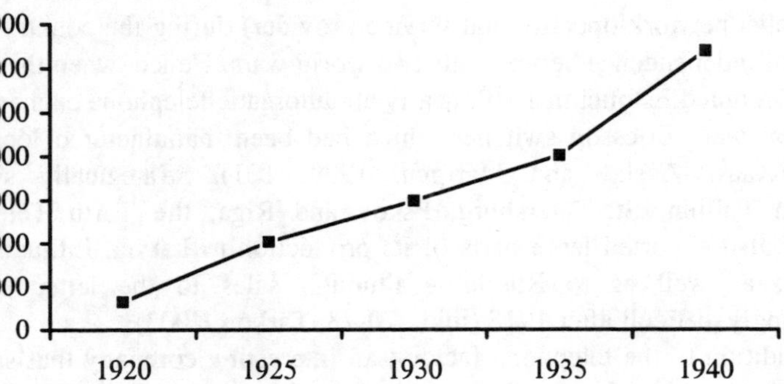

Source: *Statistical Yearbook of Estonia* (selected years).

Figure 4.1 Number of telephones in Estonia, 1920-1940

Also, it was not possible to make telephone calls to other European countries. It was only in 1919 – during Estonia's independence war – that the first international connection (to Finland) was established. By 1930, however, it was already possible to make a phone call from Estonia to nearly any European country (ibid.). The Estonian telecommunications system grew quickly during these years, and by 1939 there were 28,000 subscribers to the public networks (*Eesti Statistika*, no. 215, 1939: 507). In international comparison, on the other hand, this was still a minimal number. Sweden, for example, had by 1942 already a million telephone connections (Karlsson, 1998: 24), corresponding to a penetration ratio of some 6-7 times as high as that of Estonia. Neighboring Latvia, too, had about twice as high a penetration as Estonia (*Baltic IT Review*, no. 6, 1997: 40f.).

In the course of Estonia's first independence period, the country also became an interesting hub in the Swedish L.M. Ericsson's rapidly growing multinational telephone company. Ericsson had a strong interest in the growing markets on the Eastern shores of the Baltic Sea, and the Swedish company had even planned to move its headquarters to St. Petersburg (Jangfeldt, 1998: 182).[2] When the Bolshevik revolution in 1917 put an end to this idea, some leading engineers from Ericsson's St. Petersburg factory and from the local telephone network of the same city moved to nearby Tartu in Estonia and contributed to the development of the telephone factory that had existed there since 1907 (Tarkon, 2001). Ericsson itself invested heavily in the factory in 1929, and 'Tartu Telefonivabrik' (Tartu Telephone Factory) became the Swedish company's most important partner in Eastern Europe, producing telecommunications equipment under Ericsson licences and with the Ericsson brandname.

The Tartu factory developed close relationships to the Estonian PTO (the monopolist network operator and service provider) during the country's first period of independence between the two world wars. Hence, when the Soviet Union occupied Estonia in 1940, nearly all automatic telephone exchanges in operation were Ericsson switches which had been manufactured locally in Tartu (Kaasik-Aaslav and Maripuu, 1999: 101). Strategically situated between Tallinn, St. Petersburg, Pskov and Riga, the Tartu Telephone Factory also exported large parts of its production to Latvia, Lithuania and Finland, as well as to Russia – although sales to the latter became increasingly difficult after 1918 (ibid.: 70-73; Tarkon, 2003).

In addition to the telephone factory, an interesting company that saw the light of day in this period was the radio set producer 'Raadio-Elektroonikatehas' (RET) in Tallinn in 1935. The profile of this firm lay far outside the telecommunications sector as it was defined at that time, but it should be mentioned here due to the later convergence of radio engineering with telecommunications. RET came to play an important role in the Estonian electronics industry, and it was spurred by competition from a number of further local and foreign actors in radio engineering markets (Pihlamägi, 1999: 200f.). With respect to one further electronics-related market, it has been pointed out that the famous Minox camera, which eventually served to make neighboring Latvia's electronics industry world-famous, was originally developed in Estonia in the interwar period (Hendre, 2003).

Through Ericsson's activities in Estonia, the newly independent country began to accumulate its own competencies in telecommunications equipment. With respect to the cooperation between the network operator and the manufacturer, these relationships seem to have somewhat resembled Swedish Televerket's linkages to the Ericsson mother company in Stockholm, although the Tartu facility was not as R&D-intensive as Ericsson's main facilities in Stockholm. However, this similarity in organizational structure of the emerging telephone sector between Estonia and Sweden did not imply any *institutional* convergence: whereas Swedish and most other telephone administrations at this time were nationalized and centralized, the decentralized tsarist tradition remained in place in Estonia despite Ericsson's strong presence. This meant that Estonia followed the same path as Finland, characterized by a number of local operators in the cities in combination with one long-distance, inter-city operator.

To summarize the period up to 1940, the Estonian telecommunications sector was born through the introduction of telegraphy in the 1850s, whereby Estonia itself was directly involved merely as a user of telegraphic services. The involvement was stronger in the case of the first telephone networks, local firms being competent enough to make fairly advanced technological choices – although the statistics make plain that Estonia was a clear latecomer with respect to the expansion of telecommunications services throughout the period, and beyond, at least up to the 1960s.

On the other hand, this latecomer position did not prevent telecommunications equipment production from being established in Tartu, beginning in 1907. Interesting is the fact that in the interwar period, close relationships were built up between the Tartu factory and the local operators, in particular in Tallinn. The Tallinn telephone network had been using Ericsson equipment since the 1880s, which was presumably a good motivation for Ericsson to invest heavily in the Tartu factory. Although Tartu was not an R&D center for Ericsson, which at this time was already a global company, the Estonian factory's products were well-known (both domestically and abroad) for their surprisingly good construction and quality, which in combination with low prices led the factory to continuously win tenders not only in the Baltic Sea region, but also in China, Turkey, Persia and other countries (Kaasik-Aaslav and Maripuu, 1999: 72). The more advanced Western markets, however, never appear to have been penetrated.

Up until World War II, Estonia can thus be seen as a bridge between advanced and less advanced telecom innovation systems, and a link between Western and Eastern systems. Estonia was itself in many ways a user of technologies that had been developed in more advanced innovation systems, but at the same time it was a production site for equipment to be exported. Moreover, this production site was closely interrelated with the local telephone service sector, i.e., with the build-up and operation of telephone networks in Estonia itself. It would thus be too simplistic to state that Estonia was a mere passive receiver of foreign technology. Instead, in the interwar period, the picture emerging is rather one of intense activities aiming to use business linkages to Sweden in particular in order to achieve a successful local telephonization, at the same time as telecom production also contributed to economic growth and exports. This logic formed the basis of the Estonian telecommunications system of innovation up to World War II.

2. ESTONIAN TELECOMMUNICATIONS UNDER SOVIET RULE

World War II had seriously destructive effects on the Estonian telecommunications system. At the time of the first Soviet occupation in 1940, around 32,000 telephones had already been connected to the public networks. The destruction in connection to the events in 1944 was so far-reaching that it took around fifteen years before the pre-war level was reached again (calculated from Statistiline kogumik, 1968: 157).

After Estonia's forced incorporation into the Soviet Union in 1944, the public telecommunications system of the once free country was transferred administratively to a new republic-level Ministry of Communications, which was subordinated both to the Supreme Soviet of the Estonian SSR and to the union-level Ministry of Communications ('Minsviaz') in Moscow (Campbell, 1995). Estonia was from now on no longer free to choose its equipment suppliers, and most larger projects such as inter-republic connections were to be developed and designed by the Leningrad-based 'Lengiprosviaz' institute.

At the level of implementation-related R&D, on the other hand, the republic-level Ministry in Tallinn had also its own development organization, 'Sideministeeriumi Projekteerimis-Konstrueerimisbüroo' (Project and Construction Bureau of the Ministry of Communications, SPKB). The SPKB was created in 1968 in order to design concrete projects domestically that were connected to network expansion. It employed by the end of the Soviet period around a hundred persons and was thus an important part of the Ministry (Trapido, 1994: 4; see also Kaasik-Aaslav and Maripuu, 1999: 159). A few of its employees were also involved in developing methods for testing the quality of telephone networks (Ots, int., 2003). This appears to have been as far as the Ministry advanced in terms of innovative activities.

Due to the sensitive strategic issue of telecommunications in the totalitarian state, high-level managers and engineers were strictly controlled by the central authorities in Moscow before they could be employed by the Ministry and in particular the SPKB. Most obviously, citizens with relatives living abroad were typically prevented from employment in the telecommunication sector (Ots, int., 2003). Apart from this, however, the republic-level (Estonian) Ministry appears to have retained a certain autonomy from the central Soviet administration in the post-war development. The whole structure of the telephone system had in the interwar period gained such a momentum that a republic-based sectoral development appeared to be clearly the most efficient way of organizing republic-level network operation and service provision. The resulting relative independence of the Estonian Ministry was reflected, for example, in a much more rapid expansion and automatization of the networks than in all other Soviet republics (with exception of Latvia; see below). On the other hand, the central Minsviaz in Moscow took over the responsibility for all long-distance communications across republic or national borders. Communications to other European countries now had to go via Moscow. The special lines that went from the telephone exchanges to KGB's Tallinn headquarters were also infamous (Saarma, 1999: 131).

Whereas it was regarded reasonable at the network and connectivity layers of telephone services to administer the *operation* of telephone networks through the republic-level Minsviaz ministries, the development and production of *equipment* tended to follow a completely different Soviet logic. Estonian manufacturers in the electrical engineering industry that had survived the war were incorporated into union-level conglomerates, which consisted of a variety of production units that were located to various places all over the Union. The Estonian manufacturing units lost their independence, in the sense that they could no longer develop linkages to other firms and organizations in accordance with their own desire, but had to follow the principal plan of the relevant ministry. Before the war, as noted above, the domestic Tartu Telephone Factory had established dynamic cooperation with Ericsson and the Estonian network operator, but these relationships were now abolished, forcing the new republic-level Minsviaz to manage by itself many of the tasks that had earlier been organized in cooperation with the manufacturers (Kaasik-Aaslav and Maripuu, 1999: 110). All linkages with leading firms in Western Europe were cut off, and the subsidiaries of foreign telecommunications companies (which had typically been established in Estonia during tsarist times) were shut down or taken over by the centralized state.

A peculiar reason behind the problematic situation that followed after the war, was that the development of the telecommunications sector was not in line with the ideological claims of the Soviet state. In Marxist thought, telecommunications were identified as a 'materially unproductive' sector, and were therefore deliberately neglected in budget negotiations and five-year plans (Campbell, 1995: 1ff. and 30). In addition, point-to-point communications were seen as a potential danger with respect to the political stability of the totalitarian state. Throughout the Soviet era, the state therefore spent only about 0.1 per cent annually of the national budget on civilian telecommunications, compared to around 2 per cent for the Nordic and West European countries (*Financial World*, 15 September 1992). The operators were usually profitable, but around 75 per cent of the revenues were extracted to the national budget, thus leaving little room for re-investment and modernization. In terms of capital investment, the telecommunications sector accounted for only around 1 per cent of total investments in the national economy, which was many times below Western levels (Campbell, 1995: 47).

Moreover, from the perspective of technological change and innovation, the weak links between the organizations involved in service provision and those involved in manufacturing became increasingly problematic. In the Nordic countries and in Western Europe, it is well-known that the success of dominant firms such as Ericsson and Nokia has been very much the result of fruitful collaborative relationships between network operators and equipment manufacturers (Edquist et al., 2000). In contrast to this, the Soviet sectoral system of innovation was characterized by an artificial separation between these two subsectors. While network operation and service provision were tasks to be administered under the union- and republic-level Minsviaz ministries, the production of equipment belonged under a number of other ministries, of which the most important one was the Ministry of Communications Equipment Industry ('Minpromsredsviaz', MPSS).

MPSS was a military ministry, with a conglomerate of R&D and production facilities whose customers, however, were not primarily the operators of civilian telecommunications networks; the (civilian) Minsviaz ministries accounted for only a few per cent of MPSS's sales, and in times of output problems – which were frequent – the equipment supply to public telecommunications networks had no priority, since there were more important customers within the military.[3] This effectively prevented any organic development of dynamic relations between equipment producers and the network operators.

Given these conditions, the achievements of the Estonian Ministry of Communications and the organizations under its control were considerable. Especially in comparison to other Soviet republics, Estonia was surprisingly successful building a public telecommunications infrastructure. Estonian telecommunications representatives were successful lobbyists in Moscow, whereby the very size of the republic seems to have made it easier to acquire an over-proportional share of the produce of MPSS and other relevant ministries. The good relations with Moscow are reflected, for example, in the successful acquisition of one of the DX-200 exchanges that the Soviet Union bought from the Finnish industrial conglomerate Nokia (which at this time was moving increasingly towards telecom specialization) in 1986 and that was originally planned to be used for the telephonization of Siberia (Mets, int., 2001).

The relative success of Estonia was mirrored in a telephone density of 203 telephones per 1000 capita in 1985, which among the Soviet republics was the second highest figure (after Latvia, where the corresponding figure was 229). Telephone density in the Soviet Union as a whole in 1985 amounted to 112 telephones per 1000 capita (Campbell, 1995: 17). Also, Estonia managed to fully *automatize* its telephone system by 1975 (Statistiline aastaraamat, 1975: 192). Given the technology available, these figures and dates meant that Estonia performed exceedingly well in Soviet perspective. But at the same time, the achievements were still far below those in Western and Northern Europe, where countries such as Sweden and Finland had a telephone density about twice as high as Estonia and where the radical process of digitalization had already begun.

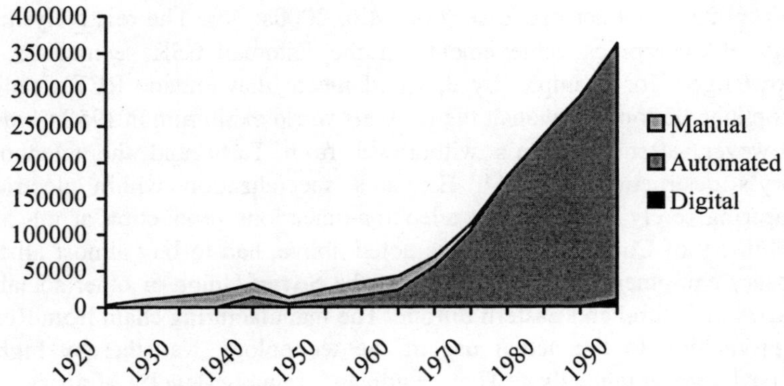

Source: *Statistical Yearbook of Estonia* (selected years).

Figure 4.2 Number of telephones in Estonia, 1920-1990 (by manual, automated and digital lines)[4]

The major problem had to do with the production and procurement of technology and equipment. Estonia had a considerable electrical engineering and electronics industry, the flagships of which in the 1930s had been the Tartu Telephone Factory and the radio set producer RET. In the first years of Soviet power, this inherited electronics industry was threatened by a centrally decided focus on high-voltage engineering in Estonia, but from the end of the 1950s the 'Thaw' under Khrushchev allowed Estonia to revive what was still left and to expand electronics production in the Estonian SSR.

Companies such as the Pöögelmann and the Kalinin electrotechnical enterprises (the latter had its own research institute in Tallinn) produced semiconductors and related devices, analogue microschemes, electronic equipment for the energy industry, etc. Electronic measurement and automation equipment was produced by two other large enterprises. The radio set producer RET was revived under the new name 'Punane RET' ('Red RET'), to which the important 'Tallinna Raadioelektroonika Konstrueerimisbüroo' (Tallinn Radio Electronics Construction Bureau, TREKB) was added as an R&D department.

Research activities at a more fundamental level were carried out within organizations answering to the Estonian Academy of Sciences: the 'Spetsiaalkonstrueerimisbüroo' (Special Construction Bureau, SKB), the 'Arvutustehnika Erikonstrueerimisbüroo' (Computer Engineering Special Construction Bureau, ATEKB), and above all the 'Küberneetika Instituut' (Institute of Cybernetics), which was founded in 1960 (see further section 4). In the vicinity of Tartu, researchers at the astronomical observatory at Tõravere were also involved in electronics-related contract research for the Soviet military and aerospace sectors (Min, 2000a: 46). The relatively high quality of electronics achievements in the Estonian SSR seemed to be acknowledged, for example, by the gold medal that Punane RET's radio-gramophone 'Estonia-2' won at the Brussels world exhibition in 1958 (ibid.).

However, after Ericsson's withdrawal from Tartu and the telephone factory's destruction in 1944, Estonia's specialization within electrical engineering rarely included any telecommunications production at all, and the Ministry of Communications, as noted above, had to buy almost all the necessary equipment from other parts of the Soviet Union or other socialist countries in Central and Eastern Europe. The manufacturing chain from R&D and production to the actual use of the technology was thereby highly dispersed geographically. The Estonian republic-level Ministry of Communications used equipment not only from 'Krasnaya Zarya' ('Red Dawn') and other Leningrad factories, but also from VEF in Riga, from the East German Robotron (which among other things produced the pseudo-digital ISTOK exchange), the Yugoslavian Iskra conglomerate and other Central and East European firms (Kaasik-Aaslav and Maripuu, 1999; Müller, 2000: 199ff.).

In general, the Estonian Ministry tried to direct its demand for equipment to the Central and East European countries (rather than to the USSR), since experience showed that equipment from these countries was technically more reliable than Soviet equipment (Saarma, E., int., 2001). However, the old Swedish-Estonian Ericsson exchanges from the interwar period were still in use in the Estonian telecommunications system, as were equipment parts from Siemens, which had been acquired by the Soviet Union as trophies during the war (Ulman, interviewed in Kaasik-Aaslav and Maripuu, 1999: 128). Hence, the Estonian telephone system resembled a somewhat confusing exhibition of telecommunications technologies from a whole array of both East and West European countries.

Soviet Estonia was thus not a center for the production of telecommunications equipment, although the Baltic rim of the USSR as a whole was an important region for the Soviet telecommunications sector. Riga was the center of gravity for the telecommunications industry in the Soviet Baltics, with the well-known Latvian VEF company, which had a long pre-war tradition (producing, among other things, the famous Minox camera) and which emerged as an important hub in the Soviet telecommunications sector (see Müller, 2000: 210).[5] VEF had also an important R&D institute directly connected to it (information from VEF-KTR and Kvant-Interkom, 2003). From an Estonian perspective, it is also worth noting the proximity of Leningrad, which was a highly important center for Soviet telecommunications, with research institutes such as LONIIS and factories such as Krasnaya Zarya – although the activities in Leningrad came to concentrate heavily on military rather than civilian telecommunications technologies during the post-war period. Also, in Estonia's immediate Western proximity, Sweden and increasingly Finland had advanced to become world leaders in telecommunications, especially towards the end of the Cold War, as mobile telephony was about to take off. Estonia was thus, from a geographical point of view, surrounded in all directions by considerable dynamism in the field of telecommunications. Under different political-economic conditions, it would have been natural for Estonia to be part of a dynamic, all-Baltic Sea telecommunications region. Under Soviet conditions, however, this did not happen.

3. DATA COMMUNICATIONS IN SOVIET ESTONIA

Apart from the traditional telegraph system, which in contrast to most Western countries continued to be widely used in Russia until the 1990s, data communications in the Soviet Union was a field where competencies accumulated mainly through experiences in the field of *industrial automation and control systems*. In contrast to R&D in fixed voice telephony, the Estonian SSR came to enjoy a privileged position in this area due to the Soviet decision to establish an Institute of Cybernetics in Tallinn in 1960. This institute, accountable to the Estonian branch of the Soviet Academy of Sciences, was created mainly in response to the increasing demand for industrial applications of what was referred to as automation and 'telemechanics' systems (Küberneetika Instituut, 2000: 7; Jaaksoo, int., 2003).

The suggestion to establish such an unusual institute in Estonia was motivated by the fact that the republic had rapidly expanding chemical and energy industries – sectors in which automation and control played a very important role (Wõrk, int., 2003). Moreover, both the chemical and energy industries in Estonia were almost exclusively based on the rich oil-shale findings in the northeastern part of the republic, and as the oil-shale industry was a somewhat unique phenomenon in the USSR, specialized research institutes focusing on oil-shale-based chemistry and energetics already existed in Estonia. This was an important circumstance, as researchers from these institutes were the individuals driving the creation of the Institute of Cybernetics, which upon formation would employ most of the personnel from the mechanics and automation departments of the Institute of Energy (Küberneetika Instituut, 2000).

Although this happened at a time when the Soviet Sputnik satellite had just shocked the Western world, Estonian researchers recognized that Western countries were already in possession of systems which, if available for import, would have been highly interesting in the Estonian context. But similar to the field of telecommunications equipment, CoCom regulations (i.e., West-East export controls) and Soviet ideology effectively hindered potentially rewarding technology transfer (Wõrk, int., 2003). This put an even heavier strain on the Soviet Union's own R&D efforts. In contrast to telecom equipment, however, where almost all R&D organizations were located *outside* Estonia, the early foundation of the Institute of Cybernetics in Tallinn meant that automation and control technologies had a fairly strong local representation in terms of R&D. The Institute emerged as an important developer of solutions for customized Local Area Networks (LANs) for the Estonian industry, and it grew by performing R&D as a subcontractor for the powerful research institutes in Moscow, Leningrad and elsewhere (ibid.). These activities led to a rapid accumulation of deep competencies in quite a diverse field of research and development – a situation that would later be of considerable importance for the development of the post-socialist innovation system.

The Institute of Cybernetics thus quickly diversified into a whole array of research fields going far beyond automation and control systems. In 1980, the fact that the Institute was both an advanced user and performer of R&D in computer hard- and software led it to activities related to data communications at a somewhat different level. This was a time when mainframes (which were as a rule IBM clones) were still the only available computers in the Soviet Union,[6] while at the same time the need for computing resources was growing rapidly. The idea then emerged to establish an ad hoc data communications network in terms of a computer time-sharing system with remote access.

A Soviet computer with 48 displays formed the heart of this project, in which half of the displays were located in the main building of the Institute itself, while the other half were dispersed around Tallinn. The users were various academic institutes but also one secondary school. Later, more powerful computers were added to the project, though the data speed remained limited to between 0.6 and 2.4 kbps (*Baltic IT Review*, no. 7, 1997: 16). Despite all limitations, the project appears to have been a noticeable achievement, and the father of the network, Arnold Reitsakas, was even honored with the IEEE computer society's Computer Pioneer Award in 1997 for the Institute's contribution to the development of computer systems and networks (Institute of Cybernetics, 2000).

Intense activities were also going on in the USSR aimed at the creation of much wider data networks. The US experience with Arpanet from the 1960s onwards seems to have inspired the Soviet Academy of Sciences to create a corresponding network in the USSR. After a series of time-consuming preliminary studies in the 1970s, the Presidium of the Academy of Sciences and the State Committee of Science and Technology in Moscow took the formal decision in March 1981 to create what became known as 'Akademset' (in English sometimes referred to as 'Academnet'). Two committees to develop and administer the project were created: the Council of Akademset (which was to handle mostly organizational matters), and, more important, a Board of Designers. The high hopes raised for the project were reflected in the election of the president of the whole Soviet Academy of Sciences, G. Martchuk, as chairman of the council. For Estonia, however, an even more important circumstance was that the Board of Designers was headed by R. Jakubaitis, director of the Institute of Electronics and Computer Science in nearby Riga. The Riga institute soon emerged as the leading actor in the whole Akademset project, providing almost all of the network's hardware and software components (*Baltic IT Review*, no. 7, 1997: 15).

Thanks to the circumstance that the 'spiritual mother' of Akademset was the Riga institute, it turned out to be possible for Estonia as a close neighbor to easily get access to all necessary components both in terms of workstations and communications equipment (Wõrk, int., 2003). This stood in sharp contrast to the enormous problems faced by the Ministry of Communications, as discussed above, in procuring up-to-date equipment for the PSTN (the public voice telephony network).

Under Latvian leadership, the Akademset project did not follow the American Arpanet's TCP/IP standard, but preferred to design the planned all-Union network on the basis of another international packet-switching standard, the at that time much more well-known X.25 protocol. The Union was divided into eight regions, one of which was formed by Estonia, Latvia, Lithuania and Belarus, and centered in Riga (*Baltic IT Review*, no. 7, 1997: 15). The main Estonian network node was located in Tallinn and was connected to the rest of Akademset by way of Riga. Later, in 1987, an additional link between Tallinn and Moscow was created (ibid.: 16).

A peculiarity of the Estonian participation in Akademset was that the republic-level Ministry of Communications was also involved. The Estonian Ministry was the only Soviet PTO (public telephone operator) that seemed to show any interest in the packet-switched data network, whereas most other republics seem to have regarded it as lying too far outside the 'true' field of telecommunications. The general profile of Akademset was that of a network built both by and for the Academy of Sciences, as a pilot network for more far-reaching data networks. The Estonian Ministry of Communications, however, saw an opportunity to extend the academic network to include data communications also at the level of public administration. Thus the X.25 node in Tallinn was not, in contrast to other Soviet republics, the only Estonian node, but no less than fourteen cities were linked to each other through the network. In 1985 the preparations and tests had reached so far that users in these cities could actually communicate with each other electronically, by sending e-mails, etc. (Wõrk, int., 2003). The Institute of Cybernetics itself used the Akademset throughout the 1980s to communicate with other academic institutes in the Soviet Union (Institute of Cybernetics, 1990: 20).

4. THE EVOLUTION OF EDUCATION AND RESEARCH

In connection with the establishment of Estonia's first national independence in 1918, an organization known as 'Eesti Tehnika Selts' (Estonian Technics Association) founded an engineering school where it was possible to study machine construction, electrical engineering, shipbuilding, hydro engineering, building and architecture. The school was renamed 'Tallinna Tehnikum' in 1919 and was made state-owned in the following year. In the 1930s, the Tallinn school came to face competition from the old university town Tartu, where a Technical Faculty was founded in 1934 – a fact that has to be seen in the context of the important and expanding telephone factory in the same town (see section 1). However, the Estonian government soon decided to move Tartu's Technical Faculty to Tallinn and merge it with the engineering school there. The result of this was the emergence of a larger educational organization that from 1938 went under the label 'Tallinna Tehnikaülikool' (Tallinn Technical University, TTÜ).

In 1944 an Electrotechnical Faculty was established at TTÜ by the Soviet regime, which renamed the university 'Tallinna Polütehniline Instituut' (Tallinn Polytechnical Institute, TPI). However, a strengthened focus on high-voltage engineering led to the shut-down of the new faculty's low-voltage specializations in 1948 (Min, 2000a: 46).

An educational infrastructure relevant to the field of telecommunications and, in a wider sense, information technologies, began to re-emerge in the Estonian SSR in the late 1950s, i.e., during the 'Thaw' under Khrushchev – and on a much larger scale than before. Courses in telegraphy and telephony technologies had earlier been included under the general label of low-voltage engineering, but from now on a structure with clearer and more specialized fields of study began to emerge. In 1957, the National Economy Council in cooperation with the republic-level Gosplan and the TPI created a new specialization of studies named 'Automation, telemechanics and electrical measurement' (TPI, 1986b: 308).

This seems to have been a response to the demand for a well-educated workforce in the Estonian industry, which since the Soviet take-over in 1944 had not only expanded very rapidly, but was now also starting to face an important technological shift due to the introduction of automation and control technologies (see also the preceding section). Thus the creation of the new Faculty of Automation with five different departments – automation; electronics; computer engineering; radio engineering; and cybernetics – which began to emerge in 1960 around the new study program, bears witness to close relationships with the local Estonian industry.

In the case of the Lenin works, for example, which expanded rapidly after the war around the oil-shale chemical industry in northeastern Estonia, the new department of automation entered into contract R&D, and the same was true for the 'Silikaatbetooni Instituut' in the same region and the radio set producer Punane RET in Tallinn (ibid.: 315). The need for well-educated radio engineers at Punane RET was clearly also the driving force in the establishment of the Department of Radio Engineering in 1966 (ibid.: 309).

A telecommunications department had actually been created at the Polytechnical Institute in 1947, motivated by the need for massive reconstruction of the whole telephone and telegraph networks in Estonia after the disastrous effects of the war (see section 2) (ibid.: 323). But with the shut-down of most low-voltage specializations at the Electrotechnical Faculty in 1948 (see above) and the radically shifting structure of Estonian industry after Soviet take-over, the department never managed to acquire the necessary momentum to retain its status as a self-standing university department when the new Faculty of Automation was created. The shut-down of the Ericsson factory in Tartu in 1944 also strongly contributed to the trend of displacing telecommunications from the curricula. In the absence of telecom equipment production enterprises in Estonia, the Soviet authorities concluded that there was no need for a domestic focus on educating telecommunications engineers. In a broad perspective, and in particular in view of later trends towards digitalization and convergence with other information technologies, the Faculty of Automation was extremely important as a supplier of engineers with deep knowledge from fields *relating* to telecoms, but after the 1950s no department of telecommunications emerged in its own right in the Estonian SSR.

At the level of secondary education, telecommunications in the traditional sense could still be studied as a specialization at a 'Tehnikum' (a technology-focused secondary school) in Tallinn. But rather than continue their studies at the Faculty of Automation, the future leading managers and engineers in the republic-level Minsviaz usually went to the famous 'Leningradskii Elektrotekhnicheski Institut Sviazi imeni M.A. Bonch-Bruyevicha' (Bonch-Bruyevich Electrotechnical Institute of Communications) in nearby Leningrad (Ulman, int., 2001; Ots, int., 2003).[7] Throughout the Soviet era, this institute remained the key educational organization from the perspective of the Estonian telecom services sector.[8]

The lack of higher education in Estonia directly addressing telecommunications technologies seems to have been a problem in the eyes of the Ministry of Communications, which pushed for introducing a telecommunications department at the Technical University. The chances for realizing this idea increased in the 1970s, when the former Minister of Communications, Bruno Saul, advanced to become Chairman of the Estonian Supreme Soviet. In cooperation with leading engineers in the Ministry, Saul managed to agree with the Polytechnical Institute that a new telecommunications teaching group was to be established within the Department of Radio Engineering by splitting the existing teaching group in half (Ulman, int., 2001). This step was accomplished in 1972, when a telecommunications (or in Soviet terminology: 'elektrosviaz', i.e., 'electrical communications') specialization was thus established (TPI, 1986a: 3). In reality, the Ministry had to supply much of the teaching staff itself during the first years (Ulman, int., 2001). From now on, around a third of the radio engineering graduates from TPI came to work with telecommunications in Estonia (Ots, int., 2003).

By the mid-1980s, teaching activities at the department, which despite its inclusion of telecommunications was still referred to as the Department of Radio Engineering, covered a fairly wide field of subjects. Apart from radio engineering and telecommunications *per se*, much emphasis was given to computer technologies and programming, relating to the shift to digital telecom technologies (TPI, 1986a: 7, 1986b: 325) – which, as has been discussed, were in reality almost never introduced into the actual Estonian networks during the Soviet era.

Theoretical teaching, moreover, appears to have been at a high level. In contrast to the problems military innovations had in spilling over to civil *industry*, Soviet innovative efforts in radio engineering and telecommunications had a beneficial impact on *basic research*, which among other things meant that excellent textbooks and manuals were available at the TPI. Furthermore, it is hardly meaningful to speak of any real difference between Eastern and Western technological trajectories in much of the teaching activities, as the style of teaching at the TPI was much more focused on basic principles that tended to have more in common with physics and other basic sciences than with any concrete technical applications (Ots, int., 2003).

The Department of Radio Engineering had also been involved in *research activities* from the beginning, in the field of signal processing since 1961, and later expanding to areas such as laser technology. In 1974, the first PhD student (in Soviet language: 'aspirant') was accepted. At about the same time, the department also managed to acquire its first computers to be used in teaching and research activities. In accordance with the dual Soviet system, however, aiming to concentrate teaching to the universities and research activities to the institutes of the Academy of Sciences, the scientific research carried out at the Department of Radio Engineering was kept at a very modest scale. Thus by 1986, only six PhDs had graduated from the department (TPI, 1986a: 2f. and 6). In the end, R&D was typically seen as a way of strengthening the financial situation of the university, whose main task, however, clearly remained the education of engineers (Ots, int., 2003).

At the University of Tartu, which had lost its Technical Faculty during the reforms of the 1930s, teaching and research activities of some relevance emerged from a somewhat different direction. In the 1960s, the Mathematical Faculty of the university diversified to include informatics, and with the rapidly broadening activities in this field, an independent programming department spun off in 1979. This programming department remained stable throughout the 1980s and until 1993, when it became the basis for the new Department of Computer Science in the newly independent Estonia (University of Tartu, 2003).

From a research perspective, on the other hand, the clearly most important organization of relevance for telecommunications and information technologies in Soviet Estonia was the Institute of Cybernetics. Because of its later key role in the emergence of re-independent Estonia's telecommunications system of innovation (which will be discussed in Chapters 5 to 7), its Soviet history should here be discussed in some length. The Institute was founded in 1960, and, as mentioned in the previous section, its creation was motivated mainly by the increasing need for industrial applications of what was referred to as automation and 'telemechanics' systems. Originally, the Institute had three departments: mechanics and applied mathematics; automation and telemechanics; and the computer center. However, the Institute soon diversified into a much wider array of activities, including fields such as mathematical methods, speech technology, instrument software, system software, physics and even biochemistry (Küberneetika Instituut, 2000: 12). By the end of the 1970s, more than 500 people worked at the Institute of Cybernetics in Tallinn, and despite the spin-off of some activities it continued to grow, reaching a peak of more than 600 employees in 1989 (ibid: 14, diagram). In the tiny Estonian Soviet republic, this was an enormous number of staff for a research institute.

The growth of the Institute bears witness to its success as an R&D organization within the Soviet Academy of Sciences. The institute became deeply integrated with the Academy system and had strong linkages through cooperation with related institutes in Moscow, Novosibirsk, Kiev and other places (Jaaksoo, int., 2003). In accordance with the centrally planned innovation system, the main customers of R&D results from the Institute were the large development departments belonging under production enterprises all over the Soviet Union, whereby the latter typically belonged to the military sector. However, the Institute had hardly ever any direct contact with the production enterprises themselves (ibid.). From the perspective of interaction in the innovation system, this was naturally a problematic feature, but on the other hand it was also an advantage, to the extent that the Institute of Cybernetics was so far away from the actual (often military) use of technologies that it was never considered necessary to keep its activities secret. As a matter of fact, this was decisive in enabling the Institute to build linkages to *international research*. In the words of the vice-director of the Institute at that time:

> In Soviet times, it was characteristic of the Estonian Academy of Science – and the Institute of Cybernetics as part of the Academy – to never do any totally closed, secret research work. For this reason, our doors were always open, and it was possible for the Institute to build good relations all over the world, even in Soviet times. In particular, we always had very good connections with Finnish universities and, to a somewhat lesser extent, with Sweden. (Võrk, int., 2003, my translation)

During the Brezhnev era it became common for researchers at the Institute to go for academic exchange to Nordic universities and academies. Popular places where Estonian scientists went were, for example, Tampere University of Technology, Helsinki University, Lund University (LTH), the Royal Institute of Technology in Stockholm (KTH) and the Swedish Royal Academy of Engineering Sciences (IVA). The stays usually lasted for three months and were formally coordinated through the central authorities in Moscow. There were several reasons for the tendency to go to Nordic rather than other countries, as stated by the Institute's last Soviet era director:

> It was clearly easier to exchange with countries near to us, where people were culturally closer . . . And it would have been very difficult to have exchanges with the US or Canada, for example, because there were so many applications for such places from the big universities in Moscow and Leningrad. But those Russian universities did not have the same degree of interest in Sweden or Finland, and they were less informed than we were about the Nordic countries. So when there appeared an appropriate place we were the ones who acted, sent an application and went there. (Jaaksoo, int., 2003, my translation)

All in all, however, the extent of the exchange was quite limited in Soviet times and seems to have been a small drop of water in the big sea of contact networks that existed *within* the Soviet Union and Eastern Europe and *within* the Western world. But still, it was to prove an important starting point for the build-up of *later* Estonian-Nordic linkages that took off from the 'perestroika' period and onwards into Estonia's new independence. The Institute's strong academic links with KTH in Stockholm and, in particular, the Tampere Technical University in Finland, were to prove crucial, for example, for the early introduction of the Internet in Estonia. This will be further discussed in the next chapter.

In 1980, the biochemistry and physics departments of the Institute of Cybernetics broke away to form the separate 'Keemiline ja Bioloogiline Füüsika Insituut' (Institute of Chemical and Biological Physics, KBFI). This organization was to play an interesting role in the early 1990s, in connection to transforming the data communications and customized software subsectors in Estonia. There will thus be reason to return to this organization.

5. SUMMARY AND CONCLUSION

In the period that has been discussed in this chapter, Estonia started out as a province of tsarist Russia. In the interwar period the country became independent (for the first time in its history), but from 1944, the incorporation into the Soviet Union meant that independence was lost again. This dramatic political history naturally came to strongly influence the evolution of the Estonian telecommunications sector.

The Soviet takeover in 1944 and the disastrous effects of the war resulted in a radical shift in system evolution. Underlying structural change was the introduction of Soviet institutions, based on Marxism-Leninism, totalitarian reasoning, central planning and new principles for interaction with the foreign environment. Technologically, this meant that Ericsson equipment and other Western technologies could no longer form the basis for sectoral innovation. Instead, Estonia was forced to orient itself towards Soviet technological trajectories. This was reinforced by dramatic organizational rearrangements, such as the shut-down of telecommunications equipment production in Tartu, and above all the total reorientation of international linkages, away from Sweden and the previously so important foreign markets towards partners and customers in the Soviet bloc. For a period of more than forty years, these structural changes came to support and reinforce each other through processes of positive feedback. At the level of activities this was reflected in the formation of a new, Soviet style of innovation.

However, the new style was not necessarily identical to the styles of other Soviet republics or East European countries in the period. Although Estonia in the Soviet period came to show much stronger similarities to the USSR than to any Western country, it would be an exaggeration to conclude that Estonia underwent a total convergence with the rest of the Soviet Union. Despite the fact that the earlier history and the republic's linkages to the West were severely weakened in their power to influence system evolution, they showed clear traces – though at somewhat more subtle levels – in the style of innovation. In particular, the Western-oriented Estonian culture, earlier inspired by countries such as Germany, Sweden and Finland, seemed to be mirrored in the people's general interest in new technological and economic developments, as well as in the Estonian will to do things efficiently (which had been so important in the capitalist period). This was a major reason for the Soviet Union to use the Estonian SSR as a test-bed for a variety of economic and other experiments. At the same time, from an external perspective, the evidence from the education and research subsector hints at the fact that contacts to Western and in particular Nordic countries were kept alive, though at a modest, almost 'sleeping' level, through which nevertheless important communication could be retained. In addition, some organizations, notably the Ministry of Communications, but also production enterprises such as Punane RET, could in many respects draw in a more direct way on their organizational histories for their Soviet developments.

Let me in the following discuss in somewhat greater detail how Estonia adopted a largely – but not purely – Soviet-type style of innovation, in terms of the activities occurring in the system.

The formulation of visions in the Estonian telecommunications system of innovation long remained focused on accomplishing basic telephonization of the republic and automatizing the PSTN. This was later followed by the vision of creating a digitalized network and a data communications infrastructure – i.e., global trends of which Estonian actors were probably more aware of than other republics, due to their relatively well-preserved contacts to Western countries. These visions later became important forces in the reorientation towards the West, as discussed further in Chapter 5. During the Soviet era, the Estonian Ministry of Communications envisioned itself a leader in service provision within the Soviet Union. In general, Estonian actors, with their Western-oriented minds and memories of the interwar (capitalist) years, collectively felt much confidence in their republic's technological and economic capacity and competencies. The Ministry of Communications and other organizations shared a belief that they could have done much better, had it not been for the often disastrous problems with procurement of equipment and components from outside Estonia. The Institute of Cybernetics, for its part, saw that their celebrated research results only rarely found their way through the Soviet innovation system to be transformed into real products and processes. At the same time, however, the Institute successfully exploited the enormous market opportunities for research that were the result of the burgeoning demand for contract research from the larger Soviet institutes.

The articulation of demand was an activity that very clearly revealed Estonia's deep integration into the all-Soviet system of innovation. The Estonian SSR itself played only a minor role in the articulation of demand, and in general the demand-side was bound to have very little influence on innovation processes, given the Soviet-type organizational and institutional structures (cf. Box 2.1 in Chapter 2). However, there were some exceptions: as noted above, there was a strong demand for Estonian R&D from other parts of the Soviet Union, especially at the Institute of Cybernetics (although it was a perceived domestic demand for R&D, related to automation of industry, that had originally been one of the key arguments for the very foundation of the Institute). Similarly, the growth of Punane RET and other electronics factories to some extent created a demand for corresponding implementation-related – and sometimes deeper – R&D, and the existence of several of the Estonian R&D organizations can hardly be understood without taking into account this effect of the local industry. The Ministry of Communications certainly articulated demand for new and better equipment, but mostly in vain, and its interest in new technologies had therefore hardly any effect on innovation, neither in Estonia nor elsewhere.

The creation of new knowledge took place in different ways. The strongest sectoral segments in Estonia were those related to radio engineering and information and computer technologies, whereas R&D in fixed voice telephony hardly occurred at all. Activities in radio engineering built on a pre-war tradition, whereas the focus on information and computer technologies represented a new Soviet-era trajectory for Estonia. R&D was performed to an impressive extent – but in highly fragmentized ways – at the Institute of Cybernetics and to a lesser extent at the universities, in particular at the Tallinn Polytechnical Institute. At the level of basic research, Estonia managed to stay integrated, albeit not very intensely, with Western academic communities – particularly with Sweden and Finland. More implementation-related knowledge was created at other R&D organizations such as SPKB and TREKB. In addition, production knowledge accumulated both in manufacturing and services enterprises – but although Estonia had a reputation of being a high performer within the Soviet Union, their product quality remained in most cases relatively low in international comparison.

With respect to *competence building*, the Estonian educational system had a very weak focus on telecommunications in a traditional sense, at least after the post-war removal of telecommunications equipment production from Tartu. Competence-building for telecommunications service provision was, instead, largely handled through Soviet educational organizations outside Estonia. On the other hand, local competence-building was strong in related areas that would be of greater importance later: radio engineering and information technology. The establishment of the Department of Radio Engineering at the Tallinn Polytechnical Institute (TPI) was a response to the Soviet decision to strengthen the RET factory, and the educational investments in computer and information technologies at TPI, the Institute of Cybernetics and the University of Tartu have to be understood in relation to the needs of the well-developed local industry. An important characteristic of this period was also that while student exchange with Western countries was hardly an issue, Estonia's being part of the Soviet Union had vast consequences for competence-building, as a large number of scientists and engineers in telecommunications-related fields earned both undergraduate and doctoral degrees at universities in other parts of the USSR.

The formation of new firms and other organizations was, not surprisingly, a very weak activity in Soviet Estonia and posed an obvious problem to the whole system of innovation. The institutional setting of the USSR hardly allowed any spontaneous firm formation or entrepreneurship at all, as new firms and other organizations could generally be created only through central planning. On the other hand, the Soviet period saw the creation of a number of key organizations. Thus the Institute of Cybernetics was created in 1960, and at about the same time the TPI saw the birth of its Department of Radio Engineering as well as other telecommunications-related departments. In 1968, the design bureau SPKB was formed as a special unit within the Ministry of Communications, although it is unclear how important this was for the system as a whole.

The adaptation of organizations was an enormously difficult activity in this period, as organizations on a number of occasions had to adapt in radical ways due to changing political circumstances. Particularly after the war, existing Estonian firms and other organizations had to restructure themselves in order to fit – at least formally – with the structure of the all-USSR system of innovation. However, further attempts to adapt organizations, for example, in response to technological advancements, were often difficult due to ideological restrictions and central planning. Thus it was only with great effort and personal links that the Department of Radio Engineering at TPI managed to broaden itself in the direction of telecommunications. The same was true for the diversification of the Ministry of Communications into data communications. The most eye-catching organizational adaptation in the latter half of the Soviet period was the phenomenal expansion of the Institute of Cybernetics, which from 1960 was built up from nothing and expanded at an impressive pace.

Networking activities were weakly developed in the Soviet period, particularly with respect to the prospects for interactive learning. The most fruitful relations seem to have had the character of informal links (including a variety of East-West links), as more formalized relations were often hindered through the centralized character of the system of innovation. It was extremely difficult to acquire resources from outside, especially in the context of new and innovative projects. Firms rarely met directly with each other, and learning through far-reaching partnerships and networks was thus seriously hindered. Although the telecommunications-related community in Estonia was small and people generally knew each other, it proved difficult to transform these informal contacts into concrete projects aimed at the development of new products and processes. An exception was probably the joint efforts that resulted in the creation of the Institute of Cybernetics. With respect to competitive relationships, these were more or less non-existent and did not play any role in the system (except for Estonia's competing with other republics for various kinds of resources and equipment).

The provision of finance was severely limited due to the overall low priority of telecommunications in the USSR. In radio engineering and information and computer technology, however, the situation was clearly more favorable. Thus R&D was largely funded through contract research for larger Soviet R&D organizations outside Estonia itself, which enabled particularly the Institute of Cybernetics to grow in an impressive way.

With regard to funding, Estonian organizations were also successful *lobbyists* in Moscow, which enabled the acquisition of a disproportionally large share of funds and modern equipment (although this was still very little in comparison with Western countries). Similarly, successful lobbying contributed to the establishment of the Institute of Cybernetics. In general, Estonia could here build on its good reputation in terms of economic development, and on its small size, which seems to have made it easier to capture a disproportional share of resources without there being any visible impact in all-Union budgets.

The creation of new institutions was, for obvious reasons, extremely difficult in the Soviet era. In particular, formal institutions were difficult to change, as such change was centrally planned in Moscow and subject to Marxist and totalitarian reasoning, which left little room for taking into account the particular circumstances and needs of the Estonian telecommunications sector. Estonians tried to compensate for this by developing informal institutions. For example, the perceived low technological level of Russian equipment led to the emergence of a rule of thumb of directing procurement towards Central Europe rather than towards Russian equipment suppliers. In a related way, the structure of international academic exchange and Estonia's proximity to the Nordic countries created the habit of primarily applying for research stays in these nearby countries rather than in the Anglo-Saxon world.

The formulation of public policy, too, was extremely difficult in the Soviet period, not least because of ideological constraints. Many policies that concerned Estonian telecommunications were formulated outside Estonia itself, and similarly to institutions, policies could seldom be moulded to fit the particular Estonian conditions. Nevertheless, the idea of making the Estonian SSR a strong location in the field of electronics in a broad sense contributed to strengthening the sector locally. Local actors here seemed to have flirted with Moscow's interest in using Estonia as a test ground for new economic developments more generally.

By the mid-1980s, then, the Estonian style of innovation had been under the influence of Soviet conditions for more than forty years. The style was not totally identical to the overall Soviet style, but components and activities of the system inevitably adapted gradually to Soviet conditions, and the Estonian system thereby also became deeply integrated with the all-Union system of innovation. At the same time, the influence of Western systems remained minimal (though not totally absent) – the infinitely much stronger external influence being that of other Soviet republics – which meant that the Estonian style was radically different from, for example, the Nordic styles of innovation in telecommunications.

These differences went far beyond a mere technological backwardness. As is clear from the evidence presented here, the Estonian system was not merely lagging behind the West; rather, innovation in Estonian telecommunications took place according to a fundamentally different logic as compared to the West. As will be seen in the following chapters, it was this different logic, or style, rather than the low technological level, that would lead to the greatest difficulties in the periods that followed.

NOTES

1. A sign of the importance of the telegraph in the decades after the Crimean War is given by the ten-fold increase in the number of telegrams sent, which according to Kaasik-Aaslav and Maripuu (1999: 177), grew from about 18 telegrams per day in the 1860s to 180 in the 1880s. In modern comparison, however, this was, of course, a very modest figure. Cf. also Pihlamägi (1999: 26).
2. The importance of Ericsson's Eastern activities was demonstrated, for example, at the World Exhibition in Paris in 1900, where the company was represented by its Russian subsidiary.
3. See Campbell (1995), where an attempt is made to map Minsviaz's complicated procurement regime in greater detail. According to Müller (2000: 196), about 80 per cent of the equipment produced in the Soviet telecommunications sector was intended for military use.
4. The disastrous effects of World War II are here clearly visible. Telephonization on a broad scale began in this perspective only around 1960. After that, however, the development was impressive. The Estonian telephone system was fully automated by 1975, the first of the Soviet republics. Digitalization, on the other hand, remained negligible throughout the Soviet period.
5. VEF was responsible for the production of the long-ditance digital Kvant exchange, which was officially based on Soviet technology but was produced in Riga under a license from Nokia during 1983-91.
6. According to Penjam (2000a: 85), however, there were also a few Apple II computers available in Estonia in the early 1980s, a fact that relates to the argument that the Institute of Cybernetics had relatively good connections, for Soviet conditions, with Western computer technology firms, as pointed out by Enn Tyugu in Küberneetika Instituut (2000: 40). Tyugu also argues that this enabled Estonia to move forward with the Western developments, relatively independent of Russian technology.
7. The Institute has been renamed Sankt-Peterburgskii Gosudarstvennyi Universitet Telekommunikatsii (St. Petersburg State University of Telecommunications). It is interesting to note that the Russian language has replaced the traditional notion of 'electrical communications' with the Western term 'telecommunications'.
8. Among the Estonians who graduated from the Leningrad institute were Jaak Ulman (Director of Tallinn Telephone Network; later Deputy Minister of Communications), Vladimir Heinrichsen (Director of the Department of Radio Engineering at TTÜ) and Arnold Reitsakas (researcher and inventor at the Institute of Cybernetics).

5. Estonia's Reorientation Towards the West (1985-1993)

This chapter investigates the transformation of the Estonian telecommunications sector from the 'perestroika' years to the re-establishment of Estonia's national independence. The main focus is on the emergence of new organizational and institutional configurations and in particular the de-linking of the Estonian system from the Soviet Union, as well as on the emerging new links in Western directions. The chapter reconstructs in depth the emergence of an organizational and institutional landscape which in the ensuing chapters will be seen to form the basis for innovative activities in re-independent Estonia.

1. NATIONAL INDEPENDENCE AND TELECOM INSTITUTIONS

During the 1980s, two broad trends can be discerned that were of critical importance for the fundamental institutional changes in the Estonian telecommunications sector that followed the reestablishment of national independence. The first was the trend towards *increasing autonomy* of the overall Estonian economy from the Soviet Union, a development that took off in the latter half of the 1980s. In September 1987, an influential newspaper article was published, in which four Estonian politicians – two of whom would later become prime ministers in the re-independent Estonia, and one of whom would become Estonia's first EU commissioner – began to officially advocate the economic autonomy of their republic within the USSR. This idea became known as 'Isemajandav Eesti' ('Selfmanaging Estonia'), the high expectations of which were reflected in its abbreviation, 'IME', which is the Estonian word for 'miracle'. Implicitly, this notion signified the perceived stagnation of the whole economy, and the possibility of detaching itself from the rest of the Soviet economy as a strategy for overcoming the stagnation.

In May 1989 the Supreme Council of the Estonian SSR approved the IME plan with an overwhelming majority and adopted the law on 'Fundamentals of the self-financing of the Estonian SSR'. The documents foresaw, among other things, a national banking system, including the establishment of a central bank and the introduction of a national currency. A breakthrough came when a central Soviet law issued in Moscow acknowledged the application of the principles of financial independence in all the three Baltic countries in November 1989 (shortly after the fall of the Berlin wall). Although the law was rather declarative in many ways and seemed to lack any real backing, it still granted a certain legal support to the aspirations of the Baltic republics (Mandel, 1999: 10; information from the Bank of Estonia).

A second and related important trend was the emerging possibility of establishing *cooperation with foreign firms*. During 1987, a series of Moscow decrees made it possible, for the first time, to legally establish joint ventures between Soviet and Western companies (Bell, 1997: 63).

These (non-sector-specific) institutional trends remained open to interpretations, but in practice they meant that new and interesting possibilities opened for reorganizing the republic-level telecommunications sector and introducing new institutions. The new economic independence implied that Estonian organizations such as the Ministry of Communications gained considerably increased power to reorganize themselves and open new channels of procurement without permission from the central authorities in Moscow. Similarly, the possibility of establishing joint ventures offered a potential source of finance for far-reaching *modernization* projects. It should be noted that these trends were evident a few years *before* the actual achievement of full political independence in Estonia. The latter is usually dated at August 1991.

Liquidating the Ministry of Communications

Towards the end of the 1980s, far-reaching institutional changes were felt to be so imminent that the Estonian Ministry of Communications deliberately limited and delayed investments in new physical infrastructure, believing that the anticipated institutional changes would soon allow for much more advanced technologies to be installed in the Estonian telecommunications networks (Sõmera, 2003: 1; Sõmera, int., 2003).

In 1990, the Ministry initiated an extensive series of changes by proposing to the Estonian government the liquidation of itself (!), a step that would be followed by the creation of two separate business-like organizations which were to be responsible for postal and telecommunications services. According to the proposition, only a minor division for handling communications issues would be retained within the government sector, preferably joining the Ministry of Transport (Sõmera in Kaasik-Aaslav and

Maripuu, 1999: 142). A separation of telecommunications from postal services was seen as a necessary prerequisite for attracting foreign investors (Saarma, E., int., 2001).[1]

The suggestion led to institutional change in the form of a new Communications Law that took all of 1990 to prepare. The main organization involved in this was the Ministry itself, which worked out the draft legislation without involving any Western experts on telecommunications legislation (Sõmera, int., 2003). However, a wide contact network (which would later turn into much more concrete cooperation) had at that time begun to be built up with Swedish and Finnish telecom organizations, and it is probable that these contacts became an effective source of inspiration for the new law.

The draft legislation was introduced to parliament in fall 1990 and was formally adopted on 16 January 1991 (i.e., seven months before Estonia's political independence from the Soviet Union). On 24 January, the government also approved a decree addressing the 'Reorganization of the Administration of Communications Services', which followed the above-mentioned suggestion of the Ministry of Communications to separate economically and financially telecommunications from postal services, starting 1 July 1991, as well as to liquidate the Ministry itself (Sideamet, 2001a). It was decided that the new enterprises replacing the Ministry were to be called 'Eesti Telekom' and 'Eesti Post', respectively (*Vestnik Sviazi*, no. 6, 1992: 33). The remnants of the Ministry were transferred to the Ministry of Transport, which was now renamed the Ministry of Transport and Communications, with only a small telecommunications department (Campbell, 1995: 134; Cunningham, 1998: 234).

The Estonian Ministry of Communications also took the brave decision to officially secede its telecommunication functions from the rest of the USSR, a step which was accomplished in 1990 and which at that time was seen as unique for a Soviet republic (*Telecommunications*, October 1991: 170f.). An immediate motivation for such an organizational decoupling from the rest of the Soviet Union was the desire of the Estonian Ministry to establish a *mobile telephony network* in Estonia in cooperation with Sweden and Finland. Finland was particularly anxious not to irritate its smooth relations with Moscow, and the Estonian Ministry therefore set out to acquire a document from Moscow that confirmed Estonia's formal rights to make autonomous decisions. The desired document was acquired, and shortly afterwards the trilateral joint venture 'Eesti Mobiiltelefon' was launched (see section 3) (Sõmera, int., 2003).

The above-mentioned decree also formally emphasized the necessity of creating an independent Estonian 'Telecommunications Inspectorate', the basis for which should be formed by the already existing State

Telecommunications Inspectorate that had earlier been part of the Ministry. The corresponding law was issued on 6 June 1991, and during the following two months the Estonian State Telecommunications Inspectorate ('Eesti Elekterside Inspektsioon') started its work (Sideamet, 2001a: 3). The main activity of this organization was to lie in the distribution of radio frequencies – work which proved not as simple as initially thought, in view of the fact that the Soviet military stayed on Estonian territory from 1991 until 1994, and demanded the utilization of a number of radio communications systems.

Towards a Concession Agreement

The issue of market regulation during these turbulent years was in practice overshadowed by the working out of a *concession agreement.* Telecommunications markets were at this time even in the West only starting to be liberalized, and more commonly countries still regarded them as natural monopolies. Thus, the issue of market regulation was outweighed by negotiations about a concession agreement for telephony in Estonia, and as such it belonged *de facto* under the new Ministry of Transport and Communications.[2]

Estonia did not have any legal experience in this field, and the issue was further complicated by the intention to transfer all existing telecommunications networks to an international joint venture – 'Eesti Telefon' – that was to be formed by Eesti Telekom and the Swedish and Finnish state-owned operators, Televerket and Tele, respectively (as discussed in section 2). This was a somewhat unique idea also in international perspective. Eesti Telekom, which seems to have been the driving force in this process, selected the US firm Debevoise & Plimton as its main legal advisor, as there was no corresponding competence available in Estonia itself (Sõmera, 2003: 2). The requirements were that all legal, economic and technological issues had to fit with international institutions as well as the national legislations of the three countries to be involved in the joint venture, i.e., Estonia, Finland and Sweden. A problem was that Estonia's national legislation was in a highly fluid stage. In particular, the country had neither a constitution nor a convertible currency until June 1992, and this uncertainty tended to delay negotiations and preparations.

During the preparatory period, from 1990 to the end of 1992, three different Estonian governments also held office. On 30 September 1992, the creation of the Nordic-Baltic joint venture based on the concession agreement was approved by the Estonian government (*Postimees*, 1 October 1992). The project was in danger, however, after the elections in October the same year, after which a new government proved more skeptical to the idea of the (temporary) 'exclusive rights' that Eesti Telefon was to acquire. Intense lobbying characterized the last months of the year, with potential competitors to the Nordic PTOs appearing behind the curtains (Sõmera, 2003: 2).

On 16 December 1992, when the concession agreement was to be signed ceremoniously, Prime Minister Mart Laar, who had been expected to participate at the event, remained absent. However, this did not prevent the agreement from being signed by the relevant parties, which were the Estonian Minister of Transport and Communications (Andi Meister), the CEO of Eesti Telekom (Toomas Sõmera, the earlier Minister of Communications) and the general directors of Televerket and Tele (Tony Hagström and Pekka Vennamo, respectively). The event took place in the medieval City Hall in Tallinn.

In the concession agreement, Eesti Telefon was awarded a monopoly license by the government for fixed telephony in Estonia until the end of the year 2000 (Kontsessioonileping, 1992).[3] An interesting aspect, however, was that the concession agreement did not address the fields of *mobile telephony* and *data communications*. These fields were left untouched by the dramatic institutional rearrangements in these early years of transition, a fact which was to prove critical in the years to come.

Creating an Institutional Environment for the Broader Field of ICT

Apart from institutional change in the field of telecommunications in a traditional sense (i.e., fixed voice telephony), substantial institutional and policymaking activities were also directed at reforming the broader field of information and communications technology (ICT). Since this included issues related to data communications, these activities should be mentioned here.

Interesting discussions started during the short IME period but took more concrete shape in 1990, when the Estonian Informatics Council was formed as an advisory body to the government. Its tasks were broadly defined in terms of working out an ICT development strategy for the republic, especially with respect to IT in public administration and the coordination of this development, but also, interestingly, the formation of a 'uniform technology policy' (*Riigi haldusinfosüsteemide arendamise ja riikliku juhtimise alused*, 1992). The field of ICT thus appears to have been regarded as the link to a new and technologically advanced national economy. The early period of the council's activity, however, was full of confusion, and its members later acknowledged that any clarity in this early period was more or less missing, and that there was no clear idea of the Estonian society's needs or opportunities in the field of ICT (Jaaksoo, 1993).

At the end of 1992, an investigation that had been ordered by the government concluded that the objectives defined in the formation of the

Informatics Council in 1990 had not at all been fulfilled. In particular, the situation with respect to IT in public administration was, according to the report, chaotic: ministries and other government bodies had largely been dealing with IT-related issues in an *ad hoc* manner, using only their limited in-house competencies, and acting in almost complete isolation from each other. Interaction among government agencies as well as between these and other organizations was thus identified as a bottleneck. The committee of eleven that prepared the report, largely consisting of people with a background from the Institute of Cybernetics, now demanded a much more systematic approach (*Riigi haldusinfosüsteemide arendamise ja riikliku juhtimise alused*, 1992).

The report led to the creation of a new government organization, the 'Department of State Information Systems' ('Riigi Infosüsteemide Osakond', RISO), in 1993, the director of which for many years was a former researcher from the Institute of Cybernetics, Arvo Ott (information from RISO; Ott, int., 2001). After long negotiations and disputes, it was decided that this organization would be established directly under the State Chancellery. While alternative suggestions included, for example, the idea of locating the new organization directly under the Ministry of Transport and Communications, the actual decision to place it under the State Chancellery seems to have given RISO an interestingly neutral position, thereby preventing its developmental activities from getting drawn into potential inter-ministerial conflicts (Siil, int., 2003).

With respect to the general technological strategy, the development of information technology in Estonia would in the words of the Informatics Council have to

> take the systems of other countries as a model, but basically we have to design and realize the systems by ourselves, because they are our systems, which have to fit our cultural environment. We have a unique opportunity to start almost from scratch, we have no old working systems, we have no bonds that prevent us from seeking and accomplishing the best solutions. (Jaaksoo, 1993, my translation)

The council also noted that Estonia had valuable experiences from the field of automated control systems development and design (ibid.).

A related organization, which was similarly created at the end of 1990, was the Informatics Fund. In its founding document, the Fund was defined as a 'state budget organization, which is formed for the development of informatics and which through financing implements the state information policy' (Agur, 1993: 10, my translation). The financing was to take place in accordance with decisions of the Informatics Council. In 1992, the money was allowed to cover the administration of systems, but from 1993, it was to go only to *developmental* activities (ibid.). The Informatics Fund thus became the main organization with responsibility for financing innovation in the broad field of IT.

However, it was difficult to speak of any 'financing' at all in the early 1990s: in 1992, the budget of the fund amounted to a mere EEK 1 million (€64,000) and in 1993 EEK 1.7 million (€109,000). According to the Fund, this corresponded to about 10 per cent of the necessary work in the field of information technology. Nevertheless, in accordance with the decision of the Informatics Council, three target programs were launched in 1992, addressing the areas of ICT in public administration; data communications and computer networks; and ICT in national security (ibid.; Jaaksoo, interviewed in *Arvutimaailm*, 1993: 6). From an innovation policy perspective, it can be noted that these target programs did not relate to the build-up of a domestic ICT industry with the goal of boosting economic development; rather, the focus was on satisfying quite specific needs related to the activities of public administration (Siil, int., 2003).

Apart from its activities with respect to financing innovation, the Informatics Fund also became the Estonian representative in a number of international standardization bodies, which involved far-reaching networking activities. It started with participation in the standardization committee of the Nordic countries, INSTA/IT. Other extensive cooperation activities were developed with the same countries: in seminars and consultations organized by the Fund, Swedish and Finnish partners were frequent participants. For example, in August 1992, the information technology department of the Swedish 'Statskontoret' carried out a seminar in Tallinn focusing on the application of IT in public administration, with the participation of more than a hundred Estonian representatives. Similar seminars focused on issues such as the development of state registers, the exchange of electronic documents through data communications networks and information standards (Agur, 1993: 10).

The Estonian *procurement policy* was anything but clear until August 1993, when a first Order of Public Procurement of Information Technology Means was enacted. This appears to have been a more or less direct response to an event that raised loud protests earlier that year. The problem began when one of the public agencies organized a public tender, and by following its own isolated rules, awarded the contract to a foreign firm, an outcome deemed unfair. The Law on Public Procurement later enacted included a paragraph that explicitly allowed special preference to domestic products, even in cases where they might not prove to be the most favorable (Agur, 1996a: 41).[4] According to the Department of State Information Systems, which came to be in charge of IT procurement, the Estonian rules that were created were 'based on the corresponding documents from the EC and

Sweden and Finland, which have been adapted to the present domestic economic and legal conditions' (Ott, 1993: 10, my translation).

From around the end of 1993, the general situation with respect to institutional change and policymaking in the broad field of ICT thus seems to have been clarified somewhat. A number of legal acts had been created, and a number of key state organizations – which were to play important roles for the sectoral system of innovation during the following years – had been established. The activities of these organizations, as well as the further process of institutional evolution, will be discussed further in Chapters 6 and 7.

2. RETURN TO THE WESTERN WORLD: THE CREATION OF EESTI TELEFON

In the field of fixed voice telephony, the union- as well as the republic-level Minsviaz ministries in the Soviet Union had for decades tried to make telecommunications a high priority in the budget negotiations with the government, suggesting ambitious plans for the expansion and modernization of the whole sector. These plans were crafted mostly in vain, as shown by the figures for actual public spending on telecommunications (see Chapter 4). However, as mentioned above, Estonia still performed surprisingly well, given the minimal resources.

At the same time, the general political changes in the Soviet Union during the 1980s worked in favor of the communication ministries. For decades, the evolution of civilian telecommunications had been severely hindered not least by the peculiarities of Marxist reasoning in connection to the telecommunications sector and by the interests of the totalitarian state. The regime regarded the counter-revolutionary dangers of public telecommunications as much greater than any social or economic benefits. With 'glasnost', however, interests were gradually redefined, and the opening-up towards the West following Gorbachev's call for 'perestroika' also broke through the isolation of the Soviet and Estonian telecommunications sector. As mentioned above, cooperative projects with West European firms and organizations became possible and easier from around 1987, at the same time as the IME program for economic autonomy gained momentum.

In these years, it became clear that Estonia was in fact striving not only for economic autonomy within the Soviet Union, but ultimately for full political independence. For such a separatist republic, telecommunications were of strategic importance, especially in view of the fact that the republic was striving for *democracy* and a *market economy* system, where communication and openness belonged to the very symbols of success. The opportunities from 1987 for East-West joint ventures indicated that foreign firms in many industries were planning to enter the gradually opening East European markets, and a well-functioning telecommunications systems was obviously a precondition for many of these foreign actors to invest. Estonia's mere fifteen (!) existing outgoing international channels, for which customers typically had to wait several days, were clearly totally unacceptable in this respect, and the fact that the lines were connected to the KGB and went exclusively through Moscow did not make the situation brighter. It was in these circumstances that the far-reaching institutional changes described in the preceding section were initiated. As outlined in the following, these institutional changes then paved the way for far-reaching organizational changes and the build-up of new partnerships, which became the basis for technological upgrading.

Searching for Strategic Partners: Entering Cooperation with Sweden's Televerket

Apart from the struggle to institutionally break out of the USSR system, intense efforts also focused on the technological issues related to the creation of new international lines and in general the modernization of the whole telecommunications infrastructure. These efforts in practice were directed towards the search for contacts with Western partners (an activity in line with the 1987 decrees as well as the IME program), and the origins of cooperation with Western European countries can be traced at least as far back as 1988. During 1988 and 1989, the Estonian Ministry discussed cooperation possibilities with a wide array of well-known PTOs, including, among others, Deutsche Telekom, France Telecom, US West, Televerket in Sweden and Tele in Finland. It turned out that several of these were indeed willing to invest in the Estonian telecommunications network, but their main interest was typically in international traffic, not in supporting the more financially risky modernization of local networks (Sõmera, int., 2003).

Two important exceptions from the general pattern of foreign attitudes were the Swedish and Finnish PTOs. In principle, these were prepared to accept the Estonian Ministry's general line of thinking, with the fixed network viewed as a whole (i.e., including both local and long-distance traffic). Apart from this, Sweden and Finland were also recognized by the Ministry as geographically and culturally close countries, whose histories were intertwined with Estonia's and whose mentality also seemed similar. Tens of thousands of Estonian emigrants had lived in Sweden since the war, meaning that almost every Estonian had one or more relatives or friends in Sweden. Finland was familiar because of linguistic proximity – many

Estonians had, at least in the northern part of the republic, no problem understanding the Finnish language, as they had been watching Finnish television since the Brezhnev era. This also meant that Estonians were familiar with developments in Finnish society, and more generally the Western world, to an extent that was seldom possible in other parts of the Soviet Union.[5] Also, the economy which was starting to open up for cooperation with Western firms was already dominated by Swedish and Finnish partners.[6]

From the perspective of the Ministry, the Swedish and Finnish PTOs thus seemed to be the ideal partners for long-term cooperation aimed at a radical modernization of the whole telecommunications infrastructure in Estonia. However, the Finnish PTO, Tele, while agreeing with the general philosophy of such a project, in practice turned out to be reluctant to actually start cooperation with the Estonian Ministry. There were two basic reasons for this: on the one hand, the *financial* risk was seen as considerable for the relatively small Finnish PTO, and on the other hand, there was also a *political* risk involved, as it was not at all clear what the reaction of the central administration in Moscow would actually be. The latter aspect was far from negligible for Finland, whose economy was much more closely linked to the Soviet Union than any other Western country and whose political traditions had set smooth relations with Moscow as its absolute priority since the tsarist days.

A somewhat similar political stance could be sensed in Sweden, where well-functioning relations with central Soviet authorities were the basis for its formal political and military neutrality in the Cold War, a circumstance that tended to overshadow sympathy for the liberation movements in the Baltic states. This was driven to absurdity in 1991, i.e., two years after the fall of the Berlin wall, when the Swedish Minister of Foreign Affairs during a visit to the Baltic states caused a scandal by making overly Moscow-friendly statements. However, in contrast to Finland, the Swedish economy was not as closely linked to the Soviet one, and in addition, the Swedish PTO, Televerket, was a considerably larger and financially more powerful organization than the Finnish Tele – a fact that was taken into consideration by the Estonian Ministry (Mets, int., 2001).

It was therefore with Televerket that the Ministry held its first serious talks in the late 1980s. In 1988, the Estonian Minister of Communications, Toomas Sõmera, met for the first time with Tony Hagström, General Director of the Swedish state-owned operator (Sõmera, interviewed in Kaasik-Aaslav and Maripuu, 1999: 141; cf. *Dagens industri*, 9 November 1988). Sõmera, obviously aware of the difficulties in creating interest in a holistic approach to modernizing an outdated East European telecommunications network, presented his Ministry's vision to Hagström and a number of his colleagues.

Interestingly, these also included representatives from Televerket's most important partner on the equipment side, Ericsson (Sõmera, int., 2003).

The first concrete result of the meeting in fall 1988 was the creation of a joint working group, consisting of five Swedish and five Estonian representatives. It met for the first time half a year after Sõmera's visit to Stockholm. The rapidly changing political situation in the Baltic states made the work somewhat irregular and there were times when it stood still for several months. After the formal decoupling of the Estonian telecommunications system from the Soviet Union in 1990, the situation became a bit easier. Soon afterwards, the joint working group presented a formal report, a hundred-page document analyzing Estonia's general economic and political situation, the general legislation, the existing telephone infrastructure, available skills within the Ministry and elsewhere and, in conclusion, the opportunities for further development and modernization (ibid.).

Finland Entering the Stage

As mentioned above, Finland had early on been interested in far-reaching cooperation with Estonia, but the financial and not least political risks had been considered too high. The Estonian Ministry's early cooperative activities with foreign operators had, therefore, instead been focused on joint work with Televerket. However, apart from the extremely ambitious project aiming at a total modernization of the Estonian networks, the contacts between Estonian and Finnish operator organizations were still intense, although they did not yet involve any business relations. This was true both at the levels of the Ministry and of the many local telephone networks that existed in Estonia and Finland.

With the formal document confirming Estonia's autonomy in the telecom sector in 1990, it proved easier to persuade Finland's Tele to enter larger business projects. As a first concrete project with Finnish involvement, the creation of new international connections between Estonia and Finland came under discussion, and in September 1990 the first direct lines to Helsinki (a microwave link across the Gulf of Finland, using the Tallinn television tower) as well as to Stockholm (via Helsinki) were put into operation (Hultén and Mölleryd, 1997: 172; *Dagens industri*, 17 September 1990). For the first time since before the war, Estonia now had an independent international connection. In January 1991, this was followed by a contract between Tele, Televerket and the Estonian Ministry that presaged the creation of a *mobile telephone network* in Estonia with the help of Nordic investments (Sõmera, int., 2003). As discussed in further detail in section 3 below, the network was launched commercially in July 1991.

With respect to the more ambitious modernization plans, however, developments were slower, and Sweden was still the only foreign partner involved until quite a late stage. After the creation of the new state-owned enterprise, Eesti Telekom, in July 1991, and the official recognition of Estonia's political independence following the events in August 1991, negotiations were intensified. The report produced earlier by the joint Estonian-Swedish working group functioned as a basis for more concrete plans.

In October 1991, the international business press reported from the Telecom '91 Fair in Geneva that Estonia and the other Baltic states were aggressively looking for Western partners for telecommunications ventures (*Telephony*, 21 October 1991). During the meeting, a schooling agreement was signed between Estonia and Sweden, with the purpose of letting Estonian telecommunications engineers and administrators acquire a more thorough knowledge of West European – and in particular Swedish – telecommunications technologies and telecommunications administration (Kaasik-Aaslav and Maripuu, 1999: 142). Toomas Sõmera and Tony Hagström also signed a Memorandum of Understanding, based on the earlier report, foreseeing 'very far-reaching' cooperation between Eesti Telekom and Televerket. Televerket announced that it was to become a major owner of the Estonian telephone administration and that a new company was to be created for this purpose. The intention was to launch the firm within six months (Televerket, 1991).

During the following months, the partners discussed the conditions under which Swedish organizations would take part in the restructuring and modernization. However, the national liberation process had proceeded much more rapidly than had been expected, and after the collapse of the whole Soviet Union in the fall of 1991, Estonia's economy faced an enormous crisis. The whole situation in the country was extremely unstable, and the falling ruble caused unexpected problems in the calculations being made in preparation for Televerket's investments.[7] Economic stabilization and recovery in Estonia seemed impossible before the introduction of a convertible currency in June 1992. Politically, in addition, there was an institutional vacuum due to the absence of a new constitution (formally, political life still proceeded in accordance with the old Soviet institutions). In summary, the risks inherent in a long-term modernization of the telecommunications infrastructure were thus seen as higher than before, which contributed to delaying the results of the negotiations.

Against this background, Televerket decided to turn to Finland, proposing Tele join the modernization project in Estonia and share the risks (Männik, int., 2001; Nurk, int., 2001). Although Finland itself was at this time facing economic recession of a kind that the country had never experienced since the war, Tele had by now become very active in the Baltics. It was deeply involved in negotiations with the Latvian incumbent operator concerning a similar far-reaching cooperation agreement, with Latvia's attractiveness including a larger market and well-developed expertise in services and manufacturing.[8] From the Estonian perspective, the new interest of Finland was welcomed, as it was regarded as a way to limit Swedish dominance in the future Estonian telecommunications sector.

From spring 1992, official reports on how preparations for modernization were proceeding no longer spoke of a Swedish-Estonian joint venture, but of a *trilateral* project, involving both Televerket and Tele as foreign partners.[9] The original idea had been that Eesti Telekom would control 51 per cent of the shares in the joint venture, while Televerket would hold 49 per cent. After Finland joined the project, Televerket's wish was to retain a 40 per cent share, thus making Tele a considerably less powerful partner with 9 per cent of the shares (Sõmera, int., 2003). In the end, however, the ownership structure was agreed to be the same as in the mobile project, in which the contract, signed more than a year earlier, stipulated that Televerket and Tele held 24.5 per cent each.

The somewhat unique outcome of Nordic-Estonian negotiations was thus the creation of a three-country joint venture, Eesti Telefon, between Televerket (Sweden), Tele (Finland) and Eesti Telekom (Estonia). This joint venture would take over the functions that had until then been under the responsibility of the Ministry of Communications. The new company would also take responsibility for the expansion and modernization of the old Soviet Estonian network. The two Nordic operators together held 49 per cent in the new company, and with the help of Nordic finance, a ten year investment program worth $25 million per year was launched, to cover the installation of 30,000 digital lines annually. The goal was to reach a public telephone penetration of 45 per cent (*Telecommunications*, November 1993).[10] In the concession agreement that was signed in December 1992, Eesti Telefon was awarded a monopoly license by the government for fixed voice telephony services in Estonia until the end of the year 2000 (Kontsessioonileping, 1992).

As noted in section 1, there was considerable dispute over the agreement in the government and parliament, as well as within the Ministry itself and the local network organizations. However, Estonia was clearly in need of both capital and new technological and organizational knowledge, and the experience of earlier cooperation with the same Nordic operators tended to reduce the perceived uncertainty. It was believed that Televerket and Tele would have the capacity to enhance and accelerate the transfer of technology and administrative knowledge.

The Nordic partners, however, saw a unique opportunity to enhance their knowledge of the former Soviet telecommunications sector, which could be a strategic advantage for further expansion in the East. For example, Telia would in 1994 refer to its prior Estonian experience as a major reason for winning a Lithuanian GSM license. The foreign partners also hoped that the venture would turn out to generate profit, although this was regarded as highly uncertain – a circumstance that was demonstrated by Televerket's hesitation in entering Estonia as its single foreign partner.

3. THE BIRTH OF EESTI MOBIILTELEFON

In summer 1980, the Soviet Union hosted the Olympic Games. These took place mostly in Moscow, but a number of water-sport events were located in Tallinn and the Baltic Sea. Preparations for these events provided Estonia with an extra source of funds from the central authorities in Moscow, which enabled a new television tower to be built in Tallinn. In addition, Estonia experienced its first major encounter with a somewhat exotic technology: mobile telephony. During the Olympics, terminals for the 'Altai' system for analogue mobile communications could be observed in Tallinn. The system was a Soviet development based on manual switching and served first and foremost the needs of important government representatives, but during the Games the mobile service was also offered to journalists. The terminals consisted of huge boxes that were mounted onto the luggage boots of cars (Viira, int., 2003; Ots, int., 2003).

In the Nordic countries, the main public telecom agencies had been experimenting with somewhat similar mobile solutions since the end of the 1940s, though with limited success. Similar to the Soviet Altai system, the Nordic solutions suffered from complicated, heavy and expensive terminal equipment. By the time of the Moscow Olympics, however, the Nordic countries were about to adopt a considerably more advanced mobile communications system, which was ascribed dramatically brighter future prospects. The Nordic Mobile Telephony (NMT) system, as it was called, had emerged largely out of Nordic cooperation, in which both operators and equipment producers from several countries took part.

Towards the end of the 1980s, NMT had become a striking commercial success in the Nordic countries, with the millionth subscriber joining the networks in January 1991. In Sweden, the penetration rate was at this time reaching 6 per cent of the population, a remarkable figure in view of the fact that it was still a matter of *analogue* mobile telephony. At the same time, further development of NMT to the digital standard GSM proceeded rapidly in the latter half of the 1980s, and in 1991 the Finnish private operator Radiolinja put into operation the world's first GSM network.

As discussed in the preceding section, when the Soviet grip of Estonia started to loosen towards the end of the 1980s, it became common for local telephone networks and large PTOs to initiate contacts with each other across the Baltic Sea. From this perspective, it is hardly surprising that the Estonian Ministry of Communications was well aware of the Nordic experiences in the field of mobile communications and of the emerging success of NMT. Nevertheless, the Nordic partners were often surprised by the enthusiasm for the future of civilian mobile telephony that the Ministry of Communications of the tiny Soviet republic showed (Makkonen, int., 2003). The Nordic world's advanced technological and market developments inspired the Ministry to investigate the possibilities of introducing civilian mobile telephony in what was still the Estonian SSR.

The issue was discussed during an informal visit of the Estonian minister, Toomas Sõmera, to Finland in 1990. When Sõmera was introduced to leaders of mobile telephony developments at Tele, he suggested an interesting and unconventional idea: Estonia would adopt NMT as the mobile technology, but the Estonian Ministry would not build its own NMT network proper; instead, the small Estonian territory – and initially only Tallinn – would function technically as an extension of Tele's own NMT network, with Estonian calls being routed through the NMT exchange in Helsinki (ibid.). For the Estonians, the technical demands would then be relatively low: what was needed were in principle only a couple of radio base stations, the most important one of which could be mounted on top of the television tower in Tallinn. The financial risk inherent in the project seemed low.

The potential establishment of an NMT-450 mobile telephony system also seemed to form a somewhat unconventional but clearly interesting solution to the problem of reliable business communications – especially at the international level. The demand for smooth international telephony was growing as a result of the new economic policy in the 'perestroika' years, during which a substantial number of foreign investors and entrepreneurs had established or were about to establish themselves in Estonia, especially in its capital Tallinn. Most of these foreign investors came from across the Gulf of Finland and the Baltic Sea – from Finland and Sweden – where the NMT-450 mobile standard was already a well-established technology experiencing exponential growth in the number of users.

One of the interesting aspects of the NMT standard was its *roaming* function, letting, for example, Swedish NMT subscribers use their mobile telephones in Finland. The system was designed for international use from the beginning, and the extension of the Nordic roaming area to Estonia, using the same mobile standard, was thus from a technical point of view a logical step (Tee, int., 2001). If realized, such an extension would enable Nordic (primarily business) visitors to use their mobile handsets in Tallinn in the same way as in their home countries. Conversely, Estonians would be directly accessible while abroad in the Nordic countries. This had an obvious attraction from the perspective of building a receptive environment to foreign business in Estonia and re-establishing East-West trade across the Baltic Sea.

Two main problems appeared: how to get permission from Moscow to use the radio frequencies, and how to organize and finance the project. Although Sõmera during the discussions in Finland signaled a strong interest in joining forces with Tele, the minister was also, as a matter of fact, in close contact with the mobile department of Televerket in Sweden. Thus the Estonians appear to have wished to involve *both* Finland and Sweden in the project from a very early stage (Makkonen, int., 2003).

Before the end of 1990, at a time when it was anything but clear that Estonia would be able to achieve the political goal of independence from the USSR, the NMT project began to take shape. The Estonian Ministry hired the general and technical directors for the future company. In January 1991, a basic agreement was signed, creating Eesti Mobiiltelefon as a joint venture between the Ministry of Communications and the two Nordic state-owned operators, Televerket (Sweden) and Tele (Finland). The Estonians held 51 per cent of the company, while the two Nordic partners each held 24.5 per cent – i.e., the same ownership structure as in Eesti Telefon, which was founded almost two years later. On 28 April 1991, the new firm was formally registered (EMT, 2001a).

EMT rented a few rooms in the Ministry's main building in Tallinn, and employees were hired, including Estonians, Swedes and Finns. From the beginning, however, the Nordic partners tended to act more as 'consultants' than actual directors at EMT. This was a deliberate policy that had been insisted upon at an early stage by the Estonian minister, and the driving personalities from the side of Televerket and Tele seem to have accepted this idea without much hesitation (Makkonen, int., 2003). Tele and Televerket agreed that the Finnish partner would provide *technical* consultancy, whereas the Swedes would be responsible for the development of the *management* side. The main Swedish representative was one of thousands of people with Estonian origins living in Sweden, a fact that in view of common language problems at this time was not to be underestimated (Tee, int., 2001).

The technologies that were used in the new network were based on the NMT-450 standard and the equipment was supplied exclusively by Ericsson – except for the handsets, which in most cases initially came from Nokia. The choice of Ericsson as main supplier appears to have been somewhat controversial in Estonia, but once Ericsson had been chosen, this turned out to have far-reaching consequences for the further development of the whole mobile sector, as will be discussed further on.[11] Similarly, in view of the fact that Ericsson Finland had been doing business directly with the Estonian Ministry for more than a year, during which an obvious mutual confidence had developed (Lehtimäki, int., 2001), it is hardly surprising that the final choice fell to Ericsson, as will be discussed in section 4. This is the more obvious in view of the problems that arose between Nokia and its Estonian customers over the issue of fixed networks (see section 4).

During the spring and early summer of 1991, the first few radio base stations were installed in Tallinn, and the first client joined the system on 5 July, i.e., a few weeks before the dramatic August events. Within the first month of operation, eight more clients joined the network (*EMT Infoleht*, no. 2, 1996: 1). The system became technically part of Tele's NMT-450 network in Finland, all calls being channeled to the mobile exchange in Helsinki across the Gulf. The first mobile phone numbers in Estonia had thus a Finnish prefix, and subscription contracts were even written in the Finnish language (EMT, 2001a).

Despite the very low number of subscribers in the early stage of EMT's activity, the network came to play an important role sooner than anybody had imagined. When Soviet troops entered Estonia in August 1991, all long-distance fixed lines were cut off, isolating the republic from the rest of the world – but the NMT system kept working both domestically and internationally. The separatist Estonian government was at this time spread out, with the prime minister temporarily residing in Stockholm and other ministers in Helsinki and Tallinn. Through the NMT network with its international roaming function, the government's activities could nonetheless be coordinated (Makkonen, int., 2003).

By the end of 1992, EMT could be said to have concluded its formative stage. The company had at this time around 2,500 clients, a figure which one year later had already grown three-fold, making Estonia, with 4.8 mobile phones per 1,000 inhabitants, the leading mobile telephone country among the former socialist states in Central and Eastern Europe (EMT, 1993). As a step in preparing for further growth, the supervisory council of EMT, consisting of representatives from the Nordic partners, initiated a number of far-reaching changes in the management board. A new general manager entered the company in December 1992, and in mid-1994, the new technical and R&D director arrived. The company had by that time acquired a structure that proved stable for many years to come. The developments that followed are discussed in detail in Chapters 6 and 7.

4. THE REORIENTATION OF THE EQUIPMENT SUBSECTOR

The Collapse of the Soviet Electronics Industry in Estonia

The effects of Gorbachev's 'perestroika' policy, which explicitly emphasized the role of innovation and hi-tech production for the future of the Soviet Union, seemed – at a first glance – to have a beneficial impact on the Estonian electronics sector. The financial situation of the state-owned research organizations tended to improve, and the links between the academic community and local industry were strengthened. For example, in 1985 a new research laboratory was created jointly by the Tallinn Polytechnical Institute (TPI) and the radio set producer Punane RET (Min, 2000a: 46).

The IME program, too, which aimed at re-establishing economic autonomy for Estonia within the USSR, reflected a great Estonian interest in hi-tech production as an important source of future economic development. It aroused great interest in the republic when EKTA, the Special Construction Bureau of the Institute of Cybernetics, presented its in-house development of a school computer, and Estonian industry impressed the whole Soviet electronics sector with Punane RET's new high-quality version of the 'Estonia' radio, which included the Soviet Union's first CD player (*Tehnikamaailm*, January 2003).

This upturn in the Estonian electronics sector seemed to be in line with the strong general focus on innovative activities in the USSR during Gorbachev's administration, as shown, for example, by the recovery of Soviet patents received in the US from 1988 onwards after a long period of stagnation (see Figure 5.1 below). Among the organizations filing foreign patent applications in these years belonged both Punane RET and several TPI departments (Min, 2000a: 46f.).

This positive trend, however, was abruptly broken by the erection of totally new national borders between Estonia and the other Soviet republics and the collapse of the Soviet Union itself that soon followed in late 1991.

Since 1944, Estonian industry had been designed almost exclusively as an integrated part of the all-Union economy and had been largely governed from Moscow. The tiny Estonian republic had seen a spectacular growth of immense production enterprises such as Punane RET, the Pöögelmann works and the Kalinin electronics factory, all with markets that lay almost totally outside the borders of the Estonian SSR (but out of reach of the Western world). Similarly, although implementation-based R&D organizations such as the Tallinn Radio Electronics Construction Bureau (TREKB) of Punane RET had been formed in the backwater of the Estonian enterprises' successes on Soviet markets, their main R&D base was still heavily concentrated on organizations located *outside* Estonia. The same was true for the supply of components, where an enterprise such as 'Tartu Apparaaditehas' (which was the successor firm of the old telephone factory) could have more than 500 Soviet suppliers, most of them in Russia (Noorem, int., 2001).

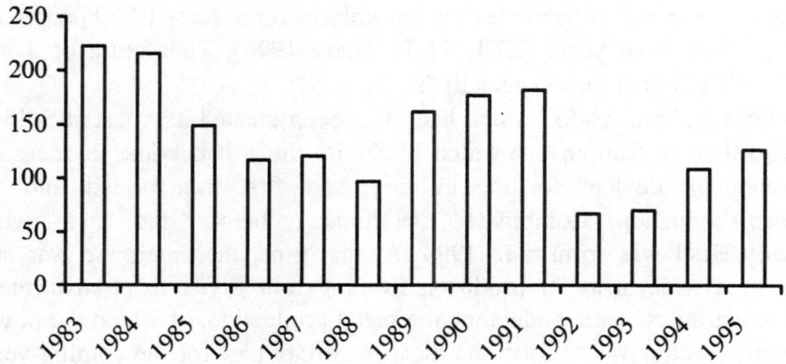

Note: Starting from 1991, the numbers are counted as the sum of Soviet and successor states' patenting.

Source: US Patent Office.

Figure 5.1 Soviet patenting in the US, 1983-1995 (number of patents annually)

In short, the Estonian electronics sector was deeply part of the all-USSR system of innovation. Even the IME program, while seeking to redirect economic decision-making towards local enterprises, had sought to profit from a retained all-Union market, with the possibility of gradually entering Western markets. The dramatic collapse of the Soviet market, however, meant that the very basis for this vision was removed long before Estonian enterprises had been able to redirect themselves towards the West. The result was disastrous.

Following the formal Estonian re-independence in August 1991, the introduction of a new convertible currency in June 1992 and new legislation, the former Soviet enterprises were transformed into Estonian state-owned companies, which were later to be privatized. In particular, it was hoped that *foreign* investors could be found. The situation seems to have been most critical for those enterprises whose main market had been within the Soviet military sector. To these belonged, for example, the Pöögelmann works, whose main production consisted of microprocessors for military use. The enterprise was transformed into the state-owned firm 'Tondi Elektroonika' in October 1992, but by then the profile of the firm had already changed dramatically: as a result of a massive reduction in production, the main product group now consisted of electronic hearing aids (formerly only a small part of its production). The goal was to enter West European markets with this consumer product (*Äripäev*, 31 October 1992). However, the company did not succeed in attracting a foreign investor, and the shares were instead sold to an Estonian holding company, 'AS Mikrel', in February 1994. The new owner had to guarantee the workplaces for at least 1,050 people for the following three years (*ETA*, 21 February 1994). The future of Tondi Elektroonika did not look especially bright.

The old Kalinin works, which had once been created as a tsarist railway car workshop in Tallinn but which in Soviet times had come to focus on semiconductor devices for the military, was first transformed into the 'Science-Production Combinate Electronics', before the state-owned company Estel was formed in 1994. At that time, the enterprise was still active in a wide array of products, forming four divisions: transformers, power electronics, semiconductors and metal production, of which the power electronics sector was to form the most important part for the coming years (Toomas Kleement, Chairman of the Board of Estel, in *Äripäev*, 15 August 1996). However, similarly to Tondi Elektroonika, Estel did not succeed in attracting a foreign investor, and the company was in 1996 instead sold to 'Estel Pluss', a firm created by Estel employees. By that time, 420 people were still working in the company (*Äripäev*, 9 February 1996 and 15 August 1996).

In Tartu, the remnants of the old telephone factory had in Soviet times formed the foundation for another military-focused production enterprise, which after the war became well-known – despite its secret status! – for its mass-production of flight recorders ('black boxes') for Soviet aircraft. In 1991, 2,500 people were still working in the Tartu factory, relying on the well-developed Soviet aerospace sector as the main market. The enterprise was transformed into the state-owned company Tarkon in December 1992 (Tarkon, 2003). The big contracts with the Soviet aerospace sector were gradually lost, however, as Russian aircraft producers reoriented themselves towards domestic suppliers of flight recorders. The company realized that it had to search not only for alternative markets, but also alternative production. In its search for strategic Western partners, contacts were first established with Swedish and Finnish firms. In particular, close cooperation began with the Swedish mechanics firm Hallberg-Sekrom in 1993, which would later turn out to have important consequences.

Tondi Elektroonika, Estel and Tarkon were examples of Estonian electronics firms that continued to exist after Soviet collapse, although massive reductions in personnel and turnover proved unavoidable. The first two continued to be bound mainly to the former Soviet market, whereas

Tarkon, as will be discussed in further detail in the next chapter, reoriented itself radically towards the West, integrating itself with the Nordic countries and focusing more and more on the booming mobile telecommunications equipment industry. However, there were also numerous electronics firms that did not manage to survive at all. Around 1988, a total of more than 25,000 people had been employed in the Estonian electronics sector. In 1993, there were only around 3,000 left (estimation by Rang, int., 2003).[12]

The Establishment of Elcoteq

One of the former Soviet enterprises that faced seemingly insurmountable problems after 1991 was Punane RET, the former flagship of the Estonian electronics industry. It had strong symbolic importance due to the fact that it had been founded in the first period of national independence, and its old name RET – 'Raadio-Elektroonikatehas' – was assumed by the company after Soviet collapse. Consumer electronics, however, was a branch that faced particularly strong competition from Western and Asian firms immediately after the opening of the former socialist markets, and it soon became obvious that RET would have very little chance to survive in this market in the absence of a foreign investor.

In early 1992, a newly founded Finnish electronics company, Elcoteq Network OY, became interested in RET. Elcoteq Network originated from within the Finnish Lohja Corporation (located in the city with the same name close to Helsinki), as a unit specializing in the manufacture of densely mounted control electronics for Lohja's ambitious project to develop flat displays. This project, however, was never completed, and the company therefore came to look for external customers, discovering that both Ericsson and Nokia were interested in their expertise as a way to solve their problems of assembling small components in mobile phones. This led to the spin-off of the unit from Lohja in 1991, forming the independent firm Elcoteq, which was to act as a subcontractor to large telecommunications producers (information from Elcoteq, 2003).

Several trends in the global telecommunications system of innovation seemed to come together here: on the one hand, the mobile telephony equipment subsector was growing rapidly, increasing the demand for Elcoteq's expertise. On the other hand, the overall trend during these years towards outsourcing in electronics manufacturing was beginning to reach telecommunications, although this was still a quite unusual phenomenon at the time of Elcoteq's creation. In addition, the emerging transformation processes in the former socialist economies indicated, somewhat unexpectedly, that a massive, relatively skilled but low-cost labor force was now available just 80 km to south of Helsinki. This was in Elcoteq's eyes an interesting alternative to the potential expansion of their production that the company had originally envisaged for China or South-East Asia (Petersen, int., 2003).

Elcoteq was thus aware of the existence of an inherited Soviet electronics sector in Estonia. After an investigation of the current situation in electronics manufacturing in the country, Elcoteq reached an agreement with RET in March 1992, according to which the Finnish company was to rent a smaller part of RET's facilities, including a limited number of the work force. With a staff of ten workers, operations thus started carefully at a test-scale (producing electronics parts for the sewing machines of Sweden's Husqvarna) with the goal of investigating opportunities for a larger investment (ibid.). The decision to initiate large-scale production in Tallinn was taken one year later, which gave rise to the creation of the Estonian subsidiary Elcoteq Baltic (which in 1995 was renamed Elcoteq Tallinn). The company took over many of the employees from RET as well as from other old electronics firms, such as the former Töögelmann (i.e., Tondi Elektroonika) and Kalinin (i.e., Estel) works (information from Elcoteq, 2004; *Äripäev*, 28 April 1997).

Elcoteq's Tallinn facility thus grew to a full-scale production plant with several hundred employees, in which production came to be directed towards the Finnish-owned company's most important customers in the telecommunications sector, Ericsson and Nokia. By 1995, the Tallinn subsidiary had grown to become Estonia's largest exporter. The developments of the company from 1994 and onwards will be discussed in detail in Chapters 6 and 7.

Reorienting Equipment Procurement

From a demand-side perspective, the Ministry of Communications was a key user of electronics-related equipment in the Estonian SSR. In the interwar period, as pointed out in the preceding chapter, the Estonian operator had been an important partner of the local telephone factory in Tartu, but since 1944 no telecommunications equipment production existed in Estonia, and the demand of the Ministry of Communications had instead been forcefully channeled exclusively towards non-Estonian equipment suppliers.

Moreover, as discussed in Chapter 4, the inter-organizational linkages that connected the Estonian network operator with foreign (Soviet and East European) equipment suppliers were generally very weak and, from the perspective of technological change, highly troublesome. The re-independent Estonia therefore did not inherit any dynamic linkages between the equipment and service sectors. The positive side of this was that it was not difficult to break the existing procurement regime of the Ministry, which, as pointed out in section 2, even deliberately postponed the installation of new Soviet equipment because of the expected emergence of new procurement opportunities.

It was not self-evident, however, from where the equipment was to be procured, given the new independent position of the Ministry, which from July 1991 acted as the self-standing state-owned company Eesti Telekom. The interwar tradition of telecommunications production in Tartu, and the image of the Estonian SSR as a hi-tech producer of electronics in the Soviet Union made it natural for the Ministry to investigate the opportunities to re-establish telecommunications equipment production in Estonia itself. In the years around 1990, this was seen by the Ministry as a way to use its own demand to stimulate the transformation of Estonia's electronics industry – a strategy that was still used in many Western countries, where for example Sweden's Televerket was deeply involved in Ericsson's successful entrance into the mobile equipment industry. The formulation of this strategy, usually referred to as 'public technology procurement', can be regarded as a policymaking activity (cf. Edquist et al., 2000).

In 1988 and 1989, the Estonian Ministry started to look for strategic partners among the large Western telecommunications producers, the idea being to transfer Western technology and knowledge to Estonia and use this to re-establish a modern telecommunications equipment sector. Talks were held with Nokia, Ericsson, Alcatel, Northern Telecom and other large manufacturers. In 1989, through mediation of a German friend, the Estonian Minister together with the head of the Tallinn telephone network had the opportunity to speak directly at Siemens' headquarters in Munich, whereby an agreement for the purchase of PBX stations was concluded.

However, it soon became clear that the realization of the idea of an Estonian telecom manufacturer presupposed that such a factory would be able to *export* a large part of its produce, in order to achieve economies of scale – and it appeared unrealistic that Estonia would be able to establish itself as a telecom manufacturer on the global market (Saarma, E., int., 2001; Mets, int., 2001; Sõmera, int., 2003). The reasons, as interpreted by the Ministry, were diverse. Opinions differed as to the suitability of the existing knowledge base in Estonian electronics, but in general it appeared to the Ministry that the gap between the Soviet-Estonian technological capabilities and those asked for by the Nordic operators would require expensive and time-consuming efforts to adapt the local firms to the modernization project. This was obvious to international manufacturers, who in the end preferred to establish themselves in countries where the local markets were larger (such as Russia and Poland), expecting the domestic producers to be able to satisfy the domestic Eastern markets but hardly able to export anything to the West.

The prospects were more promising with respect to the *import* of ready-made equipment from international suppliers. Such activities started in 1990. Nokia had, as a matter of fact, already in 1986 delivered a DX-200 exchange to the Estonian SSR, but this was an agreement that had been signed in Moscow between Nokia and the central Soviet Minsviaz, and was thus not a direct contract between Finland and Estonia. During the following years, however, it became popular among the local Estonian telephone network organizations to establish contacts with local telephone networks in Finland. Estonian representatives visited various local Finnish networks, and in some cases Finland sent parts of its old equipment as development aid to Estonia (Mets, int., 2001; Saarma, E., int., 2001).

It was not until 1990, however, that the first *business* relations were established in the equipment field. In that year, the Estonian communications minister visited Ericsson's Finnish subsidiary in Helsinki. Contacts had existed at personal levels long before, and the more official meeting that now took place combined personal and business interests. Representatives of Ericsson Finland, in their turn, were invited to Estonia, and in June 1990, the Estonian Ministry signed its first business agreement with Ericsson Finland, for the delivery of an X.25 data communications system (see the section on data communications below) (Lehtimäki, int., 2001).

The agreement was controversial within Ericsson and came as a surprise to the company's headquarters in Sweden, where the general attitude towards Eastern Europe at this time seems to have been much more hesitant than at Ericsson Finland, as a result of the political and economic instability in the region: the financial status of the Estonian Ministry of Communications was anything but clear, and it would take another two years before a convertible currency was introduced in Estonia (ibid.). Ericsson Finland, more eager to take the risk of entering the Estonian market, 'solved' the problem of currency by offering the Estonian Ministry a postponement of the payment for the X.25 network for three years (Sõmera, int., 2003).

With the benefit of hindsight, it is clear that these very early contacts established with the Swedish telecommunications giant through its Finnish subsidiary were significant for more far-reaching developments, which were to follow in the coming years and which will be discussed in detail in Chapters 6 and 7. The smooth relations at personal as well as organizational levels made Ericsson a natural choice of partner and equipment supplier by the Estonian Ministry of Communications in its first large-scale projects.[13]

Estonia's relations to Ericsson's competitor Nokia were totally different. Nokia suffered from its reputation as collaborator and ally with Moscow, an impression strengthened by its failure to deliver a second DX-200 exchange to Tallinn in the summer of 1990. The equipment had been ordered long in advance, but in the meantime Estonia had, as pointed out above, seceded its telecommunications sector from the USSR. In line with the established model of Nokia's business in the USSR, the transaction was organized via Moscow, whereby the central Minsviaz, given the new situation, decided to redirect the order so that the DX exchange was not sent to Tallinn but to a Siberian network. The Estonian Ministry was not informed, which created tensions with the Nokia headquarters in Finland (Sõmera, int., 2003).

Ericsson, as well as other international equipment vendors, (re-)opened offices in Estonia, whereby historical ties between the Swedish company and Estonia were deliberately used to increase its attractiveness. For example, in December 1992, i.e., at about the same time as Eesti Telefon got its concession, Ericsson's sales office in Estonia, the firm Baltimore, organized a seminar in Tartu on Ericsson's technologies and their potential future use in Estonia, with the motto being: *'You remember what Ericsson used to be; come and see, what Ericsson is now!'* (advertisement in *Postimees*, 9 December 1992, my translation).

5. DATA COMMUNICATIONS: FROM AKADEMSET TO THE INTERNET

Chapter 4 discussed data communications in Estonia in the context of its emergence from automation and control systems. Advanced R&D was carried out in this field in Estonia, especially at the Institute of Cybernetics. As pointed out, the Institute was an 'open' R&D organization in the sense that its activities were not subject to the secrecy of organizations more closely related to the military complex, and it had thus been able to build a fine – though thin – international network.

In addition to academic East-West cooperation, researchers from the Estonian Institute of Cybernetics were sporadically working directly with Finnish business firms, a trend that appears to have resulted from Finland's unique economic ties to the Soviet Union in general. For the emerging field of data communications, the most notable example of such cooperation was the Estonian participation in the Soviet-Finnish joint venture Elorg Data in Helsinki. The main shareholder in the joint venture was the Finnish oil company Teboil, and the actual work of Elorg Data was focused on developing software for Teboil's petrol stations in Finland. This had in itself little to do with data communications, but the participation of the Tallinn-based Institute of Cybernetics in the venture suggested possibilities of establishing electronic communication between the Institute (in Tallinn) and Elorg Data (in Helsinki) (Wõrk, int., 2003; Institute of Cybernetics, 1990: 20).

As late as in the mid-1980s, it had been virtually impossible to get permission from Moscow to establish a direct data communication link between the Estonian SSR and a foreign country, but with the emerging 'perestroika' it was attempted anew and with more successful results. Thus a leased line could now be acquired, which went into operation at the end of 1988 and through which e-mails and data files could be sent at the (admittedly modest) speed of 9.6 kbps (ibid.).

Behind this project was the Estonian wish to establish a connection to international databases and the emerging *Internet*. So far, the only possibility for Estonia to access Western networks was through a single minimal 9.6 kbps channel in Moscow, which was shared by *all* (!) Soviet users and in which usage was strictly controlled by authorities in Moscow. Until 1988, this 9.6 kbps channel was the microscopic window Estonian data communications had to the Western world. Using the leased line to Elorg Data in Helsinki, an alternative connection to the outside world became technically possible. However, Elorg Data did not manage to get the corresponding permission from the Finnish government, which, as discussed above, was extremely careful in its relations with the Soviet Union (Wõrk, 1997: 17). Nevertheless, some actors involved argue that the connection was *actually* used for accessing foreign data networks (Institute of Cybernetics, 1990). In any case, the connection was closed in April 1990, officially because of its high cost (Wõrk, 1997: 17).

On the more academic side, Estonia's cooperation with Tampere University of Technology was more fruitful in bringing international data communications to the Estonian SSR. In fall 1988, Soviet reforms and Estonian-Nordic academic links had proceeded so far that it was possible to create an academic Estonian-Finnish Committee on Informatics, the goal of which was to build direct cooperative links between Estonia and Finland (i.e., not via Moscow) in the field of informatics. The first meeting of the committee took place in Tallinn in November, drawing participants broadly from the universities of Helsinki, Tampere and Turku and, from the Estonian side, the Institute of Cybernetics, the University of Tartu and Tallinn Technical University (*Arvutustehnika & Andmetöötlus*, no. 3, 1989: 38).

The need for better electronic communications was discussed at several committee meetings, and activities in this field were intensified after the shut-down of the leased-line to Finland via Elorg Data in 1990. The Finnish academic partners acted as mediators between the Institute of Cybernetics and the relevant Finnish authorities, who, given the Moscow-friendly politics in Helsinki, were not necessarily willing to allow a Soviet republic to use Finland as a gateway to international data networks.[14] With the help of the Technical University in Tampere, the Institute of Cybernetics managed to obtain permission to be linked via a dial-up connection to the FUUG and

FUNET networks in Finland. The Finnish side delivered the necessary equipment – Nokia modems and an auto-dialer – to the Institute in September 1990 (Wõrk, 1997), but it was the researchers at the Estonian institute who put the technology into operation (Tammet, 1996). The service went online in December 1990. For 400 rubles, it was also possible for non-academic Estonian users to link to the Institute of Cybernetics and send and receive e-mails as well as read news from the emerging Internet newsgroups (ibid.).

From this moment Estonia was basically part of the Internet in the wider sense of this concept, albeit still using the Soviet domain '.su'. Some observers argue that this Internet connection, similar to mobile telephony, played an important role as a channel to the outside world during the dramatic events of August 1991; while the normal telephone lines were cut off, it was still possible to use data communications and send e-mails to the West.

At the same time, however, both the Institute of Cybernetics and the Institute of Chemical Physics and Biophysics (KBFI) worked intensely to introduce the Internet into Estonia in its more narrow sense, i.e., building upon the TCP/IP protocol. This project was much more expensive, as it required sophisticated hardware (such as Internet routers). The dominant suppliers in this market were American firms, and due to the CoCom regulations aimed at hindering hi-tech exports to the Soviet Union, the availability – not to mention the affordability – of such equipment was far from self-evident. In June 1991, however, representatives from the three Baltic countries – in reality the same people who had been the national representatives in the Soviet-era Akademset project (see Chapter 4) – took part in the first worldwide Internet conference (INET '91), held in Copenhagen. During the conference, the Baltic representatives met with Nordic colleagues, who in turn introduced them to people from UNESCO. It turned out that the Baltic republics had a good chance of obtaining considerable financial support from UNESCO for establishing a common TCP/IP backbone network for the three countries. For UNESCO, it was important that the three republics act together, and such cooperation was natural for the Baltics, having already been involved jointly in connection with the creation of Akademset's X.25 network.

KBFI, for their part, in cooperation with the Biocenter at the University of Tartu, also sought to introduce the Internet in Estonia, though starting from a totally different ideology: they had close traditional collaboration with the Royal Institute of Technology (KTH) in Stockholm, and together they had the idea of connecting the Estonian university towns, Tartu and Tallinn, via *satellite* to the Swedish academic network, SUNET. At the Institute of Cybernetics, this idea was not considered viable due to the enormous relative costs of using satellite connections to such a nearby country as Sweden –

especially as the Institute was aware of Eesti Telefon's ongoing realization of new fiber-optic links to the Nordic countries.

In the volatile fall of 1991, representatives from the Institute of Cybernetics and KBFI flew together to Paris to present their applications to UNESCO. In March 1992, it was decided that the suggestion of the Institute of Cybernetics was to be supported, but not the one of KBFI. However, the latter managed to acquire finance from other sources: the Soros Foundation granted money for a six-month satellite connection to Sweden, and the Swedish BITS Foundation granted another six months. As it was much easier – though more expensive – to establish a satellite link, KBFI was the first organization to establish a true Internet connection between Estonia and the West; it was opened in April 1992 at a ceremonial meeting with representatives from Sweden and Estonia (Tammet, 1996: 93). This made it natural for KBFI to become the general administrator of the Estonian educational and research network. KBFI created the Estonian domain '.ee' and formed the organization EENet as a counterpart to the Swedish SUNET and the Finnish FUNET (ibid.).

The Institute of Cybernetics, for their part, would soon establish its own Internet connection to the Nordic world when Eesti Telefon opened a new powerful microwave link to Helsinki, which went online in May 1992, i.e., only a few weeks after the establishment of KBFI's satellite connections. However, the opening up of EstNet, as the Institute's network was called, followed only in January 1993, as there were problems getting the necessary permission to import the Cisco equipment. When a new link between Tallinn and Tartu was put into operation, and the grant for the satellite connection depleted by spring 1993, the satellite was shut down. The links between Tallinn and Helsinki were also strengthened considerably through the opening of a new fibre optic cable across the Gulf of Finland at about the same time (information from EsData, 1994).

The Institute of Cybernetics had earlier regarded itself as the natural administrator of the Estonian educational and research data network, but after the 'battle' with KBFI it changed direction in an interesting way, emerging as the first *commercial* Internet access provider in Estonia, thus entering a new market. This started with contacts between the Institute and the Finnish FUNET. FUNET was the organization that arranged the international connections to Finland for the Institute of Cybernetics, but when the new microwave link opened, the Finnish Tele, which distributed the international lines, initiated contact with the Institute and proposed more far-reaching direct cooperation between Tele and the Institute. The idea was to establish an Estonian-Finnish joint venture aimed at offering high-quality data communications services to large firms and other organizations in Estonia, and the rationale was the articulation of demand from a rapidly increasing number of subsidiaries of large Finnish and other foreign firms in Estonia. Tele had for some years been offering the Finnish-developed protocol-free so-called FASTNET service in Finland, and with the establishment of the new microwave link across the Gulf of Finland it seemed natural to extend the service to Estonia. The main idea was to enable smooth linkages between Finnish firms' headquarters and their Estonian subsidiaries, but the rise of the Internet was also taken into account (Wõrk, int., 2003).

At this time, the newly established Eesti Telefon was also gaining interest in data communications. The involvement of the Ministry of Communications in creating an Estonian X.25 network linked to Akademset was continued in 1990 through the application of more modern X.25 technologies from Ericsson (Eripax), as noted above. The network was called Estpak, and in connection with the formation of Eesti Telefon in December 1992, a wholly-owned subsidiary, Estpak Data, was created dedicated to the X.25 network (*Baltic IT Review*, no. 1, 1996: 57). In 1993, Estpak Data started to show some interest in the Internet, but more from the perspective of Eesti Telefon's internal use than for commercial application (Martens, 1997a: 67). However, although data communications services provision was not defined as a core business of Eesti Telefon, it was interested in keeping an eye on rapid developments in this field.

As the front line in Estonia in this respect was represented by the Institute of Cybernetics, Eesti Telefon expressed interest in joining the collaboration between the Institute and Tele. Thus, in the beginning of 1994, the new firm EsData was formed as a joint venture between the Institute of Cybernetics, the Finnish Tele and Eesti Telefon.[15] It was mainly directed towards the Estonian FASTNET extension, but it also offered dial-up Internet access. However, the market for the latter was still close to zero; it would require the appearance of the world wide web and the Netscape browser before the Internet appeared on a larger scale in the former Soviet republic.[16] But when these decisive global innovations arrived, Estonia was well-prepared to quickly take them into use – and even to engage extensively in their further development. This will be elaborated in detail in Chapters 6 and 7.

6. THE EMERGENCE OF SMALL TELECOMMUNICATIONS-RELATED FIRMS

The preceding sections outlined the development of network operators, equipment suppliers and their partners. Up to the mid-1990s, these actors may in a global perspective be regarded as the core of any telecommunications sector. The paradigm shift in the 1990s in connection to the emergence of the Internet, however, implied that a number of whole new layers of actors have become relevant for innovation in telecommunications. Such actors are smaller IT and related companies, which in the course of the 1990s gradually emerged both as partners and competitors of traditional actors in the telecom sector. This section discusses the origins of these new actors in Estonia.

The failure of Gorbachev's new program for the computerization of Soviet society, which was adopted in 1987, resulted in a disastrous undersupply of computing facilities. Critics argued that the program was much too isolationist, that its main weakness was the total reliance placed upon domestic innovation capacities and Soviet technological trajectories, and that the program should have been much more open to international collaboration, incorporating the Eastern countries into the ICT industry's international division of labor (Adirim, 1991: 665). Obviously, even if general stagnation meant that the Soviet bloc fell further behind during the 1980s, the spotty achievements of the East promised some potential mutual benefits from such an international division of innovative labor.

Internationalization was exactly what followed from the late 1980s, though perhaps not in the way that Soviet observers had expected. In particular, the undersupply of domestic computers stimulated an increased trade with ready-made Western computer and electronic component manufacturers – i.e., as far as the CoCom regulations allowed – and around 1990 it was estimated that the imports accounted for around 40 per cent of PCs in the USSR (ibid.: 662). This trend was further strengthened by the decrees issued in 1987 that allowed international joint ventures. Joint ventures became a way for Western manufacturers to enter Eastern markets, or to outsource software production to Soviet subcontractors (who were attractive due to their advanced skills and low labor costs).

In Estonia, both these developments – foreign firms seeking to enter Eastern markets and Eastern software firms doing subcontracting work for customers in the West – contributed strongly to the rise of a completely new subsector within the field of information and communications technologies. This subsector was characterized by new small firms engaging in the import and retailing of Western equipment and software; computer assembly based on imported components; service activities such as installation and maintenance; the creation of a wide array of software mainly for foreign private or local public customers; etc. A massive wave of entry followed in the years around 1990, which also involved far-reaching networking activities with both foreign and domestic firms.

This trend was important for later developments in the telecommunications system of innovation. It is not within the scope of this study to reconstruct the development of the entire small-firm segment relating to IT in Estonia, but a few cases – to which there will be reason to return in the ensuing chapters – should be discussed briefly here to illustrate the sectoral landscape in which certain actors were formed that would later play important roles for innovation in telecommunications services and equipment in Estonia.

MicroLink, a company that would later grow to become a dominant player among Baltic IT firms, was founded in August 1991, i.e., in direct connection to Estonia's re-independence. It was based on domestic capital. The business idea was to produce personal computers through the assembly of imported components. The first MicroLink computer was put together in March 1992 (information from MicroLink, 2001). Despite the general hi-tech profile of the computer sector, the company could at this time hardly be seen as innovative in any creative sense; the production of personal computers through assembly did not necessarily have anything to do with knowledge- or capital-intensive activities, but was rather a business where firms could become 'national champions with a screwdriver' (Bitzer, 2000: 257).[17] The most important knowledge was highly codified and directly available through the purchase of foreign components, and the Estonian PC industry could thus emerge without being dependent upon any important links to Soviet era R&D facilities (such as the Institute of Cybernetics – see below). To the extent that innovation did occur, it was a matter of minor adaptations.

Baltic Computer Systems was established in January 1989 with aims to provide IT solutions for local governments in Estonia. The initiator was the Tallinn city government, which in the preceding years had established contacts with its counterparts in Stockholm and Helsinki. After the Soviet decrees in 1987 paved the way for creating Soviet-Western joint ventures, it became possible to use existing Estonian-Nordic links in more concrete ways in the field of governmental IT systems. It was with the dominant Swedish municipal provider of IT systems, Kommundata, that the Tallinn city government established a business joint venture in 1989. It was owned 50 per cent by each partner and was one of the first international joint ventures created in Estonia. The idea was to use and adapt (to local conditions) the solutions Kommundata was providing to the municipal government in Sweden. The close relation that Kommundata had with its counterpart in Finland, KT-Tietokeskus, enabled BCS in June 1991 to initiate its first foreign project, the development (in cooperation with KT-Tietokeskus) of a software system to be used in municipal environment protection in Finland. In connection with the reorganization of Kommundata in the beginning of the 1990s, through which the municipally-owned Swedish firm was sold to private actors (eventually ending up within the Finnish-Swedish TietoEnator group), Baltic Computer Systems became a wholly Estonian-owned company, although one owner was a Swedish Estonian, through whom tight linkages with Sweden continued to be influential. Thus when Baltic

Computer Systems later emerged as an important actor in the field of IT education, an important element in the firm's evolution was the mutual training project with the Swedish computer association Dataföreningen and the Swedish development aid organization SIDA (see Chapter 7) (Sild, int., 2003; Kirme, 1993: 18f.).

Abobase Systems was formed in 1991 as a spin-off from the somewhat unique Soviet-era Estonian IT company Mainor Parvi, which was part of the Soviet initiative to allow the emergence of small businesses in the 'perestroika' period. In 1989, still under Mainor Parvi, Abobase started activities in the Nordic IT markets and gained valuable experience through projects such as the development of a software system for a Finnish bank and the booking system for Nordström & Thulin's Tallinn-Stockholm ferry line, Estline. As an independent company, however, Abobase started as a reseller of Hewlett & Packard, whose Finnish subsidiary was at that time itself thinking about how to enter the new Baltic markets. As for most other Estonian companies acting on the domestic software market, the relationships to HP and other global players became more important in the early 1990s; Abobase had begun as an aggressive innovator in database systems, offering its own alternatives to the then globally dominant DBase products. Such an offensive innovation strategy was possible as long as the systems were based on DOS operating system, but with the emergence of more advanced operating systems the production of database software based on own systems proved so complex that it was found more efficient to rely on partnerships with global software platforms offered by firms such as Oracle. Abobase was owned 100 per cent by Estonian private capital (Saarma, K., int., 2001).

Regio was established in 1990 and became known during the following years mainly for its production of maps. It was a spin-off from the Department of Geography at the University of Tartu, whose scientists anticipated the liberalization of map production in Estonia – an area that had been subject to far-reaching secrecy and strict regulations during Soviet times. Preparations for entering the emergent map market were initiated in 1988, whereby an important feature was that Regio involved computer technology from the very beginning, in particular with respect to databases of place names. For a country where the field of map production had hardly existed at all prior to 1988 and where computing resources had been severely limited, the new political and economic conditions from around 1990 seem to have motivated Regio to immediately switch to more modern technologies as compared to those that had incrementally evolved in Western countries. Thus by 1994, Regio had switched 100 per cent to digital map production, i.e., there was no manual work at all carried out. This was somewhat unusual in international perspective, where, for example, Finnish map producers continued to use semi-manual techniques (Jagomägi, int., 2003). A few years later, Regio's competencies in map technologies and geographical information systems (GIS) would turn out to be of key importance for innovative activities in the emerging field of mobile telecommunications positioning services.

Assert was established in December 1991 by five former employees of the Institute of Cybernetics together with the Swedish biotech-related IT consultant Bionix, which at that time was still a young firm, having been founded in 1987. Assert's area of activity was extremely broadly defined, dealing initially with 'informatics and computer technology; consulting and education in social and economic areas; export, import and renting of computer, measurement and office machinery; and purchase and sale of securities and real estate'. However, it soon became clear that the actual focus was to be computer hardware and software. In 1992, Assert managed to get the Baltic states' first agreement with IBM for the reselling of its products in Estonia. A similar agreement was concluded with Fujitsu and later on with further international vendors. In 1993, a first *software* contract was acquired with the help of Assert's Swedish contacts, focusing on programming in the form of subcontracting for a Swedish firm. In the following year, the company entered the US software market, but it would turn out that Assert's main focus in software would be the Nordic countries. By 1995, the company had 25 employees, which at that time meant that it was one of the largest Estonian IT companies (information from Cellnetwork [formerly Assert], 2003).

Univoice was created in fall 1990 as a sister firm to a Finnish company with the same name that had been established in the beginning of the same year. The founder was an Estonian entrepreneur and former researcher from the Institute of Cybernetics, Olev Künnap, who had emigrated to Finland in the early 1980s. There he gained experience from a number of projects, such as the automatic measurement of water level in water reservoirs and its connection to computer networks. A former colleague from the Institute's speech technology department, Arvo Ott, became the director of the Estonian Univoice, whose activities aimed at designing – in cooperation with its Finnish sister company – telephone systems to be used in banking and related areas. In 1993, the firm was split into two different companies, Univoice and Voicecom. The former concentrated on the development and sales of small and medium-sized private telephone exchanges (PBXs), whereas the latter continued to focus on the emerging banking sector in Estonia. In 1994, Voicecom thus developed the first automated telephone banking system in Estonia, the main client of which was the Tallinna Pank (information from Voicecom, 2003; *Äripäev*, 26 September 1996 and 12 November 1996).

Crebit was a software firm which, similarly to Univoice, early on developed close relations to the banking sector, which was built up from nothing in Estonia beginning around 1989. Crebit's founders had earlier been employed in an Estonian electronics-related R&D institute as specialists in mathematical modeling, and around 1990 they had been working for a few years in Finland, in a firm focusing on communications solutions for conference centers and related customers. In 1991, Crebit started a close and, as it would turn out, crucial cooperation with the new Estonian bank Hansapank, for whom Crebit emerged as the key supplier of IT solutions. Crebit was thereby also responsible for the creation of Hansapank's inter-office data communications network. In 1993, Crebit created for Hansapank an electronic banking services system, Telehansa. This was an 'off-line' solution, but it was nevertheless an important exercise for the upcoming Internet banking later on in the 1990s (*Baltic IT Review*, no. 1, 1996: 57; Hansapank, 1999; information from Hansapank, 2003; Luštšik, 2003; Sildmäe, int., 2003; see also Ideon, 2001: 74ff.).

Reveko was established in January 1990 with plans to import and resell office machinery in Estonia. The founders went to nearby Helsinki to look for opportunities in this respect, where they found it surprisingly easy to establish an agreement with Canon's Finnish subsidiary for the reselling of Canon's products in Estonia. This first contract was followed by similar agreements with other international office machinery vendors, which eventually included telecommunications equipment in terms of small PBX systems. The rapidly growing telecommunications market led to an increasing emphasis on the latter, and in 1993 Reveko was split into two firms, of which one, Reveko Telekom, focused exclusively on the telecommunications market. At about the same time, the firm was contacted by Tele, the Finnish PTO, whose Erinox telephones Reveko had been selling through an agreement with one of Tele's Finnish resellers. Tele thought it more effective to use Reveko directly as a reseller, and this became the beginning of a long-term cooperation with the Finnish PTO. Reveko also diversified its sales by entering more complex segments of the telecom equipment market, of which Nortel's Meridian systems became Reveko's most important brand. However, the firm was still intensely focused on sales, and there hardly existed any in-house technical know-how at all until around 1996. This would soon become a problem (Peets, int., 2001).

Traba Projekt was a firm with a somewhat different background. It was founded in the end of 1991 as a spin-off from Eesti Telekom, i.e., the former Ministry of Communications. More precisely, it was the successor of SPKB, the Ministry's design bureau, i.e., the implementation-related R&D organization that had been active since 1968. Of around 100 employees, only ten engineers-designers were retained. The director, Garri Trapido, similar to many of the former employees, had a background as a graduate from the Department of Radio Engineering at the Polytechnical Institute. Traba's main customer was after its creation still the old partner Eesti Telefon, for whom the firm designed a number of projects related to the build-up of an optical-fiber backbone network in the country (Trapido, 1994: 4; information from the Department of Radio and Communications Engineering).

These small firms had as a rule no parallels in the earlier history of the Estonian telecommunications sector. In the course of the 1990s, however, they were to play important roles and develop in interesting ways both internally and externally. These developments will be discussed further in Chapters 6 and 7.

7. THE REORIENTATION OF EDUCATION AND RESEARCH

Tallinn Polytechnical Institute (Tallinn Technical University)

In Chapter 4, it was pointed out that tertiary education relating to telecommunications in a traditional sense (fixed voice telephony) was largely missing in Estonia in the Soviet era. On the other hand, programs in radio engineering, electronics and information technology had been strongly supported and well developed in order to support Estonian industry (though the corresponding curricula were largely designed in a centralized way in Moscow).

The unfavorable conditions for telecommunications education in a traditional sense were further emphasized by strict academic boundaries drawn in the USSR between telecommunications ('side' in Estonian; 'sviaz' in Russian) and radio engineering ('raadiotehnika'; 'radiotekhnika'). There were even specialized universities which dealt exclusively with 'sviaz' or with 'radiotekhnika'. To traditional telecommunications belonged the Bonch-Bruyevich Institute in Leningrad, which had been an extremely important educational organization for the supply of Estonian telecommunications engineers (as discussed in Chapter 4). At the Tallinn Polytechnical Institute (TPI), on the other hand, the strict division between the two related specializations theoretically meant that it should not have been possible to introduce telecommunications education within the Department of Radio Engineering. Nevertheless, political pragmatism and an Estonian government head (Bruno Saul) who himself was a communications engineer did in the end quietly enable the introduction of telecommunications courses at the TPI from the mid-1970s.

This action turned out to be important towards the end of the 1980s. With the striving for national independence, it started to be realized that Estonia might in the future have to educate its own telecommunications engineers instead of sending them to Leningrad, and at the same time it was observed that the field of telecommunications was gaining in importance socially and economically in global perspective. When the Soviet Union finally collapsed in 1991, at which time the Polytechnical Institute assumed its old name, Tallinn Technical University (TTÜ), the already existing, however minor, specialization in telecommunications ('side') meant that the Department of Radio Engineering was fairly well prepared to adjust to the new circumstances and that the transformation of study programs could be carried out in a relatively smooth way. Despite the fact that from 1991 the Bonch-Bruyevich Institute in Leningrad was suddenly practically closed to Estonians (who were now foreigners in Russia) and that basically all telecommunications tertiary education was from then on the responsibility of the Tallinn department, not more than around a third of the curricula were considered necessary to overhaul (Ots, int., 2003).

More serious were the effects of the Soviet collapse with respect to the department's R&D activities, which had so far accounted for about half of its financing. As pointed out in Chapter 4, this R&D had been carried out either for local industry (mainly in terms of contract R&D for the TREKB design bureau of Punane RET) or for various industrial branch R&D institutes in Moscow, Leningrad, Kiev and other Soviet cities. Under Gorbachev, the financial situation initially improved for the Estonian R&D organizations, and a new joint laboratory was also created in 1985 between Punane RET and the TPI (Min, 2000a: 46f.). But this situation changed dramatically in connection to the collapse of the Soviet Union. Although the R&D contracts were not lost over night, but rather continued to decrease until as late as 1996, the effects were disastrous for the financial situation of the TPI, as shown by its decrease in personnel: in 1986, 52 people had been employed at the Department of Radio Engineering; a decade later there would be only around 20 (calculated from TPI, 1986a, and present statistics; also confimed by Ots, int., 2003).

The orientation towards telecommunications in a broader sense was in this situation seen as a possible strategy for survival in the R&D arena. Notably, the large Nordic investments in Eesti Telefon (see above) meant that this new operator seemed a potentially promising supporting force for telecommunications education and R&D. Although it was hardly expected that Eesti Telefon would be able to replace the whole spectrum of telecom-related organizations that had been supporting the department through R&D contracts in the Soviet era, the Nordic-Estonian operator was simply the only large organization with reasonable financial health in the Estonian telecommunications sector in the early 1990s.

Not surprisingly, therefore, a new professor of telecommunications was formally installed at the department in 1992, financed directly by Eesti

Telekom. The name of the Department of Radio Engineering itself was changed in 1992 to the Department of Radio and Communications Engineering ('Raadio- ja sidetehnika instituut', RSTI). Later on, Eesti Mobiiltelefon (EMT), which in the early 1990s was still a very small firm, would similarly emerge as an interesting partner for the department. Such a link, focusing on mobile communications, was even more natural from the perspective of the department's strong historical focus on radio communications.

In *electronics* education, the response to Estonian economic autonomy seems to have come somewhat earlier and more forcefully than in radio engineering. The old curricula were kept until the fall of 1989, when the last 'Soviet-style' engineers – whose specialization was always referred to as 'industrial electronics' – were accepted. In the following year, 1990, the Department of Electronics used the new Estonian autonomy to reorganize the education. Starting from 1990, a new study direction 'Electrotechnics' appeared, with a possibility to specialize in electronics and automation. The study program now departed from its earlier content, which had been imposed from Moscow and which had followed an all-Union pattern. However, the dramatic collapse of the Estonian electronics industry and the country's full re-independence caused further changes, and in 1992 the study direction was renamed 'Computer and Systems Engineering' (Tamm, 2002: 6f.).

The last 'Soviet-shaped' electronics engineers graduated in 1994 (some even later); altogether there had been around 600 graduates since the start in 1962. These graduates long remained the basis of engineering competencies in the Estonian electronics industry, a circumstance that was strengthened due to the declining number of applicants to electronics engineering in the early and mid-1990s (ibid.).

There were radically varying opinions about the desired future of electronics engineering education in the newly independent Estonia. With the shut-down of many former Soviet electronics production enterprises, after which only slightly more than a tenth of the earlier workforce could be kept (see above), some argued that electronics education be simply removed from the Technical University (Min, 2000a: 47).

International contacts had been extremely limited for the TPI in the Soviet era. From the early 1970s to the late 1980s, for example, only three people from the Department of Radio Engineering ever went to a Western country. When guests arrived from the West, controls were rigorous. Exchange within the Soviet bloc was more common, and scientists from the Department of Radio Engineering went, for example, to Moscow, Kiev, Kharkov, Leningrad, Novgorod, Riga, Vilnius, Kaunas and Belarus, as well as to Hungary, Poland and East Germany. When one of the staff at the department was invited to Helsinki in 1987, this still had to be arranged through Moscow.

This situation naturally changed with the opening of the Estonian SSR in the years around 1990. From the end of 1991, it proved impossible to retain any meaningful formal contacts with the earlier so important Bonch-Bruyevich Institute in Leningrad, where guests from Estonia were from now on not allowed to enter any further than the special room for foreign visitors. This did not prevent the maintenance of personal links, but at the level of university departments the relations were almost totally lost. At the same time, scientific exchange with the West tended at least initially to remain at a low level (probably due to the enormous work-load in Tallinn after the severe reduction of staff). However, new opportunities with respect to *student exchange* increased dramatically, largely due to financial support from the Nordic side. In particular, Chalmers University of Technology in Gothenburg played an important role: of the 6-7 students that the Department of Radio and Communications Engineering sent to Chalmers to carry out their master's theses, several would later arrive at key positions in the Estonian telecommunications system of innovation (see Chapters 6 and 7) (ibid.).

The Institute of Cybernetics

As discussed in the section on data communications above, the Technical University was also involved in the academic Estonian-Finnish Committee on Informatics, which held its first meeting in Tallinn in fall 1988. All of Finland's technical universities were represented at the event, and representing Estonia were the two main universities, the Tallinn Polytechnical Institute, and the classic Tartu University. The meeting also took place at the Institute of Cybernetics, which was the driving organization in Estonia behind the creation of the committee. Apart from research cooperation *per se*, the committee was strongly involved in the introduction of new data communications technologies in Estonia, as discussed in section 5.

The Institute of Cybernetics was at this point in time an extremely powerful R&D organization, with impressive continuous growth from its creation in 1960. The peak of its development was reached in the same year as the Committee on Informatics was formed, 1988, when the number of staff for the first – and only – time reached above 600 people (Institute of Cybernetics, 1990: 14).

By Estonian measures, this was an enormous number of staff. Towards the end of the 1980s the staff were involved in a wide array of projects, which often had interesting connections to the field of telecommunications in both a narrow and a broad sense. As discussed in section 5, the Institute was a key actor in the development of data communications in Estonia and the introduction of the Internet. The growing demand for computing resources in an environment of problematic hardware supply also motivated the Institute to engage in making the computing resources of its own computer center available to other institutes of the Estonian Academy of Sciences, as well as to other organizations, by using telecommunications networks. Research was also being carried out by the Institute on transmission errors in the data communications system (ibid.: 19f.).

Much stronger, however, were the Institute's accumulated competencies in the field of industrial local area networking, the demand for which was seen to be growing rapidly and where its R&D continued to focus on everything from protocols to implementation software. The fast real time reaction and high reliability industrial LANs that were developed included a focus on the somewhat exotic use of fiber-optic systems (which in reality seldom left the R&D complex for actual implementation in industry) and were developed according to international standards. The Institute also developed distributed CAD software for heterogeneous LANs (ibid.: 23).

Up to the latter half of the 1980s, these R&D activities continued to be largely based on demanding Soviet customers, mostly in Russia and the Ukraine, resulting in a situation where around 90 per cent of the R&D was carried out for the large industrial branch R&D organizations in Moscow, Leningrad, Novosibirsk, Kiev and other cities (estimated by Jaaksoo, int., 2003). Estonian R&D thus involved considerable networking activities. Starting from the official recognition of Estonia's economic independence in 1989, these contracts started to be cancelled. The natural reaction was that the various departments of the Institute began to look around for alternative customers.

At a first glance, there should have been rich opportunities *in Estonia itself*, given the well-developed electronics industry in the republic. But in reality, as discussed above, these electronics production enterprises were severely hit by a massive loss of customers and were therefore hardly capable of acting as demanding buyers of the Institute's advanced R&D. Similarly, it was far from self-evident that the Institute's broad field of specializations 'fitted' the character of the local industry, as the R&D complex and industry had been built almost exclusively in relation to customers outside Estonia rather than in relation to each other.

In this situation, the various departments at the Institute started to look for business relationships with *Western* customers. This was not easy, given the totally different structure of Western systems of innovation, but to a certain extent it was possible to build on already existing linkages to the West. For example, when in the early 1980s the Institute of Cybernetics had designed and developed all the software for a new control system for a water cleaning plant in Tallinn, the actual realization of the project had been carried out not by Soviet production enterprises, but by a *Finnish* firm, Valmet Automation (Wõrk, int., 2003). At that time, such a project had still been a somewhat unconventional solution. However, with the new political and economic situation emerging in the last few years of the 1980s, the possibilities for establishing similar cooperation with foreign firms became easier to realize.

The cooperation between the Institute of Cybernetics and Finland through the Soviet-Finnish joint venture Elorg Data has already been discussed above. However, the late 1980s also saw the emergence of further joint ventures and small firms, the goals of which were explicitly formulated in terms of possibilities for the *commercialization* of the Institute's research results. The development firm Kompakt, which was founded in May 1988, was wholly-owned by the Institute, but it found markets also outside Estonia, more specifically in Finland and to a lesser extent in Hungary. The Finnish projects consisted of a control system for a central heating plant in Kuusamo in Northern Finland, and software for industrial LANs for Finnish industrial firms (Institute of Cybernetics, 1990: 25).[18] An Estonian-Finnish joint venture, in addition to Elorg Data, was Ektaco, whose activity was very broadly formulated in terms of 'development, small-scale manufacturing and marketing of computer systems and electronic equipment; consultation and training' (Institute of Cybernetics, 1990: 35).

The crucial need for international cooperation with foreign firms and universities was one of the main explicit emphases in the regular five-year report that was published by the Institute in 1990. In reality, out of the 80 publications listed in this report, 45 were English-language publications, while 35 remained Russian-language (there were thus no Estonian-language publications at all!). As pointed out in Chapter 4, research contacts between the Institute of Cybernetics and Western academic organizations had been fairly strong in the Brezhnev era, although the research stays abroad (mainly in Sweden and Finland) had been somewhat sporadic and mostly limited to three months. The continuing importance of the contacts in the late 1980s was demonstrated by, for example, joint Estonian-Swedish scientific publications, but the actual exchange and international conference activities were now intensified. For example, already in 1985, a 'Soviet-Finnish international symposium on mini, micro and personal computers and their applications' was organized by the Institute in Tallinn (Institute of Cybernetics, 1987a), and was followed two years later by a similar Soviet-Finnish symposium on 'Computers in Education' (Institute of Cybernetics, 1987b). In 1989, the Institute even arranged a Soviet-French symposium on informatics in the Estonian republican capital (Institute of Cybernetics, 1989). In the outgoing direction, the Institute took part in activities such as the VTT symposium on software development trends in Finland in 1989 (Institute of Cybernetics, 1990: 18).

The most ambitious attempts to exploit the political situation of the late 1980s in a creative way seem to have been made by EKTA, the Institute's Special Design Bureau. The already mentioned joint venture Ektaco was initiated by EKTA, but the Bureau also created two joint ventures with firms from Switzerland and the US, initiatives which were quite unusual in Estonia at this time and which seemed to reflect the Institute's wide international contact network, and the respect as an R&D organization it had gained in both East and West. EKTA had gained a leading position in the USSR in the field of process control systems, resulting in a number of special awards from Moscow, but also international fame in the West for theoretical developments (ibid.: 27f.). It was also at EKTA that the Estonian 8-bit school computer, 'Juku', was developed.

However, by the end of the 1980s, a clear dissatisfaction was voiced loudly within EKTA concerning the current situation in Soviet IT and electronics. Complaints were raised concerning the low level of electronic components available within the USSR, and the fact that Estonian researchers could not see their R&D results implemented in industry due, for example, to the lack of certain integrated circuits (ibid.: 29.). The general view seems to have been that much more could have been achieved given a different set of political and economic rules, and that the current situation represented a tragic waste of powerful intellectual resources and technological competencies. The dissatisfaction went so far that the attempts to influence the political system became a field of activity in its own right at the Institute of Cybernetics, and among the Institute's list of publications in 1989 one could thus find studies of innovation policy as well as debate articles on economic development (ibid.: 21 and 27).

Both the increasing openness with respect to foreign universities and private business, and the rapid global development of computer technology, came to motivate far-reaching organizational changes within the Institute of Cybernetics in the late 1980s. Fundamental reshufflings of various departments took place in 1986 and 1988 and a new director arrived in 1989. The large computing center which had been built for central computers with corresponding software development was found to be largely useless as Western equipment became available. The computing center was therefore transformed into a department dealing with the exploitation of existing Western computer technologies. It was within this department that the activities took place to facilitate the introduction of modern data communications as discussed in section 5, and it was on the basis of the totally transformed computing center that the Internet access provider EsData finally emerged in 1994 (Wõrk, int., 2003; Küberneetika Instituut, 2000: 12ff.).

The creation of EsData was one of many new firm formations that began around 1988 with Elorg Data, Kompakt, Ektaco and other joint ventures or wholly-owned firms. Formally, the Institute of Cybernetics experienced a dramatic decline in personnel: by nearly 75 per cent between 1990 and 1994 (calculated from Küberneetika Instituut, 2000: 14, diagram). In reality, however, a substantial share of this loss was the result of new firm creation, a process the Institute encouraged, seeing it more effective to retain existing competencies than to attempt reorienting itself towards Estonian and Western markets (Jaaksoo, int., 2003; Tani, 2000: 49). While it proved impossible to prevent substantial unemployment, a significant part of the scientific potential did manage to survive in this radically new way. Moreover, most of the new firms that spun off from the Institute stayed in the physical premises of the Institute of Cybernetics, which by the mid-1990s therefore had come to resemble something like a science park. Thus in the building of EKTA, for example, which had accounted for about half of the Institute's 600 employees in the last Soviet years, around 200 former employees of the Institute were still working in 1994 (Tani, 2000: 49).

In addition to these developments, many former employees of the Institute started their own firms in other parts of Tallinn and Estonia, such as the founders of Assert (see section 6), or were appointed to positions in new government bodies such as the Department of State Information Systems (RISO) or the Informatics Council. Other researchers who left the Institute went abroad. For example, the former head of the Department of Programming, Enn Tyugu, became professor of computer science at the Royal Institute of Technology (KTH) in Stockholm, and one of the PhD students, Tanel Tammet, chose to finish his thesis at Chalmers University of Technology in Gothenburg, where he later stayed as a lecturer.

The formal organizational changes in the Institute itself took place mainly in 1993, when a majority of departmental units were liquidated. On the other hand, the new Department of Information Technology saw the light of day in the same year, and in 1995 a new Data Communications Department was created. At the same time the Institute collaborated with the Technical University in the establishment of a the 'Informatics Center of the Universities' (Küberneetika Instituut, 2000: 16). The increasing focus on data communications and, in particular, information security would later turn out to be of vast importance for the future. The Institute's main transformation, however, leading to its split into an institute that joined the Technical University and a private, business-oriented R&D organization, was to be postponed to the late 1990s.

8. SUMMARY AND CONCLUSION

This chapter has covered a period from the mid-1980s to the first years of the re-independent Estonian Republic. The starting point was the stagnating telecommunications sector in Soviet Estonia, which then, through a series of radical changes, was dramatically transformed. The situation that resulted by the early 1990s was contradictory: a number of interesting and completely new opportunities had emerged, while at the same time, there was a severe downturn in the economy. Service providers of fixed and mobile telecommunications as well as providers of data communications had found strong foreign partners and investors and appeared to face a relatively smooth transformation, while the electronics sector, which in the late 1980s had employed around 25,000 persons, seemed almost to have ceased to exist by the end of the period.

The primary focus of the chapter has not been on innovation as such. Rather than addressing innovative outcomes, the chapter has mapped the transformations of the entire organizational and institutional landscape and the Estonian style of innovation in response to the collapse of socialism and the reestablishment of national independence. It was a period in which actors were forced to make very important choices, and whereas the actual creation of new products and processes stayed somewhat in the background, a radical reconfiguration of the whole sectoral system took place and thereby redefined its very logic, setting the stage for further developments in re-independent Estonia. It was thus an important formative period, whose dynamics are important to understand if we want to grasp the system's development in the following periods.

The structure of the system was radically altered. The overall institutional changes in the 'perestroika' and national liberation periods opened up new opportunities for very far-reaching structural transformation. The new institutions paved the way for important organizational changes (leading to joint ventures, foreign investment projects, new inter-organizational relations, the formation of small firms, new government agencies, etc.). The new organizational arrangements became important vehicles for exploiting technological opportunities relating to the digitalization of fixed networks, the introduction of mobile telephony, the emergence of the Internet in Estonia, etc. A variety of foreign actors were involved in this, and the relations between the Estonian system and its foreign environment were thus of great importance.

However, the acquaintance with new technologies, such as the Internet and mobile telephony, tended to raise further questions as to how the acquisition and further development of telecommunications technologies were to be organized (internationally or domestically? with or without domestic production and R&D? etc.) and what new institutions would be necessary for that (what role for competition? protect local R&D or not? what type of support organizations? etc.). Deciding upon these issues required considerable knowledge and specialized competencies, and the Estonian system secured these both by drawing on foreign advisors and by building on domestic, inherited competencies from the Soviet era. This enabled the Estonian system of innovation to respond to the new situation through the creation and shaping of new institutions (e.g., participation of the Ministry of Communications and the Institute of Cybernetics in the working-out of new legislation and policy), and by making the technological choices that steered the system in the direction of certain technologies rather than others (Estonian operators choosing NMT and in particular Ericsson technologies; strong interest in the Internet rather than other data communications technologies; etc.). While several issues remained unresolved and uncertain, the Estonian telecommunications sector had by the end of the period made surprising progress in resolving the structural tensions to which the new political situation had given rise.

The Estonian style of innovation, too, was in this period undergoing radical transformation. Let me in the following discuss the main elements of this style for the period covered in the chapter.

Clearly, the 'perestroika' period and the emerging Estonian independence were closely associated with radically new opportunities for *formulating visions*. As noted in the preceding chapter, advanced visions had not been absent in Soviet times, but it was only now that they actually seemed possible to realize. Discussions were in this period therefore very intense concerning future development strategies in all sectoral segments. Ideas relating to digitalization of networks, establishment of data communications facilities and the creation of a mobile telephony system could now be transformed into real-world projects. Close contact with Nordic operators and equipment vendors, both among the absolute world leaders in their fields, thereby appears to have functioned as a great inspiration, not least because Estonian actors – despite the long Soviet influence – considered themselves to be culturally close to the Nordic world. At the same time, future prospects were intensely debated with respect to the usability of the inherited resources in a totally new political and economic era. In the short IME period, a popular vision was to gain economic independence from the USSR but at the same time benefit from its huge markets. When this vision lost its relevance after 1991, alternative ideas included the establishment of an Estonian 'national champion' in telecommunications equipment, but this, too, was soon given up in favor of the emerging trends towards acting as subcontractors to Western and in particular Nordic equipment producers.

The articulation of demand underwent a total reorientation in the years around 1990. In terms of consumer demand, there was a strong perceived need for improved telecommunications services in relation to the gradual establishment of a market economy. In particular, many foreign firms that were establishing themselves in Estonia had a need for high-quality long-distance connections. At the level of operators, the Ministry of Communications initially restricted its own demand for equipment from the Soviet Union and East European countries in a deliberate way, as it was expected that new opportunities would soon open up for procuring higher-quality equipment from the West. This came true from around 1990, after which demand was directed towards suppliers such as Ericsson. At the same time, it was being realized that the size of the republic made domestic demand much too limited for a local equipment factory to be created, and the collapsing Soviet demand for Estonian industrial output and R&D was a decisive shock. It was hoped that domestic or foreign demand would be able to compensate for this, and emerging foreign investment seemed to give some hope here with respect to production – but not with respect to R&D. Regarding data communications, the Soviet experience with Akademset and close collaborative activities between Estonian and Finnish universities and firms played an interesting role in stimulating early demand for Internet access.

With respect to the *creation of new knowledge*, the 'perestroika' period was associated with a temporary upturn, as mirrored in, for example, patent statistics. Estonia produced the Soviet Union's first CD player, and R&D was initially boosted through the joint TPI-RET laboratory and international events such as the Estonian-Finnish committee on informatics from 1988 and other international conferences. However, from the end of the 1980s, research contracts from other parts of the Soviet Union started to vanish, and there was hardly any new demand domestically or from the West that could compensate. On the other hand, some inherited knowledge was further developed within the new business community, where researchers sought to build on their expertise and develop it further for commercial purposes. The impact inherited knowledge would have on future innovation is addressed in the next chapter.

Competence-building was an activity that had an unusual character in much of this period, as Estonia during the perestroika years remained part of the all-Soviet educational system, while at the same time the new era opened up opportunities for exchange and cooperation with the West. With independence, however, the opportunities for Estonia to profit from competence-building at Russian universities and institutes decreased radically. Towards the end of the period, competence-building was further hit by a loss of staff at universities and institutes, due to the collapse of contract research which had earlier been an important source of funding for universities. At the same time, the opportunities for going to the West were still quite limited, mainly due to lack of funding, but interesting opportunities seemed to open up, for example, with respect to student exchange with the Nordic countries. Outside universities, important training took place in collaboration with Nordic operators and Ericsson, as well as through Nordic guidance in IT activities in public administration.

New firm formation became, for the first time since almost half a century, an important activity during this period. A wave of entry of small Estonian firms followed in the years around 1990. They included a variety of equipment retailers, but also software firms and computer assembly companies, which often built on the competencies and ideas of highly skilled researchers from Estonian universities and academic institutes. A disproportional share of these new telecommunications-related firms seems to have been created by persons with close contacts to Sweden and Finland, countries that were also dominant in the formation of international joint ventures. In the government sector, a number of new bodies were also created in connection to 'perestroika' and the policy of IME. In academia, new committees, often involving foreign (especially Finnish) participants, were formed, as was the joint laboratory between the TPI and Punane RET. On the other hand, the idea of establishing a telephone factory ended in failure at an early stage. In the new segment of mobile communications, the first Estonian operator, EMT, entered successfully, based on Nordic investments and having the character of a spin-off from the incumbent fixed operator.

The adaptation of organizations focused on the difficult transition to an emerging capitalist market system, and above all on surviving this transition. The firm-internal adaptations involved a completely new organizational culture, with the introduction of hard budget constraints and decentralized (firm-based) decision-making. Firms lost their earlier security and had to learn to act in a more risky and uncertain environment, but they also acquired greater freedom. This meant that specialized suppliers and a wide array of small firms could be created. At the same time, production enterprises and R&D organizations had to adapt to collapsing Soviet demand, which often enforced dramatic reductions in employment and not seldom total liquidation.

With respect to *networking activities*, a fundamental shift occurred as a result of the new freedom to build linkages according to firms' own desires and needs. This was the case in both domestic and foreign inter-organizational relations. Transnational networking activities were substantial and ubiquitous: Estonian actors developed important links with foreign operators, equipment suppliers, IT firms, universities, standard organizations, municipal IT organizations, financial actors, etc. A pattern crystallized in which almost all relations to foreign partners went in the direction of Sweden and Finland. This was so in all segments of the sector, from fixed and mobile telecommunications to data communications and education, as well as in research and politics. Many of these international links focused on foreign direct investment arrangements, but an important issue was also the build-up of new procurement regimes, such as between operators and equipment producers. Domestic collaboration appeared to be less strongly pronounced.

The *financial* situation seemed initially to improve for innovation-oriented activities in connection to Gorbachev's emphasis on information technology as an instrument for 'perestroika'. With the emerging economic and then political independence of Estonia, however, both production enterprises in telecom-related fields and research institutes and universities faced a disastrous crisis in terms of funding, as the once important Soviet resources disappeared almost completely. A partial compensation for this was the new investment activities from the side of foreign firms, such as in connection to the formation of EMT, Eesti Telefon, Elcoteq and some small firms, with the help of Nordic capital. International funders were also involved, such as UNESCO and the Soros Foundation. This funding, however, had a very different character from that of the Soviet period, being oriented towards the restructuring of organizations and investments into physical infrastructure, rather than towards R&D.

Consultancy, advice and lobbying activities evolved in an interesting way during the period of investigation here. Domestically, the Institute of Cybernetics emerged as an important advisor to the Estonian government in matters related to information technology. In fixed and mobile telecommunications, far-reaching input came from the Nordic incumbent operators, who later also entered Estonia as investors. Nordic consultations were facilitated by cultural proximity to the Nordic world and in particular by linguistic proximity to Finland and the participation of exiled Estonians living in Sweden. On the manufacturing side, Ericsson appears to have entered successfully both as an advisor and a lobbyist for the Ministry of Communications.

With respect to *creating, changing and abolishing formal institutions*, this period covers the establishment of totally new rules and legislation, which generally aimed to replace the old socialist system with a capitalist market economy. This took place in different stages, whereby economic autonomy was followed by political independence. Estonia started to build Western-inspired institutions out of nothing, with the most important driving actor being the Ministry of Communications itself, which saw this as part of its reorientation towards and new collaboration with the West. Far-reaching regulation was introduced in fixed telephony, but the data and mobile communications market as well as the equipment side were liberalized. This was to prove important for the years to follow.

The *formulation of public policy* was dominated by the desire to build a liberal market economy in Estonia. Intense debates therefore followed when it came to introducing a temporary monopoly policy in fixed telephony. The overall laissez-faire politics, in combination with the general lack of funds and the only weakly developed policy capabilities, also meant that the government rarely intervened in any direct way to prevent the dramatic collapse of production enterprises and research institutes. This was despite the fact that representatives from the academic community were strong participants in the working out of new policies and institutions.

As can be seen from this discussion, Estonia's Soviet and earlier history played an important role in shaping this style. Soviet competencies were a key resource in rearranging activities such as the formulation of visions, new firm formation, policymaking and institution-building – all of which would clearly have looked completely different had it not been for the inherited competencies. Similarly, existing links to the Nordic countries – particularly within the academic community – turned out to be an important historical resource, for example, for the introduction of the Internet in Estonia as well as for early student and research exchange.

The influence of the 'new' foreign environment was, of course, considerable. An interesting characteristic is the gradually crystallizing concentration of international linkages in the specific direction of Sweden and Finland. In light of the strong cultural and geographical proximity to these countries, this development was hardly surprising – the least not in the telecommunications sector where the Nordic countries were the world leaders. Relationships with these two countries therefore tended to show up in practically every activity in important ways. The Nordic countries became sources of inspiration for the formulation of Estonian visions; demand for equipment was largely directed towards these countries; new firms were formed with the help of Nordic funding, consulting and networking; and Nordic actors appeared in the creation of new institutions, in competence-building through student exchange, in training in cooperation with Nordic firms, in knowledge-creation through joint conferences, etc.

What the roles of history and interaction with the Nordic systems of innovation would mean for the further transformation of the Estonian system are investigated in great detail in the upcoming chapters. The Estonian system as it appeared in the years around 1990 signaled interesting future opportunities as well as possible threats. By the end of the period covered in this chapter, the situation was still highly uncertain and fluid, but many important decisions and steps had already been taken – steps that were to prove crucial for the dynamics of innovation during the following decade.

NOTES

1. It can be noted that the separation of postal from telecommunications services was a pan-European trend during these years, i.e., it was not limited to Eastern developments. For example, Deutsche Bundespost was split-up according to a similar pattern in 1989, thus creating Deutsche Telekom.
2. Thus the concession agreement in 1992 was signed by the Minister of Communication, Andi Meister.
3. The concession agreement was signed by A. Meister and K. Saar (representing the Estonian government) and T. Sõmera, T. Hagström and P. Vennamo (representing Eesti Telefon); see also Cunningham (1998: 234). The agreement mentioned a possible extension of the monopoly rights for another six years, but this was never realized.
4. The preference for domestic products was later removed, as negotiations from 1997 onwards for EU accession required this.
5. These arguments are put forward in striking agreement by almost all interviewees.
6. In 1990, for example, the foreign participation in Estonian firms was dominated by Finland (50 per cent), Sweden (15 per cent) and Germany (7 per cent) – see Mandel (1999: 29).
7. The problem especially concerned the calculation of the value of fixed assets in existing Estonian networks. The Nordic investments were to be of a similar size as the value of the networks.
8. As noted in Chapter 4, Latvia was the leader in telephonization within the Soviet Union, followed by Estonia, and Latvia also played a key role in the creation of the first Soviet data communications network, Akademset. With respect to equipment production, Latvia hosted the famous VEF company. A Scientific Research Institute belonging directly under VEF contributed to making the company not only a producer but also a key developer and designer of equipment to be installed in telephone networks all over the Soviet Union. Towards the end of the Soviet period, VEF even gained considerable experience in the emerging segments of digital long-distance and business exchanges. Semi-digital exchanges were launched by VEF in 1978 and fully digital exchanges in 1989. See, e.g., Müller (2000: 210); information from VEF-KTR (2003) and Kvant-Interkom (2003).
9. At the pressconference held on 21 January 1992, the Nordic interests were represented by Televerket and Ericsson, Tele being absent and not mentioned. In May, however, it was reported that Tele was taking part in planning an Estonian-Swedish-Finnish joint venture. See *BNS*, 21 January 1992 and 18 May 1992.
10. The sum $250 million in ten years was enormous for the small country. But a comparison of undertakings of Swedish and Finnish operators in Estonia with those of the West German operator in the ex-GDR at the same time, in the $36,000 million 'Telekom 2000' project, indicates the Estonian money was more like a small development aid to the Baltic neighbor.
11. Some participants involved in the tendering procedure for the NMT system strongly argued that Nokia in fact made a better offer than Ericsson, but that Ericsson was still chosen for a number of other, more controversial reasons.
12. Agur (1996b: 6) estimates the workforce in the Estonian electronics sector at 28,000 before Soviet breakdown; Tani (1993: 15) estimates the number of employees at 26,000 in the Estonian electronics sector in 1991, representing 12 per cent of the total industrial employment.
13. However, see also the section on the birth of Eesti Mobiiltelefon below, where the choice of Ericsson as an equipment supplier was very controversial.
14. Cf. the state-owned PTO Tele's hesitation to enter collaboration with the Estonian Ministry of Communication at about the same time (section 4.2).
15. Formally, the ownership also included the state-owned Estonian holding company Eesti Telekom, which owned 51 per cent of Eesti Telefon. The shares divided among Tele (35 per cent), the Institute of Cybernetics (35 per cent), Eesti Telefon (15 per cent) and Eesti Telekom (15 per cent).
16. At the time of EsData's launch of commercial Internet access, there were around 400 computers in the whole country connected to the Internet. Two years later, however, there

were 4,000 computers and another two years later 16,000 computers connected to the Internet. See further Chapter 6.

17. This is also confirmed by Martinson (int., 2001).
18. The Hungarian customer of software for industrial LANs was the well-known firm Videoton.

6. Imitation and Experimentation (1993-1998)

After the introduction of the Estonian kroon (EEK)[1] and the adoption of the new national constitution in June 1992, the stage for economic and political transformation of Estonia had been set. In September 1992, the first free elections took place in the re-independent country, resulting in the installation of a pronouncedly liberal government headed by the new prime minister Mart Laar, a mere 32-year-old historian. Soon thereafter, Estonia also got a new President, Lennart Meri.

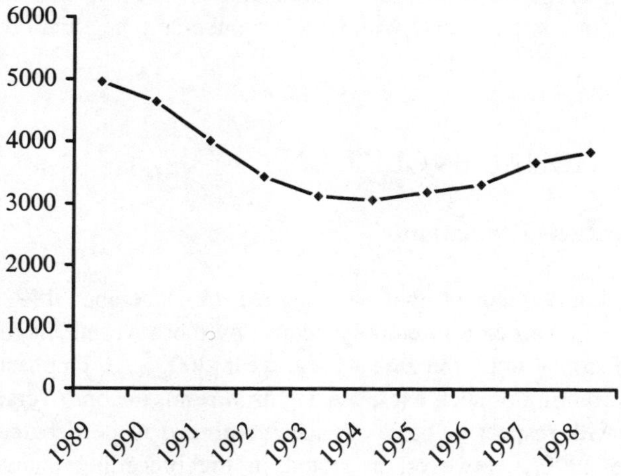

Source: *Transition Report Update*, April 2001: 15.

Figure 6.1 Development of real GDP per capita in Estonia, 1989-1998 (€, constant 2000 prices)

From an economic perspective, what followed in the mid-1990s was a radical restructuring of the whole Estonian economy, hitting in particular the oversized Soviet-era industry. As late as 1989, the industrial sector's share of the Estonian gross domestic product had equaled nearly 40 per cent.

Agriculture's share was 22 per cent, while service sectors such as wholesale and retail trade, transport and communications and business services were severely underdeveloped (as was typical in the Soviet-style economies). By 1998, the share of industry in GDP had almost halved, corresponding to 22 per cent, whereas agriculture had declined even more, now equaling 7 per cent (including forestry and fishing). These numbers already corresponded roughly to typical Western economies. Similarly, the above-mentioned service sectors had almost doubled their importance in GDP.

The combined effect of the collapsing Soviet-era Estonian economy and the emergence of new economic fields of activity resulted in the economy starting to grow again after having reached a low-level mark in 1994.

With respect to the international reorientation of the Estonian economy, an overwhelming majority of both privatized and newly established firms were based on foreign direct investment from Sweden and Finland. By 1999, these two countries stood for around two-thirds of the cumulative foreign investments in Estonia. Sweden and Finland also emerged as Estonia's most important trade partners, together standing for around one-third of Estonia's exports throughout the mid-1990s. At the same time, Estonia aspired to become a member of the European Union, and in 1997 it was formally endorsed as a 'candidate country' with which membership negotiations could begin.

1. INSTITUTIONAL EVOLUTION

Telecommunications Regulation

In the concession agreement that was signed in December 1992, Eesti Telefon had been awarded a monopoly license by the government for fixed telephony in Estonia until the end of the year 2000. As emphasized in Chapter 5, the granting of such exclusive rights was not uncontroversial, and the monopoly with respect to basic services continued to be debated lively throughout the 1990s. However, as noted in the preceding chapter, the concession did not address the fields of mobile telephony and data communications. These fields were actually left more or less untouched by the dramatic institutional rearrangements in the early years of transition – a fact which was to prove critical in the years to come.

As late as in the early 1990s, mobile and data communications were in practice still relatively exotic phenomena even in the most advanced Western markets, where not more than a few per cent of the population were users of mobile phones or the Internet. In the mid-1990s, however, GSM started to be established on a large scale in most European countries (the real international breakthrough year being 1995), while at the same time innovations such as the world wide web and Netscape's Internet browser became the starting point for the phenomenal growth of the Internet that followed at about the same time. The enormous market growth in mobile telephony and the Internet subsectors of telecommunications implied a shift with respect to the focus of telecommunications innovation, turning more and more to the new market segments as well as to the convergence between various technologies and markets. The lack of regulation in these new and crucial subsectors of telecommunications in Estonia seemed in an interesting way to imply that competition would be allowed to work freely as a driving force in the evolution of the sectoral system of innovation.

Telecommunications and IT were in the mid-1990s in practice still treated as two separate fields in Estonian public policymaking. Telecommunications in a traditional sense was a field organized under the Telecommunications Inspectorate and the Ministry of Transport and Communications. The Inspectorate focused mostly on technical issues (such as the distribution of radio frequencies) rather than market regulation. The Ministry, where only a small group with a staff of four worked with telecommunications (the major part dealt with transport), was responsible for preparing legislation, elaborating communications policy, planning development programs, issuing licenses for public networks and services and representing Estonia in international organizations such as ITU and CEPT (Ott, 1996a: 11). In practice, however, the most influential public body was probably the government itself, which in accordance with Eesti Telefon's concession agreement was to approve the latter's business plans each year. This possibility was frequently used, as will be discussed in section 2 below.

With respect to mobile telephony, the worldwide rise of GSM meant that competitors of EMT were becoming interested in getting licenses in Estonia. In the absence of exclusive rights in this field, new entrants could start operation surprisingly easily in Estonia – a situation which contrasted sharply to the conditions in neighboring Latvia, where attempts to introduce competition in mobile telephony were characterized by a series of corruption scandals. In Estonia, in contrast, two GSM operators entered the market at virtually the same time in January 1995. Moreover, despite the minimal size of the Estonian market, a third GSM license was soon issued, and by May 1997 three mobile providers were active on the Estonian market, resulting in a situation that was highly unusual at that time in global perspective.

A precondition for GSM to take off was also the fact that the 900 MHz frequency band became available after the exit of the old Soviet military forces from the Baltics in the summer of 1994. Here, the Telecommunications Inspectorate played an important role in terms of technical regulation (Sideamet, 2001a). The evolution of the mobile communications subsector is discussed further in section 3.

Policymaking and Institutional Change in ICT

By the end of 1993, the Informatics Council decided to begin to create a broader document, the 'Estonian Informatics Development Plan'. In September 1994, a first version of the plan was published (in the following I refer to the version of the plan that was published in *Arvutimaailm*, no. 9, 1994: 26-30). It dealt in a very broad way with the information society, in terms of overall goals such as 'raising the efficiency and competitiveness of the economy', 'better response to social needs', 'raising the quality of life' and 'deepening the country's sovereignty and cultural identity' – goals that were very similar to those put forward in corresponding development plans presented a few months earlier in the EU, Sweden and Finland.[2]

The Estonian plan complained that the state lacked an informatics budget, and that the state should make better use of its power to influence the private sector, whereby it was argued that

> the state's task is to create a favorable environment for the application of information technology in the private sector, applying to this end tax, credit and customs policy and legal acts that guarantee the security of information. A market for information services based on free competition has to be developed. (My translation)

Among the legal acts, the Databases Act and the Personal Data Protection Act were put forward as examples. With respect to the development of standards, it was argued that these would, without exception, have to be based on the standards of international associations, although it was noted that some would have to be adapted to specific Estonian conditions. The development plan also noted the connection between the process of informatization and the development of a *national industrial and innovation policy*,[3] although it remained unclear what was actually intended: the use of information technology as an instrument in the working out of these policies or, rather, selecting the information technology sector as a prioritized target for innovation policy.

With respect to the priorities (i.e., *within* the field of ICT) selected in the plans, these were concentrated in five directions, or 'programs':

1) data communications;
2) IT systems for national security;
3) IT systems for public administration;

4) state registers; and
5) geographical information systems.

A 'key issue for the information society' was seen to lie in the connection of IT appliances to telecommunications networks. The development plan here showed considerable irritation with respect to the lively debate that surrounded Eesti Telefon's exclusive rights. The plan strongly defended the existence of exclusive rights and pointed out that free competition could only be realized when a strong technological basis for that had been created:

> it is difficult to understand how, in the absence of an agreement guaranteeing exclusive rights, it would have been possible for us to get large capital investments for the renewal and expansion of telecommunications networks and exchanges. Yet even with these [exclusive rights], open competition occurs with respect to additional services, for example in data communications services, which always have to be supported. (My translation)

In 1995 and 1996, much of the policymaking in the field of telecommunications and IT took a somewhat new direction, turning to a focus on competence-building activities in the information society. In spring 1995, the Estonian ambassador to the United States, Toomas Hendrik Ilves, presented an idea that in itself sounded simple: to equip all Estonian schools with computers, viewing this as a vehicle for enabling Estonia to take a big leap into the information society. The idea was supported by the Minister of Education, Jaak Aaviksoo (who was a physics professor), and gained momentum towards the end of the year, when the use of the Internet had already started to grow unexpectedly rapidly. The project was backed in a ceremonial way by the Estonian President, Lennart Meri, who declared the 'Tiger Leap' project, as it came to be called, officially opened in an address on national television in February 1996 (information from Tiger Leap Foundation, 2003; *Baltic IT Review*, no. 2, 1996: 9; no. 3, 1996: 61).

With the coming of the Internet, the idea of computerization of Estonian schools was soon complemented by the goal to *connect all schools to the Internet*. The initial focus on hardware tended to shift more and more to 'soft' aspects such as the training of teachers in IT subjects. In a broader perspective, the Tiger Leap project also came to function as a basis for discussions on institutional aspects of information technology, as well as on the influence of information technology on mass media, culture and other aspects of life. An information society and Internet 'fever' seemed to have captured Estonia:

> The idea of the Tiger Leap project will be to prepare students and teachers for the expected changes, and all of society through them . . . The coming of the network into the countryside will reduce migration into towns . . . School buildings, as well

> as museums and libraries in the countryside, will serve as information, computer education and hobby centers – places for direct communication for people. Through them, access to 'virtual' resources will be granted for everybody. Step by step, urban people will return to the countryside . . . Network access in homes will enable people to try distance education, marginal and talented students being the pioneers: an inability to conform to formal requirements will no longer be an obstacle to getting an education; nor will one need to waste one's time on items that are too elementary. Learning must become a life-long process . . . (Ruut, 1996: 62)

It was also argued that

> paradoxically, Estonia's small size and non-developed structures represent advantages. We don't have rigid governmental structures, social welfare mechanisms haven't been developed, our infrastructure needs replacing. We will have to build without breaking. The guidability of a system is inversely proportional to its size: with respect to rapid change, small countries are ahead! (ibid.)

The committee responsible for the Tiger Leap project consisted of an interesting combination of Estonians living both in Estonia and abroad: in addition to the Minister of Education and the ambassador to the US, the committee included a professor from the Tallinn Technical University, an Estonian professor from Chalmers University of Technology in Sweden, the CEO of Baltic Computer Systems (created as a joint venture between the City of Tallinn and the Swedish Kommundata – see the preceding chapter, section 6) and a graduate from a Tartu secondary school. The Tiger Leap Foundation was founded and initially supported by the Ministry of Education and 37 firms and private persons. The firms were mostly small computer and IT companies established after 1990, including Baltic Computer Systems, MicroLink, Regio and others as well as the Estonian Computer Association (AFA). The list of private persons included, among others, researchers from the Technical University's and University of Tartu's computer sciences departments and the CEO of EsData (Estonia's first commercial Internet access provider – see Chapter 5). Entirely missing from the list of founders and supporters were, on the other hand, the fixed and mobile telecommunications operators, representatives of the Department of Electronics and the Department of Radio and Communications Engineering of the TTÜ (Tiger Leap Foundation, list of founders, 1996). This obviously signified a continuing perceived division between IT, telecommunications and electronics. The convergence of these fields was not yet a reality in the Estonian system of innovation.

The growing awareness of the importance of the Internet led to interesting discussions about the future of the Estonian ICT sector, and a number of rearrangements followed at the political level starting in late 1996. The link between the Informatics Council and the Informatics Fund was cut off, as the Council was reorganized as a separate commission (consisting not only of IT experts but also of representatives from very mixed areas of society) of the government (to whom it was to provide advice), and the Fund was transformed into the Estonian Informatics Center in January 1997. The Center became closely linked to the Department of State Information Systems (RISO), and was later to play an important role in connection with the planning and implementation of new radical IT projects. The function of financing innovation projects in the private sector, however, was handed over to the overall Estonian Innovation Fund, although the Center could still function as a financer through its ordering of projects in which private firms would take part in interesting ways during the following years (Siil 1997: 9-12; Ott, 1998).[4]

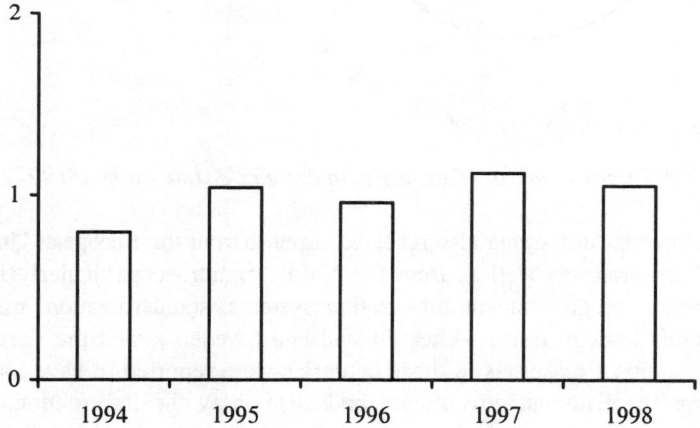

Source: *Baltic IT Review*, no. 11, 1998.

Figure 6.2 IT costs in Estonia's state budget, 1994-1998 (percentages of total state budget)

In the mid-1990s, however, the overall IT budget of the state was much too low to give room for any far-reaching innovative projects. State expenses on IT remained at about 1 per cent of the total state budget between 1995 and 1998, which in relation to the weak overall state budget was minimal: by 1998, the absolute amount of IT expenditure was $10.7 million (although this did not include the Tiger Leap project) (Odrats, 1998: 24).[5] Moreover, the structure of expenses signaled that public administration was still mostly concerned with building a basic technical environment through the simple purchase of ready-made hardware and software, rather than using its purchasing power as a demand-side policy instrument to stimulate the development of more innovative IT solutions among private firms (see Figures 6.2 and 6.3).

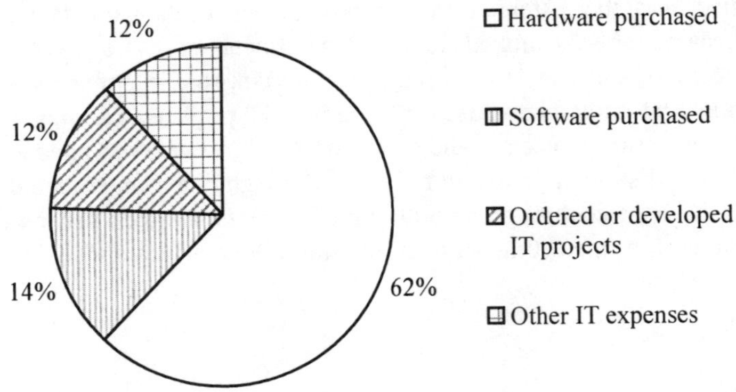

Source: *Baltic IT Review*, no. 11, 1998.

Figure 6.3 The structure of IT expenses in Estonia's state budget, 1997

IT development in Estonia also received support from the European Union's PHARE program, as well as from the Nordic countries, particularly in the development of government information systems, standardization, training and overall strategy issues. After Finland and Sweden joined the European Union, Estonia's close ties to those countries were reported to have made it easier for the Estonian government bodies to 'play the information game according to European rules' (Agur, 1996b: 9). It was also by Finland and Sweden that the largest bilateral projects were financed in the mid-1990s, a circumstance that was explained in terms of the cultural affinities of Estonia with the Nordic countries:

> The need for very close cooperation between the project manager and the object organization requires knowledge about the local educational and cultural background, legislation, etc. Familiarity with the national language or the possibility to guarantee fast and adequate translation is often a most important factor. (Ott, 1996b: 10)

The state's intense activities with respect to IT were not supported, however, by all actors in the Estonian telecommunications sector in a broader sense. As noted above, the powerful fixed telephone and mobile communications firms did not participate in the Tiger Leap project's early years, probably reflecting their strategies in which data communications did not at all belong to the core activities of these firms. A more direct criticism of the state's strong focus on IT came from the Department of Electronics at the Technical University, where it was argued that the state should also focus on the *equipment* side of the information society. Estonia, it was argued, should prevent the loss of its historically accumulated competencies in electronics, in particular in the field of *electronics design* – an area not very dependent upon expensive machinery but seen to increasingly contribute economic value worldwide (Rang, 2000: 10; *Arvutustehnika & Andmetöötlus*, no. 1, 1994: 1ff.). The electronics department contended that

> Estonia has a unique opportunity to push through a quick and decisive renewal, as the people have high qualifications and there are no old technologies [bearing down] on the throat. (*Arvutustehnika & Andmetöötlus*, no. 1, 1994: 3, my translation)

Critique was also directed towards the overall linkages between economic and industrial policy and IT activities. At the Department of Electronics, it was even argued that Estonia did not have an industrial policy at all (Min, 2000a: 47). Similarly, the Institute of Cybernetics criticized the Estonian government for not connecting its foreign investment policy with regulations that could stimulate foreign investors to take into account the domestic R&D base, as it was seen that the privatized and new firms rarely turned to the existing and highly advanced Estonian R&D organizations to solve problems relating to the introduction of new products and processes in the private sector (Jaaksoo, int., 2003).

An area in which institutional change was delayed was the legislative response to the Internet revolution. Despite eager activities in Estonia with respect to internetization and emerging Internet-based services (such as Internet banking – see section 5), government agencies remained surprisingly passive with respect to the renewal of legislation. It was only in mid-1997 that the government seemed to become increasingly aware of the new needs. In particular, this concerned the issue of digital signatures, electronic ID cards and rules regulating the use of databases. Slowly, the focus also turned to what came to known as 'e-government'. These and further developments are discussed in greater detail in Chapter 7.

2. THE FIXED NETWORKS: MODERNIZING THE INHERITED INFRASTRUCTURE

Throughout the world, the provision of fixed telephony services was a business that by the early 1990s had reached a somewhat peculiar and highly uncertain stage of development. Exciting opportunities as well as considerable threats had gradually become evident during the 1970s and 1980s, and they continued to be driven forward by the intertwined evolution of telecommunications technologies, organizational structures and the institutional environment. The digital technologies that started to be implemented in public telephone networks from the 1970s in the West opened a variety of new opportunities in terms of service product innovation, thus potentially changing the profile of actors that had so far been commonly known as 'network operators'. With the introduction of new value-added services, network operators started to regard themselves increasingly as 'service providers', a trend that grew even stronger with the ongoing convergence between telephony and data communications and which signaled an expected shift in the balance of revenues away from network operation towards new developments in the middleware and applications layers.

Organizational and institutional changes were initiated in many countries, partly driven by technological change: in view of the new technological landscape, traditional PTOs began to feel imprisoned in their old structure as public utilities and wished to become business-like organizations. In particular, it was demanded that postal and telecommunications services be organizationally separated in countries where they had belonged under a single administration. During the 1980s, in addition, the neo-liberal shift in Western politics led to questioning the telecom sector's status as a 'natural monopoly', and arguments for the introduction of competition in the sector were raised. For the telecommunications sector, which since the end of the 19th century had in fact been developing in a strikingly smooth way, these (technological as well as non-technological) developments were of a revolutionary nature.

It was amidst these dramatic developments that Eesti Telefon was created in late 1992. The new company, which started operating formally in January 1993 as a joint venture between Eesti Telekom, Televerket and Tele, was immediately thrown into a carousel that had been turning relatively quietly in the West for more than a century but which was now quickly gaining speed and seemed to be heading for an unknown world. With two of the world's most advanced PTOs as strategic investors, it was clear that Eesti Telefon was also going to be strongly influenced by ongoing developments in the West.

However, the point of departure was radically different in Estonia as compared to that in Sweden and Finland. The planned modernization of the fixed infrastructure was a long-term and highly complex project that could not be realized overnight nor be built from scratch. On the one hand, Eesti Telefon was open for a transition in the direction towards Western – and in particular Nordic – PTOs. But on the other hand, the concrete point of departure was the existing organization and infrastructure of the old Soviet Estonian Ministry of Communications. Eesti Telefon, the 'new' operator, inherited an eclectic Soviet-style telephone network, based almost exclusively on analogue technologies, as well as the entire staff of the old republic-level Minsviaz, amounting to around 5,000 employees. Eesti Telefon also inherited a waiting list of 200,000 individuals and juridical persons wishing to get a telephone connection (Eesti Telefon, 1997a).

Technologically, the general modernization plan of Eesti Telefon built upon the idea of an 'overlay network' to be built as a new digital network that would gradually replace the old analogue infrastructure. This was a common overall strategy in Eastern Europe at this time, and the idea stemmed from the depressing results of investigations that had mapped the technical strength of the Soviet network. In Estonia, modernization began with the creation of a fiber-optic backbone connecting all the major municipal centers in the country, where larger digital exchanges were thus installed. Subscribers were then gradually connected to local digital exchanges. The idea was to begin with those parts of the regions that faced the worst conditions, although the capital, Tallinn, was at the same time prioritized. All in all, the task was gigantic, but it was facilitated by the idea of using the existing system as far as possible, to the extent that the 'last mile' of Soviet copper wire was to remain in place. The plan thus differed considerably from the more ambitious modernization in eastern Germany, where Deutsche Telekom in the famous project 'Telekom 2000' invested no less than $36,000 million and thereby drew fiber-optic cables *all the way to the end user*. The Swedish and Finnish investors never dreamed about such a solution, and the planned investments for modernization as a whole amounted to about $250 million, which in terms of per capita was about fifteen times less than the East German project. In total, the ten year investment program that was launched aimed at covering the installation of 30,000 digital lines annually. The goal was to reach a public telephone penetration of 45 per cent (*Telecommunications*, November 1993).

The process of modernizing the infrastructure had begun before the formal creation of Eesti Telefon. As mentioned in Chapter 5, a digital microwave link was opened between Tallinn and Helsinki in spring 1992, and in the same year Eesti Telekom – still without any formal joint ownership with foreign partners – ordered a new international and long-distance digital exchange (AXE) from Ericsson. Therefore, the point of departure in the beginning of 1993 was not a 100 per cent Soviet-style network: Eesti Telefon had from its very launch the most modern globally available international switch at its disposal as well as good connectivity to Finland (although the increasing data traffic soon made much more powerful transmission links necessary). In early 1993, a first optical fiber cable was also installed between Tallinn and Helsinki (Sõmera, int., 2003).

Throughout 1992 and 1993, it was only in Tallinn that subscribers could get access to digital services; by the end of 1993, there were 4,400 digital lines in operation in the capital, and in the following year another 24,400 digital lines were installed (Eesti Telefon, 1998b). In 1994, digital technology also reached Tartu, the university town, where approximately 10,000 digital lines were created (Eesti Telefon, 1994: 5). In this context, the existing satellite connection that had so far given the university access to the Internet (cf. Chapter 5) could also be closed down (Wõrk, int., 2003). Also in 1994, a fiber-optic cable was drawn from Tallinn to Narva (with the plan to later connect it with Russia) and digital exchanges were installed in the towns of Jõhvi and Pärnu (Eesti Telefon, 1994: 5). The real breakthrough, however, seems to have occurred in 1995, when most Estonian counties got their first access to digital technologies, as the Western half-circle of the fiber-optic backbone was completed. New international fiber-optic connections were also opened with Russia in the East, Riga in the South and Sweden in the West (Eesti Telefon, 1995). In December 1997, the last optical fiber-link in the Estonian backbone network was completed, connecting Jõhvi and Tartu (Eesti Telefon, 1997a, b).

However, digitalization and the elimination of the long waiting list were processes that consumed enormous financial resources as well as time, and by the beginning of 1997, after four years of Eesti Telefon's operation, the number of subscribers that were connected to digital exchanges were still only 28 per cent of the total number of subscribers, that is, the PSTN (the public telephone network) was still 72 per cent analogue. The penetration ratio had increased to 30 per cent, but this, too, was still far from the goal of 45 per cent.

The continuing scarcity of digital lines was at this time beginning to be seen as an important obstacle in Estonia's striving for the 'information society', in particular because Internet and other data communications services were already starting to take off in the country. Many actors and observers were severely critical of the telephone company's ability to guarantee the provision of a modern physical infrastructure. The idea of a monopoly in fixed telephony, albeit temporary, had always been difficult to accept in a country that had just liberated itself from central planning with its ubiquitous state monopolism, and in the country's neo-liberal political climate Eesti Telefon was sometimes seen as a new manifestation of the old system. Especially in the countryside, Eesti Telefon was by no means a popular company among the population, who found it difficult to see the logical link between the continuing scarcity of telecommunications in large parts of Estonia and the image of welfare and progress that Eesti Telefon wished to establish for itself.[6]

These critical voices were accompanied by an increasingly tougher attitude towards Eesti Telefon in political circles. In August 1996, the Minister of European Affairs, Endel Lippmaa, resigned, one reason being his criticism towards what he regarded as a policy of monopolism in the telecom business. 'In telecommunications, Estonia has become a Swedish company's colony to pump profit from', he argued (*Äripäev*, 7 August 1996: 5, my translation).[7]

However, of the three countries involved, Estonia appears to have been the one which gained most from the joint activities. The very first few years especially enabled crucial assistance from the Nordic partners in planning and strategizing the modernization program. In these early years, there were around 15-20 Nordic representatives holding key positions within Eesti Telefon. In contrast to mobile telephony (see below), the main group of foreign directors came from Sweden rather than from Finland. These persons provided significant input into the build-up of the operator, though obviously more so with respect to organizational and managerial than technological issues. The key problem for Eesti Telefon was perceived to be organizational – the capacity to function as a modern Western-style network operator – rather than the learning of new technological competencies. With respect to technology, it was true that Estonian engineers had little practical experience from working with digital technologies in public networks. But at the same time the Estonians were well-prepared theoretically through their thorough Soviet education; most Estonians who were later appointed to key technical positions in Eesti Telefon had worked or trained with the local Technical University or the Bonch-Bruyevich Institute in Leningrad.

Moreover, close contacts with Telia went hand in hand with relations to Telia's traditional partner on the equipment side, Ericsson. It was to Ericsson's facilities in Finland and Sweden that Eesti Telefon's engineers were sent for training in connection to the installation and operation of the AXE equipment in Estonia, and on the technological side Eesti Telefon's relationship to Ericsson appears to have grown increasingly important relative to the relationships to Nordic investors.

By the mid-1990s, Eesti Telefon complained that Telia had too much influence on developments that could be better handled by the Estonians themselves. There followed increasing demands from the Estonian side to transfer the management of Eesti Telefon fully to the Estonians. A compromise was found in January 1997, when Eesti Telefon's Swedish general director, Guy Sundqvist, left and was replaced by a Swedish-Estonian, Jaan Männik (Eesti Telefon, 1997a).

1997 was in general a year of substantial organizational changes within Eesti Telefon. In addition to changes in the management board, the firm went through a reorganization that mainly followed reforms carried out about a year earlier in Sweden. Notably, the *regional* structure of the company was replaced by a *functional* one, whereby separate units for information systems, customer service, production and support functions were formed. For the first time, Eesti Telefon also got a special department for *new services development*, marking an organizational adaptation to the new digital era.

These developments pointed towards a new era of fixed communications in Estonia. At the same time, however, Eesti Telefon remained responsible for the far from finished 'basic telephonization' of the country, as stated in the concession agreement of 1992. The government, being generally critical of how slow Eesti Telefon's expansion and modernization of its infrastructure was proceeding, made it very difficult for the company to get state approval of its business plans. The company was forced to rewrite its three-year business plan for 1997-1999, resulting in the obligation to increase the speed of modernization from 30,000 new lines annually to no less than 50,000 lines annually for the coming four years. For the first time, it was also specified that the waiting list was to be fully eliminated by the end of the year 2000. The planned investments for the coming years were thereby also raised to EEK 655 million annually, which was almost twice as much as the annual investment amounts that had been stated in 1992 (Eesti Telefon, 1997c). In December 1995, Eesti Telefon's general director (at that time still a Swede) estimated the annual investments for the coming years to around EEK 400 million (*Äripäev*, 15 December 1995).[8]

In this connection, one of the critical problems was Eesti Telefon's investments with respect to basic telephonization of thinly populated areas. The installation of completely new copper wire to thousands of dispersed rural settlements was extremely expensive in terms of costs per line. From the early 1990s, however, Eesti Telefon had experimented with a somewhat unusual solution that seemed to have the potential of radically reducing these costs: the wireless local loop (WLL). In cooperation with Ericsson, a minor network had in 1994 been installed that was based on Ericsson's 'RAS 1000' system. This solution can be described as a fixed, wireless variant of Ericsson's NMT-450 mobile telephony system, which meant that it was an analogue radio communications network with a relatively low frequency. The network, which in reality was among the first of its kind in the world, was installed on trial for 300 subscribers in a rural area close to Tartu in southeastern Estonia, and it was jointly financed by Eesti Telefon and federal and local governments (Eesti Telefon, 1994: 5ff.).[9] In 1995 and 1996, RAS 1000 networks were put into operation in southern Estonia, based on the same financial model and with about the same number of subscribers (Eesti Telefon, 1995, 1996a, b, c).

After the new obligations arising from the government's criticism in 1996, the RAS system took hold at Eesti Telefon, which decided to implement the Ericsson solution on a massive scale. In January 1998, a contract – the largest ever in the history of Eesti Telefon – was signed with Ericsson for the delivery of no less than 30,000 rural lines, supported by about forty RAS 1000 local exchanges. With this project, to be carried out gradually over three years starting in May 1998, Eesti Telefon judged that the rural telephonization would finally be completed (Eesti Telefon, 1998a, c). For Ericsson, the contract was regarded as the real breakthrough in the company's business with Estonia (Lehtimäki, int., 2001).

The vast use of the RAS 1000 system to solve the problem of rural telephonization in Estonia was an interesting case of solutions being found far outside the generally accepted ways of dealing with telecommunications services in the Western world. Local wireless loop technologies had, as a matter of fact, their main prospects in markets where telephone penetration was still low – such as in Eastern Europe and developing countries – and where the persistence of earlier technologies did not hinder the introduction of this new paradigm. On the other hand, however, RAS 1000 was in the end acknowledged as a technology with severe shortcomings in view of the information society and the emerging Internet era: it was, as pointed out above, basically an analogue system with strict limits, and it was in principle not constructed for advanced data communications. Although Eesti Telefon was keen to point out that most of the existing digital services would also be made available for RAS 1000 subscribers, it was clear – already before their installation – that the capacity of the lines did not at all correspond to the future demands of the Internet and other data communications solutions (Männik, int., 2001).

Nevertheless, both Telia and Tele took great interest in Estonia's experimentation with new and unusual solutions. From their perspective, Estonia was an interesting telecommunications market not least because of its vast need for new solutions in combination with the smallness of the country. This, together with an increasingly obvious openness towards new technologies among the Estonians, seemed to make the country unusually well-suited as a *test market*, where future technologies could be tried out on a smaller scale before being launched in the Nordic home countries (cf. Kõstner, 1996). However, a growing problem in this respect concerned international developments, with the emerging liberalization of the European telecommunications markets, on which Eesti Telefon's two Nordic investors, Telia and Tele, had started to act as *competitors*. From this perspective, the *joint* involvement of Telia and Tele in Estonia involved substantial risks with respect to the potential leakage of valuable knowledge. Obviously, this situation may have looked different, had Televerket accepted to enter Estonia as a *single* foreign investor in 1992 (Nurk, int., 2001).

Nevertheless, Estonia and the whole Baltic Sea region was the geographical field that both Telia and Tele came to gradually define as their new home markets. Much later, in 2002, this common interest would spur the *merger* of the operators, but they had by then been cooperating in the Baltic states for more than a decade. Outside Estonia, the two operators had jointly owned 49 per cent of the Latvian mobile operator LMT since 1992, and Tele was also the main foreign investor for fixed telecommunications in Latvia – a situation that was strengthened after the withdrawal of UK-based Cable & Wireless from the Baltics. In Lithuania, Telia and Tele – the latter by then about to change its name to Sonera – won the tender to jointly privatize 60 per cent of the fixed operator Lietuvos Telekomas in 1998, a deal which according to the operators was facilitated by Telia and Tele's long experience in Estonia and Latvia. Telia and Sonera also invested heavily in the main Lithuanian mobile operator Omnitel. In addition, they were important shareholders in the Russian mobile operator NorthWest GSM in St. Petersburg. In 1999, Telia declared that its aim was to become the leading network operator in the 'Baltic Sea region', that is, on both sides of the Sea. Allusions to Swedish hegemony in the 17th century Baltic Sea region and even to the age of the Vikings were not uncommon (see, e.g., Telia, 1999: 8f.).

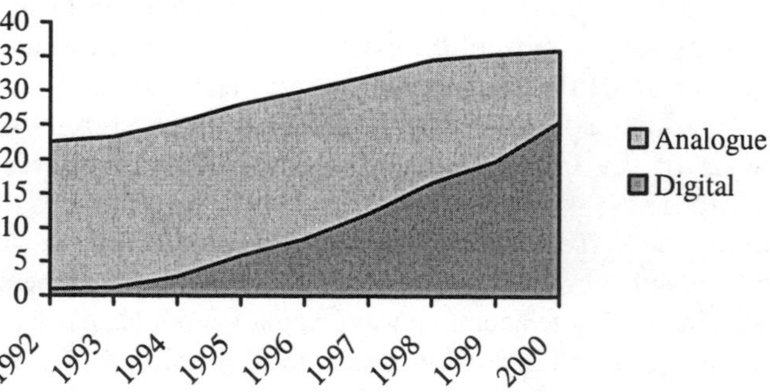

Source: Annual Reports of Eesti Telefon.

Figure 6.4 Fixed telephone penetration in Estonia, 1992-1990 (by analogue and digital lines per 100 inhabitants)

In terms of actual progress in Estonia, new lines were installed at a rate of about 30,000 connections annually, whereby fixed telephone penetration increased by about two percentage points per year up to 1998. This was about three times as fast as during the 1980s (calculated for the years 1980-85 from

Statistika aastaraamat, 1985: 137). During the latter half of the 1990s, however, the demand for new lines decreased, probably because of the shift towards mobile telephony, and the obligation to install 50,000 lines annually in order to reach a penetration of 45 per cent therefore lost relevance. However, the process of digitalization accelerated in response to increased investments enforced by the government. The overall development is illustrated in Figure 6.4.

3. MOBILE TELEPHONY: ENTERING THE DIGITAL ERA

After its launch in July 1991, the NMT-450 mobile telephony system operated by EMT grew to become an unexpected great success in Estonia. Despite the steep decline in overall economic activity in the newly re-independent country, and notwithstanding the high prices charged for the terminals and the telephony service, the number of NMT subscribers grew at an exponential rate (cf. Table 6.1). In terms of penetration ratios, mobile telephony in Estonia was in the mid-1990s at about the same level as countries such as Germany and France, and far ahead of other Baltic countries and Russia. In EMT's 1995 Annual Report, the Finnish chairman of the supervisory board, Matti Makkonen, predicted that 'the developments we see today are quickly lifting Estonia into the same league as Finland and Sweden, already world leaders in mobile penetration' (EMT, 1995).

Table 6.1 NMT users in the Baltic states, 1993-1995 (number of terminals per 100 inhabitants)

	1993	1994	1995
Estonia	0.6	0.9	1.3
Latvia	0.2	0.3	0.4
Lithuania	0.0	0.1	0.2

Source: Hultén and Mölleryd (1997: 170).

In 1991, the Estonian NMT network had been launched as an extension of Tele's NMT network in Finland, which meant that Estonia did not have its own NMT exchange at the start. In other words, all calls were connected through the NMT exchange in Helsinki – even if the two communicating parties resided in Estonia. The only substantial investment which had to be carried out at this very early stage was with respect to a few radio base stations in Tallinn. During the following years, however, EMT carried out a continuous network expansion both in the form of a higher density of base stations in Tallinn and other heavily populated areas, and through the growing territory covered by EMT's network. By the mid-1990s, the NMT network covered more or less all the inhabited Estonian territory, except areas such as the wide forests and swamps in the northeastern part of the country (*EMT Infoleht*, no. 1, 1996: 1, map). An NMT exchange, delivered by Ericsson, was also installed in Tallinn.

In December 1992, a period of organizational adaptations was initiated at EMT, as the temporary general director, Vladimir Jakovlev, was replaced by Tõnu Tee, a former director of the Soviet-era computer center of the Estonian Statistics Office. At the time of his appointment (which roughly coincided with Eesti Telefon's creation in the subsector of fixed telephony), EMT was still a very small firm with 20-30 employees and about 500-600 customers (that is, NMT subscribers). EMT rented a few rooms on the premises of Eesti Telefon and indeed appeared much like a loosely connected corporate venture of Eesti Telefon: the mobile company saw itself as sharply contrasting to existing lines of business and representing a totally different corporate culture from that of the fixed operator. From the perspective of Eesti Telefon, on the other hand, EMT was long seen as a curiosity aside the core activities of the Eesti Telekom group. Within a few years, however, the balance between the two sister companies would change dramatically.

A few Finnish representatives of Tele joined the workforce of EMT in the initial years, supporting in particular the technical activities of the new operator. Especially during the initial period when EMT's network was linked to Tele's, Finnish technical management was seen as necessary. The Finns played an important role in transferring knowledge related to mobile network design, whereby they also brought the necessary equipment to Estonia. The Swedish representatives did little here; Telia's people instead focused on consultations regarding the management and financial operation of EMT. As it was not self-evident that the employees of EMT had English language skills at this early stage of transition in Estonia, it is perhaps not surprising that the main Swedish consultant was a Swedish-Estonian.

While EMT was still a small company, networking activities with the Nordic partners took other expressions. It became a tradition in the early years, for example, that the whole company went for a Christmas cruise across the Baltic Sea to Helsinki or Stockholm, with visits at the partners' – Telia's and Tele's – head offices in Sweden and Finland. The general director of EMT was also initially sent for a two-month training period at Telia in Stockholm upon his appointment (Tee, int., 2001).

With respect to technology, EMT appears in practice to have had *two* technical directors simultaneously during the first few years: a Finn and an Estonian, the latter being expected to take over after some time. Similar to the fixed operator, EMT regarded the Nordic representatives as 'consultants' rather than directors. The Estonian technical director, however, decided to leave the company in 1994, which led to further important adaptations in the organization, as the new technical director, Tõnu Grünberg, took over. Grünberg had been studying at the local Department of Radio and Communications Engineering at Tallinn Technical University, but he had also completed a master's thesis in this field at Chalmers University of Technology in Sweden.

Grünberg's appointment concluded the formation of a stable board of *Estonian* managers at EMT. This corresponded with a structure that had been a deliberate vision of Eesti Telekom from the very beginning. The Nordic investors and consultants at EMT were satisfied with the arrangement, which from now on was to hold for many years to come. Starting from 1996, Telia and Tele no longer had any representatives on EMT's management board. The direct Nordic influence in the company was from then on limited to the supervisory board.

During the first half-decade of EMT's activities, the quality of the physical infrastructure was substantially improved. In 1994, practically all NMT base stations had still been connected to the exchange via analogue transmission lines. By the summer of 1996, 70 per cent of the transmission links were digital (Tõnu Grünberg in *EMT Infoleht*, no. 3, 1996: 1). The telephones used in the Estonian NMT network were dominated by Nokia, which had a market share of 63 per cent in September 1996. At second place was Ericsson, which in terms of infrastructure was usually the single supplier, but among the terminals had a mere 11 per cent share (*EMT Infoleht*, no. 4, 1996: 2).

Towards a GSM Network: EMT vs. Radiolinja

By the time of the Soviet collapse and Estonia's fully regained independence in fall 1991, analogue mobile telephony was no longer a radical new technology in the Western world and the Nordic countries. Digital (second-generation) mobile standards were about to emerge as full-fledged technologies in the West, and on the Western shores of the Baltic Sea the new GSM standard had been launched in 1991 by Radiolinja, a private Finnish competitor of EMT's Nordic investors, Televerket and Tele. Although Eesti Telekom was aware of these trends when EMT was founded, GSM had at this early stage never been presented as a viable alternative to NMT in the discussions; there seemed to be no need for such a complex and expensive system designed for a mass customer base.

Two years later, mobile telephony was looked upon in a completely different light in Estonia. Although penetration was still below 1 percent of the population, the accelerating growth and the encouraging experiences from the Nordic markets (the impressions from which were channelled effectively to Estonia through Telia, Tele and Ericsson) made it possible to imagine mobile telephony services as a future mass product.

In September 1993, a few GSM radio base stations were installed in Tallinn, forming Estonia's first digital mobile network, and the first calls were successfully carried out. The circumstances, however, were quite special. Estonia, through a bit of luck, had the opportunity to organize the international GSM MoU plenary meeting, and this seemed to be a perfect opportunity to demonstrate the eager technological spirit of EMT. It was agreed with Tele that the launch of GSM in Estonia was to follow the same pattern as in the case of NMT, that is, based on the idea that the Estonian network would initially function as an extension of Tele's network in Finland. The plan was successful, and GSM telephones could thus be used during the international meeting (EMT, 2001a: 9). However, it would take another year and a half before a full commercial launch of digital mobile communications could begin in Estonia.

In the case of NMT-450, EMT had often been accused of controlling a monopoly of mobile telephony in Estonia: 'From time to time the argument appears that NMT-450 is monopolized in Estonia', observed the company's technical director Tõnu Grünberg in 1996, while protesting: 'This argument is senseless, as the NMT-450 system as a standard is not created to enable several operators in one country' (Tõnu Grünberg in *EMT Infoleht*, no. 3, 1996: 1, my translation).[10]

In the digital era, however, there was nothing inherent in the *technology* that could prevent fierce competition, and in addition, there did not exist any *formal institutions* (in terms of regulations) that stopped competitors from building their mobile networks in Estonia; as discussed in Chapter 5, the concession agreement signed in December 1992 had given exclusive rights to Eesti Telefon for fixed telephony, but EMT, as a mobile provider, had never acquired any similar privileges, and the issue of market regulation in mobile telephony (as well as in data communications, see section 5 below) had in reality not been touched upon at all from a formal point of view.

This inspired the privately owned Finnish mobile operator Radiolinja, which had become the world's first commercial operator of GSM, to try its luck on the southern shores of the Finnish Gulf. In spring 1994, Radiolinja advertised in the Estonian newspapers for future Estonian employees of what would become EMT's competitor. In June, it was revealed that the position of general director at Radiolinja Eesti was to be taken by none other than EMT's development director, Ants Viira. Intense activities now focused on building the new network of Radiolinja, which in contrast to EMT chose Nokia as its main supplier of infrastructure (reflecting Radiolinja's traditional relations to Nokia in Finland). In November 1994, six months after the formal establishment of the firm, the first call could be carried out in Radiolinja Eesti's GSM network. A Nokia GSM exchange was installed on the company's premises in Tallinn, and on 10 January 1995, Radiolinja Eesti launched Estonia's first commercial GSM network (Viira, int., 2003).[11]

It is not clear why EMT, in spite of actually having established a functioning GSM network in Tallinn for the special purposes of the conference in September 1993, was initially left behind by Radiolinja in the commercial launch of the new generation mobile telephony. One might suspect that the loss of the person-embodied competence of EMT's (later Radiolinja's) development director contributed to delaying EMT's launch of GSM. But at the same time, EMT was doubtless an organization that had vastly more accumulated experience on the organizational level than the completely new firm Radiolinja Eesti.

On the other hand, it is not self-evident that an early introduction of GSM was a main concern at EMT. Estonia was in 1994 still a country in deep economic recession, which clearly had its impact on the prospects for rapid market growth in GSM telephony. In other words, it was not self-evident that Estonia was 'ready' for GSM in 1994. A third and more peculiar circumstance that tended to retard the birth of EMT's GSM network, was the perceived uncertainty with respect to the presence of the Soviet military on Estonian territory. The military was traditionally using the 900 MHz frequency band, and it was not completely clear when the radio spectrum would be free (Makkonen, int., 2003). It was only in September 1994 that the Soviet military finally left Estonia.

However, when it became clear that Radiolinja was about to launch its own GSM network, EMT could hardly wait any longer. From now on, it was clear that the company had to do everything to respond to the new competition. In September 1994, EMT signed an important supply contract with Ericsson for GSM equipment (*EMT Infoleht*, no. 1, 1996: 2), and after the official launch of its network three weeks later than Radiolinja,[12] EMT spent enormous efforts on overtaking Radiolinja in terms of number of customers, network coverage, quality and variety of services. Extensive developmental activities were thus initiated during 1995, 'in order to introduce new services and integrate them with the existing ones' (Tõnu Grünberg in EMT, 1995)

Digital mobile telephony doubtless offered radically new opportunities exactly with regard to the introduction of new services. While developmental activities in NMT on the side of the operator had in principle been limited to the improvement of the plain voice service quality, the digital era of GSM appeared to open up a whole new universe of technological and market opportunities. At the center stood the development of *mobile data communications*.

In spring 1995, data and fax communications were for the first time made possible in EMT's GSM network. In May, Voicemail was launched and in July SMS was introduced. At the fair Tele'95 in November, EMT tested 'Cell Broadcast'.[13] Although there was nothing radical in these services in international perspective, it is interesting to note that EMT at the time of their introduction was a firm that was no longer crucially dependent upon consultations with its Swedish and Finnish investors for its developmental activities. The most important external link seems instead to have been the relation to Ericsson, which in 1996 and 1997 supplied around 85 per cent of EMT's total equipment (Tõnu Grünberg in *Äripäev*, 12 December 1996).[14] The operator's investments amounted to around EEK 250 million in 1996, which was about as much as for the three previous years taken together (Tõnu Tee in *EMT Infoleht*, no. 1, 1996: 1).

From the perspective of networking activities, the growth of EMT led to interesting changes in its relation to Ericsson, as discussed further in section 4. In addition, due to the increasing number of customers, intense work for the part of the operators also had to be carried out with respect to *databases* and in particular the *billing system*. Here, the mobile operator for the first time established serious business contacts with Estonian IT-related firms. While the main data system was supplied by Oracle (EMT, 1996), many parts of the billing software had to be developed for the specific Estonian context. This meant a chance for local IT firms to do business, whereby, for example, a firm such as Aprote could be responsible for developing the function that makes it possible to set a limit to a phone call (Saarsen, int., 2001). At the same time, EMT connected its billing system to Hansapank's electronic banking system, Telehansa (see Chapter 5, section 6, and section 5.2 below) (EMT, 1996). With the rapidly increasing number of customers and the differentiation of services during the following years, the IT systems of EMT grew increasingly complex, which meant that EMT began to emerge as an important customer for the Estonian software industry and a key partner for the banks. There will be reason to discuss this in further detail in Chapter 7.

Ritabell: The Coming of a Third Operator

The GSM networks in Estonia grew relatively rapidly in terms of territorial coverage. This was so in particular with respect to EMT's network, which by the end of 1996 covered all county centers and main highways. However, GSM in Estonia was still a rather exclusive phenomenon from a customer perspective. By September 1996, the GSM networks were used by about 25,000 subscribers, which was about the same number as NMT subscribers. EMT was the clearly dominant operator, standing for 88 per cent of the subscribers, while Radiolinja had 12 per cent (*Äripäev*, 5 September 1996).

The marketing directors of EMT and Radiolinja, however, were both optimistic about the future growth of GSM in Estonia, prognosing around a ten-fold increase in the number of subscribers up to the year 2000 (*Äripäev*, 6 September 1996). This in itself optimistic prognosis would turn out to be a clear understatement of the actual future.

At this time, *competition* in GSM telephony was still a relatively underdeveloped activity worldwide. Many GSM networks around the world were still monopolies, and even in the most advanced telecommunications markets it was rare for there to be more than two operators in a country. Exceptions were Sweden and Hongkong, which both had three active operators (*Äripäev*, 28 April 1997). Estonia, for its part, could already be said to be advanced in terms of competition in GSM, thanks to its two operators which had launched their networks at almost the same time in January 1995. For a country with a mere 1.5 million inhabitants as the maximal customer base, this was in itself an interesting achievement.

In February 1996, however, a newly founded firm, Ritabell, announced its intention to build a *third* GSM network in Estonia (*Äripäev*, 1 February 1996). Ritabell was a joint venture between the Estonian firm Levicom (52 per cent), which had experience with the cable TV business and also acted as a reseller of mobile handsets for EMT, and Millicom (48 per cent), which was controlled by the Swedish Kinnevik group. The latter was behind the private mobile operator Comviq in Sweden, which had been involved in mobile telephony during the NMT era. Levicom and Millicom were also collaborating together with a number of other investors in a consortium that was at the same time bidding for Latvia's second GSM licence.

The director of Ritabell for many years to come was Toomas Peek, who in the 1970s had graduated with distinction from the Department of Radio Engineering at the Polytechnical Institute, where he had studied at the department together with, among others, the founder of Levicom and later Minister of Finance, Tõnis Palts (list of graduates, Department of Radio Engineering). In September 1996, Ritabell signed the main supply contract with Siemens, which had been selected as the main equipment supplier. The event took place through a formal ceremony in the medieval city hall in Tallinn and was seen as significant enough for the Estonian Minister of Transport and Communications to attend (*Äripäev*, 5 September 1996).

In March 1997, Ritabell's new network started to be tested, and in the beginning of May it was commercially launched under the brandname 'Q GSM'. The event was accompanied by a massive advertisement campaign of a kind that had never been seen in the Estonian mobile business before (Tõnu Tee in *Äripäev*, 11 June 1997). The dramatic reductions that were initially offered to new Ritabell subscribers can in a sense be seen as the beginning of the real 'mobile fever' that came to characterize this business in Estonia during the following years.

At the time of Ritabell's entering the Estonian market, there were about 64,500 GSM subscribers in the country. 53,000 belonged to EMT and 11,500 to Radiolinja, whose market share had thus somewhat increased. It remained to be seen how Ritabell would alter this market structure, and how the system of innovation would respond to the new competitive environment.

4. THE EQUIPMENT INDUSTRY: EXPLOITING LOW-WAGE LABOR

Procurement of Equipment

With the technological transition to digital technologies in the networks of fixed and mobile network operators, the old linkages to the Russian and East European telecommunications equipment industry that had been totally dominant during almost half a century were cut off in a way that did not allow for any gradual transition or compromises. Never again, it seemed, would an Estonian operator conclude an agreement for the delivery of telecom equipment from a Russian or East European supplier. The Soviet and East European technologies that were already installed would continue to be used for many years, as everything could not be replaced at once, but old equipment that was taken out of operation in the modernization process became a reliable source of spare parts, which made it unnecessary to keep up contacts with Krasnaya Zarya and other Eastern manufacturers. Eesti Telefon, which inherited the Soviet-era fixed network, even started to *export* some Soviet-era equipment back to the East, where modernization was a much slower process and the analogue equipment therefore was still in demand (Mets, int., 2001).

Replacing the old linkages to Soviet and East European manufacturers, the relationships that had begun to emerge with Western suppliers at around 1990 gradually gained momentum with the launch of the ambitious modernization project of Eesti Telefon and the construction of the new mobile networks of EMT, Radiolinja and Ritabell. In general, the overall procurement landscape that emerged in the course of the 1990s was characterized by EMT relying almost exclusively on Ericsson technologies, Radiolinja largely on equipment from Nokia, and Ritabell building a tight relationship with Siemens. On the fixed side, Eesti Telefon relied on Ericsson's AXE for its two international exchanges and some of the smaller switches, but in contrast to its sister company EMT, a second main supplier was Siemens, which with its EWSD platform stood for around half of the equipment (Männik, int., 2001). This was, at least partly, a deliberate policy resulting from Eesti Telefon's desire not to become too dependent upon a single supplier (Mets, int., 2001).

Geographically, Helsinki emerged as the most important point of connection for Estonia in terms of procurement. Helsinki was not only the international headquarters of Nokia but as pointed out in Chapter 5, Ericsson's business with Estonia had from 1990 re-emerged via the Swedish company's Finnish subsidiary, rather than via Stockholm. After the somewhat exotic visit to Munich in 1989, where Eesti Telefon had purchased a number of PBXs directly from Siemens' headquarters, Estonia's procurement of equipment from Siemens was also organized through the German group's Finnish subsidiary.[15]

Domestically, the early 1990s saw the emergence of smaller firms that focused on the import and reselling of smaller-scale equipment from Ericsson, Nokia and Siemens. Ericsson and Siemens offered popular PBX systems, and all three firms were through their resellers leading vendors of terminal equipment. In addition, in order to support their business with Estonian operators, Ericsson and Siemens (but not Nokia) opened their own agencies in Tallinn in 1993.

The intense activities of EMT in connection with GSM contributed strongly to further organizational changes with respect to Ericsson's business with Estonia. Up to 1995, Ericsson dealt with Estonia partly through the local reselling firm Baltimore and partly through Ericsson's Finland headquarters in Helsinki. It was the latter that had been reponsible for Ericsson's business with Estonian operators (Lehtimäki, int., 2001). In January 1996, however, the Estonian subsidiary Ericsson Eesti was created through the acquisition of Baltimore, one of the reasons being the need for better coordination of joint activities on the part of Eesti Telefon and EMT in connection with the delivery, repair and updating (reprogramming) of the exponentially growing amount of Ericsson equipment installed in Estonian networks. Ericsson's sales in Estonia increased from around EEK 100 million in 1994 to EEK 600 million in 1998, a development that seemed to motivate a more active role for Ericsson on the southern shores of the Gulf of Finland (Mets, int., 2001; Lehtimäki, int., 2001).

Ericsson Finland, however, was not satisfied with the new subsidiary's first year of activities in the rapidly growing market. The desired direction of business for the subsidiary was made clear in spring 1997 by appointing to Ericsson Eesti a new general director who had earlier been responsible for Ericsson Finland's operator business in Estonia (ibid.). The responsibility for the company's operator business in Estonia was now moved to the Estonian subsidiary. In addition, as the new director, while in Finland, had also been responsible for Ericsson's operator business with Latvia, Ericsson Eesti now took over the responsibility for sales to the neighboring Baltic country (Lehtimäki, int., 2001).

The Evolution of Elcoteq

In Chapter 5 it was mentioned that Estonia through the Ministry of Communications and later Eesti Telekom had taken an interest in establishing telecommunications manufacturing *in Estonia itself*. The idea had been to create a joint venture together with one of the well-known international equipment producers, and thereby build upon existing competencies and traditions – for example, from Tartu, where the pre-war telephone factory was still a legend and very much alive in the collective memory. Relatively soon, however, Eesti Telekom concluded that this vision had no future, and it was instead decided that the necessary equipment was to be procured directly from the large international producers (see above).

However, Eesti Telefon's decision to rely on foreign equipment suppliers did not necessarily mean that the domestic equipment industry was doomed to death. On the contrary, after the disastrous collapse in the years around 1990, domestic activities in the Estonian electronics sector showed interesting signs of revival. Although it appeared that the huge Soviet-era electronics sector – in which nearly 30,000 people had recently found employment – had been lost forever through the broken linkages with the East and the collapsing markets there, one could by the mid-1990s point to a number of encouraging examples of new or privatized firms that had begun to orient themselves successfully towards the Western markets. Moreover, although telecommunications represented only a negligible share of the republic's electronics industry in the Soviet era, the extremely rapid worldwide growth of the telecom sector and the geographical as well as cultural proximity to two of the Western world's leading telecom countries, made Estonia an interesting potential location for the *outsourcing* of telecom manufacturing – in particular for subcontractors to the large Swedish and Finnish telecom vendors, Ericsson and Nokia.

The most obvious example of this was the development of Elcoteq Tallinn, whose creation was discussed in Chapter 5. At the time of the formal establishment of the company's Estonian branch in 1993, 20 people had been employed, a number that grew to 840 by December 1995. Many of these people were earlier workers of RET and other old electronics production enterprises such as the former Pöögelmann (i.e., Tondi Elektroonika) and Kalinin (i.e., Estel) factories. Elcoteq Tallinn's production was now focused on around 30 products for seven different foreign – mainly Nordic – customers. The main products were computer display electronics, accessories for mobile phones (battery chargers) and audio-video components.

Production processes included considerable manual work, although three industrial robots were also used. The Estonian plant conducted all testing activities itself (information from Elcoteq, 2004). All goods were exported, and with a total export value of EEK 438 million in 1994 and EEK 1,361 million in 1995, Elcoteq Tallinn had by then already grown to become Estonia's largest overall exporter (*Äripäev*, 8 December 1995 and 6 March 1996).

During the following years, Elcoteq Network developed in an interesting way, and Elcoteq Tallinn as part of it, through an increased concentration on the production of telecommunications equipment, with the two Nordic telecom giants, Ericsson and Nokia, as increasingly important customers. In 1997, 62 per cent of the company's turnover originated in the production of mobile phone components and accessories, while industrial electronics (where the main customer was ABB) represented 14 per cent. During the first half of 1998, the share of telecom equipment had increased to 71 per cent, whereas industrial electronics had decreased to 13 per cent. Ericsson and Nokia's share of the turnover at the same increased from 66 per cent to 76 per cent (*Äripäev*, 7 August 1998).

By 1997, in addition to mobile phone components and accessories, production also included parts for GSM radio base stations. In the same year, however, a breakthrough occurred through an agreement with Ericsson that made Elcoteq Network's Tallinn plant responsible for the *manufacture of a mobile phone from start to finish*. Elcoteq Tallinn had at a very early stage gained experience not only in the assembly process itself, but also in related testing activities. It had also begun to procure components on its own, and although the finished goods were not sent directly to the end customer, the Tallinn plant had been involved in packaging and logistics, too. All in all, this meant that Elcoteq Tallinn had the necessary ability to provide the 'full' electronics manufacturing service. An important factor that increased Elcoteq Tallinn's reliability in the eyes of the customer was the existence of a Phone Repair Center, which had been established one year earlier. It was a joint project with Ericsson, which sought a means to gain expertise in repairing returned and defected products (information from Elcoteq, 2004).[16]

Thus, the new joint Ericsson-Elcoteq plans stated that the new phones were to be manufactured in 'high volumes in a *box build* mode, which means that Elcoteq delivers finished, quality controlled, tested and packaged products to Ericsson [or its customers]'. Elcoteq Tallinn was thus now required to procure the required electronics components and other parts for the phone in question – the new Ericsson GA-628 model – as well as manufacture and test the products, and pack and dispatch them directly to the customer's distribution center (Elcoteq, 1997a). Such responsibility for the whole production process had never been carried out anywhere in the global mobile telecommunications sector. Interestingly and perhaps surprisingly, it was thus in a former Soviet republic that a subcontractor for the first time in mobile telephone history had full responsibility for the manufacture of a phone from start to finish. The manufacturing of Ericsson GA-628 started in the summer of 1997.

Elcoteq's Tallinn plant was extended several times through green-field expansion of its territory on the outskirts of the Estonian capital. This growth, which followed Western standards and which was accompanied by the installation of advanced Western machinery, contrasted sharply to the vast empty facilities of the former Estonian electronics production giants, whose sad ruins at that time could be seen in all industrial areas in Tallinn and elsewhere. Elcoteq emerged as a new symbol of hi-tech – though still labor-intensive – production in the new re-independent Estonia. The factory was frequently shown to visitors when profiling Estonia as a country of modern industry. The perceived importance of the company for the entire Estonian economy as well as for the Finnish-Estonian relations were explicitly demonstrated at a number of occasions. In June 1996, for example, none other than the President of the Republic of Estonia, Lennart Meri, and the President of Finland, Martti Ahtisaari, ceremonially laid the foundation stone of the Tallinn factory's expansion. At the occasion, Elcoteq Tallinn's general director argued:

> Thanks to our factory in Tallinn, we have been able to take up the challenge of intensifying international competition by manufacturing electronics in large series run cost-effectively. Thus we can successfully compete with the cost level of the Far East, where the toughest competitors of European electronics manufacturers operate. (Elcoteq, 1996)

In 1996, the firm was chosen as the 'foreign investor of the year'. In the celebration that took place on this occasion in March 1997, the jury underlined the overall 'vitalizing effect' of Elcoteq's success on the development of Estonia's total economy. It was noted that 'Elcoteq [Tallinn] has introduced new modes of operation and created possibilities for other Estonian companies to work as part of its network'. A speech was made on the occasion by the protector of the contest, President Lennart Meri. In 1997, Elcoteq Tallinn received an award for having successfully developed Estonian Production Quality and Technology (Elcoteq, 1997b). In October that year, around 1,900 people worked in the Tallinn factory (*Äripäev*, 28 April 1997).

Clearly, the ultimate rationale for the company's operation in Estonia was the availability of low-wage labor. Elcoteq's Finnish headquarters calculated in 1997 that production costs were no less than ten times higher in Finland as compared to Estonia (ibid.). At the same time, Estonia offered, in the eyes of

Elcoteq, a number of additional advantages, such as an open-minded economic environment; low energy costs; a suitable logistical location; far-reaching customs exemptions; flexible labor legislation; and educated personnel (information from Elcoteq, 2004). In this perspective, it was hardly surprising that the mid-1990s saw a decrease in the number of employees in Elcoteq's Finnish plants while the number of employees in Estonia grew rapidly. As a consequence, the Tallinn plant became with time relatively more important for the whole Elcoteq corporation. An interesting aspect of Elcoteq's Tallinn branch, however, was the fact that it consisted almost exclusively of local staff – not only at the workers' level, but also with respect to the board of managers. Moreover, the competencies in terms of the overall management and build-up of an 'electronics manufacturing services' (EMS) plant that had been accumulated practically from zero starting in 1992 proved to be of considerable value for the corporation's further expansion and globalization that started to take off towards the latter half of the 1990s.

So far, Elcoteq's operations had been limited to Finland and Estonia, with the main customers located in Sweden and Finland. In 1997, Elcoteq investigated the idea of expanding its activities to Latvia or Russia (ibid.). The slower growing labor costs in Russia motivated Elcoteq to first choose St. Petersburg as a location, where in October 1997 pilot production was launched. The responsibility for the build-up of the St. Petersburg plant was delegated to Elcoteq Tallinn, whose representatives were thus in charge of educating Russian managers and generally building-up production there (*Äripäev*, 10 October 1997). The Russian plant reportedly made 'encouraging progress in terms of quality and productivity', but bureaucratic difficulties arose in the negotiations with city officials in St. Petersburg and with federal Russian authorities in terms of 'unsolved problems relating to customs and logistics procedures'. Thus in November 1999, there had still been no decision for taking the step to increase production capacity from a pilot plant to a full-size production enterprise in the Russian city (Elcoteq, 1999).

Of greater importance for the corporation seems to have been the opening of a new plant in Hungary in 1998, which was then the first full-scale factory outside the Gulf of Finland. This marked the breakthrough in Elcoteq's globalization strategy, which had been indicated by the company's introduction on the Helsinki stock exchange at the end of 1997. Newly employed Hungarians of Elcoteq now came for training to Elcoteq Tallinn, which thus functioned as the basis for knowledge transfer to Hungary, where the new plant was opened in the end of 1998. A similar process took place in the following year, when Elcoteq prepared its launch of a production plant in China (Petersen, int., 2003). Within a couple of years the company had increased its capacity many times over. In 1999 the network of plants covered more than ten countries in Europe, America and Asia; however the core was still undoubtedly on the shores of the Finnish Gulf: in Finland and Estonia.

In a striking way, opinions differed widely in Estonia about Elcoteq and its role in the Estonian economy. While it often won awards in terms of 'best foreign investor', 'best exporter', etc., some critics argued this was due merely to its size. While the company's proponents pointed out that Elcoteq's employees acquired high-level qualifications that could later spill over to other Estonian companies, its critics noted the low wages paid and the profits leaving Estonia. The company itself drew attention to the modern equipment in the Tallinn plant, including component placing automats, wave-soldering equipment and automated testing equipment. Because of the advanced machinery most personnel were not involved in production itself, but rather dealt with testing and related activities. In this sense it was not completely clear to what extent Elcoteq Tallinn represented a way to a more innovative future, or was rather becoming locked into a position where competitiveness remained determined by low-wage labor and other cost advantages. In the following period, somewhat less ambiguous signs of where the company was heading would be given (see Chapter 7).

Other Subcontractors

As noted in Chapter 5, Elcoteq's growth in Estonia had originally started with Finnish representatives of the firm looking for potential enterprises that could function as a basis for an Estonian production facility. However, although test production had begun in cooperation with RET in Tallinn, the firm grew mostly through the establishment of completely new facilities. Elcoteq thus never took part in the privatization of the old Soviet Estonian electronics enterprises.

But Elcoteq was not the only foreign firm that took interest in Estonia. Other potential investors – in particular from Sweden and Finland – similarly visited the country in search of investment opportunities in the early 1990s. One of the more interesting investment objects in the inherited electronics sector was Tarkon, the Soviet successor of the legendary telephone factory, which in Soviet times had mainly been producing flight recorders for Soviet aircraft (see also Chapters 4 and 5). Some Estonian observers envisioned Tarkon would be able to-collaborate with the well-known university in Tartu, and by commercializing research results, transform the enterprise into an innovative hi-tech producer. However, while Tarkon was probably able to manufacture advanced electronics products, the management team did not see any basis for product innovation based on findings from the university:

> We are ready to bring into production products or components derived from research at universities and other scientific organizations, if there is a market for

> them. Unfortunately, our [Estonian] inventors have not been able to offer products which could be industrially produced. (Toomas Noorem, CEO of Tarkon, in an interview with *Äripäev*, 11 November 1996, my translation)

Domestic inventions originating in the universities were thus seen to be non-existent. Furthermore, Tarkon had practically no in-house product development competence at all (Noorem, int., 2001).

Instead, the future seemed to point in a completely different direction. In 1993, the factory, which until recently had employed 2,500 people but whose Eastern markets were now collapsing, saw a number of Western visitors and potential investors. Among these belonged the Swedish fine mechanics firm Hallberg-Sekrom. A collaborative agreement was initiated in which the Swedish firm started to outsource part of its production to Tarkon. The components outsourced for manufacture included telecommunications equipment that Hallberg-Sekrom produced for the Swedish firm Allgon, which specialized in parts for radio base stations. Tarkon thus supplied Allgon via Hallberg-Sekrom. The early relationships to these Swedish firms would turn out to be of vital importance for the company's future (Carlsson, int., 2003; see further Chapter 7).

In the years around 1990, as noted above, Tarkon had also been of interest as a potential factory for telecommunications equipment production in connection with plans for modernizing the Estonian telephone networks. The basis for this idea was the historical role played by the Tartu factory before the war, in which far-reaching cooperation took place with domestic network operators (as discussed in Chapter 4). The idea soon turned out to be unrealistic, but in 1994, Tarkon revived its cooperation with Eesti Telefon – though at a somewhat different scale: based on a license from an Israeli company, an agreement was reached between Tarkon and Eesti Telefon for the production of around 1,000 new *payphones* to be installed in Estonia, replacing the old Soviet ones (ibid.). Thus Tarkon made its come-back in the Estonian telecommunications business.

However, cooperation with Hallberg-Sekrom would play the most important role for Tarkon's future. Collaboration with the Swedish company gradually deepened, which in 1996 resulted in the Swedish firm buying 60 per cent of the shares in the then still state-owned Tarkon (Tarkon, 2003). The Estonian privatization agency had been trying to sell the company since March 1994, but negotiations with different potential investors had not led anywhere (possibly as a consequence of the problematic debts that the factory had accumulated in the early 1990s). In 1996, the state decided to liquidate the whole company, and it started to sell all inventories, from real estate to production machinery, in a public auction. In the summer, however, an agreement was struck with Hallberg-Sekrom, which agreed to buy the part of the factory that was still operating for the modest price of $0.5 million

(Toomas Noorem, CEO of Tarkon, in an interview with *Äripäev*, 11 November 1996).

Investments by the new owners enabled more far-reaching reorganization of the firm, in particular with respect to the introduction of Western machinery. Tarkon's production initially focused on different kinds of mechanical parts, which were later complemented by labor-intensive assembly of cable systems. The customers were found in widely varying cable markets. However, with the rapid growth of the mobile telecommunications industry, the later 1990s saw a gradual concentration on the assembly of parts for radio base stations and other areas within telecommunications equipment. In addition to Allgon, an interesting customer was Ericsson, once the major owner of the old telephone factory (Carlsson, int., 2003; Noorem, int., 2001). With growing orders from Swedish customers, who proved satisfied with the Swedish-owned Estonian factory, Tarkon left its low-level mark and grew from around 225 employees in 1996 to 327 two years later. In 1998, the turnover of the company equalled EEK 65 million, an increase of almost 100 per cent as compared to the preceding year (Noorem, 2002). This remarkable growth continued until the Nordic telecom crisis beginning in 2000.

Another venture that sought to build on an existing plant was the collaboration between Nokia Cables and the former Soviet Harju Elekter plant, forming the joint venture Keila Kaabel in 1992. The goal was to supply the Estonian, Latvian and Lithuanian markets with cables for power supply and communications. The firm developed successfully during the 1990s and became the dominant supplier of cables in a number of market segments in the Baltic states. It also got recognition through an ISO 9000 certificate. However, the joint venture came to focus almost exclusively on cables for power supply, whereas the communications business remained in the shadows (Enno, int., 2003; information from Keila Kaabel, 2003; see also *BNS*, 26 October 1993).

The inherited Estonian *semiconductor industry* faced huge problems throughout the mid-1990s. Opinions differed with respect to the actual level of inherited competencies and equipment in this industry, but the main characteristic of the Estonian semiconductor sector in the mid-1990s was the almost complete lack of foreign investment. The lack of investment, in turn, made it virtually impossible for the old firms to acquire the production equipment necessary to compete on the advanced international electronics markets. Those few firms that produced semiconductors therefore, not surprisingly, had their focus on markets in the former Soviet Union, China and developing countries. In 1998, Tondi Elektroonika (the former Pöögelmann works) eventually terminated its semiconductor production, shifting to the import of necessary components used for its core activity, which was now the production of electronic hearing aids (Tani, 1999: 57). Also Estel Pluss (the former Kalinin works) drastically reduced its semiconductor production, and it seemed unrealistic that the production could continue for long (ibid.).

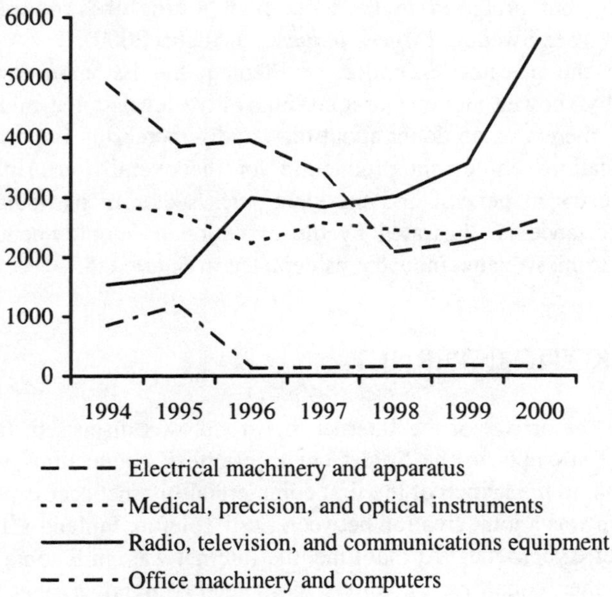

Source: Raagmaa et al. (2003: 11).

Figure 6.5 Employment in electronics-related sectors in Estonia, 1994-2000[17]

A big problem was that many firms, in particular those that did not acquire foreign investments, continued to use extremely outdated equipment from Soviet times. In many firms work operations were performed manually, and many products were produced in very small quantities, which often meant that it was not viable to use modern production technologies to any greater extent. From an innovation perspective, a serious problem in firms that did not manage to acquire foreign investment or establish themselves as subcontractors to powerful Western firms, was the lack of modern *testing equipment*. In many Estonian electronics firms testing was therefore limited to visual control, which in many cases proved unacceptable (ibid.). Elcoteq was in reality a rare exception to this pattern, as it acquired such equipment directly from its customers. In view of the fact that testing is often a first step towards capturing more innovative value-added products and processes, this hindered a movement towards more innovative activities. Thus a firm such as *Glaskomponent*, a Swedish producer of, among other things, glass components for radio base stations and broadband networks, established an Estonian subsidiary (employing 20 people) for labor-intensive volume manufacturing, but preferred to keep the final testing and control at the mother company in Sweden (*Dagens Industri*, 3 August 2000).

Thus with the possible exception of Elcoteq, the Estonian equipment industry hardly showed any encouraging innovative level in the mid-1990s. Nevertheless, there was no doubt about the rapidly increasing importance of telecommunications equipment production for the overall performance of Estonian industry in general and the electronics sector in particular. This growing importance is illustrated by the evolution of employment in the Estonian electronics-related industry, as depicted in Figure 6.5.

5. THE EXOTIC INTERNET

In Chapter 5, the arrival of the Internet to Estonia was discussed, from the difficult early attempts in the Soviet era to establish connections with the Western world, to the launch of the first commercial Internet access provider, EsData, which was a joint creation between Eesti Telefon, Finland's Tele and the Institute of Cybernetics. At that time, the Internet was, in Estonia as well as in most other countries, an almost completely unknown phenomenon among ordinary consumers of telecommunications services. In 1994 and 1995, however, important global innovations – the world wide web and the Netscape browser – entered the stage on a broader basis, which proved decisive for the Internet boom in the West during the latter half of the 1990s. As it turned out, Estonia was more than prepared to ride this trend (typically in a much more enthusiastic way than countries such as France and Germany), and the mid-1990s can thus be seen as the period in which the Internet found its way out from its exotic status to a widely used technology among broad layers of the population.

5.1 Provision of Internet Access

The period under discussion here represents the crystallization of Internet access provision as the most important market with respect to data communications services. This is perhaps best illustrated by the evolving strategy of Eesti Telefon as a data communications provider during the 1990s, as discussed in the following.

Eesti Telefon as a provider of Internet access

As noted in Chapter 4, the Ministry of Communications (i.e., the predecessor of Eesti Telefon) had been involved in data communications in Soviet times in connection to the Akademset project in the 1980s. The responsible division for the Ministry's participation in the project had been the computer center of the Ministry. Similar to other computer centers belonging to different Soviet-era Estonian organizations, the Ministry's center faced radical change from the end of the 1980s, as modern Western computers started to become available, which tended to make the computer centers obsolete. As a matter of fact, the most relevant function of the center that remained by 1990 was data communications, and the center was therefore transformed in 1990 into an organization called Estpak Data, which belonged under the Ministry and later became a subsidiary of Eesti Telefon (Wõrk, 1993: 26).

Estpak Data thus became Eesti Telefon's provider of data communications services. Estpak provided first and foremost access to the X.25 network – its customers were mainly banks and related firms – as the provision of Internet access was hardly defined as an area to focus on in any commercial sense. Up to 1994-95, Eesti Telefon seems to have regarded the Internet as an area outside their core business and therefore preferably handled by other organizations, in particular EENet, which was responsible for the Internet access of the educational, research and cultural sector (cf. Chapter 5).

The formation of Estpak therefore did not resemble a 'corporate venture' in a Western sense, i.e., a new department aimed at a promising future business area. Rather, it reflected the inherited responsibilities of the old Soviet Ministry of Communications, which had begun constructing the X.25 network long before the new institutional situation had become clear in December 1992 (when the concession agreement was signed). The core activities of Eesti Telefon were seen to lie in the provision of fixed telephone services, with nearly all energy channelled to the modernization program for physical infrastructure. In the course of the 1990s, however, this situation was bound to change in an interesting way.

In 1993, Eesti Telefon decided to use its Estpak subsidiary to construct a TCP/IP network, covering all of Estonia's major cities, but this was regarded more as a response to the operator's own, internal demand. Still in mid-1994, Eesti Telefon did not have any intention to start offering commercial Internet services. In the midst of creating the internal network, however, the operator recognized that there was great public interest in the project. As a result, the subsidiary Estpak started to provide Internet services in a number of cities in 1995 (Martens, 1994: 26, 1997a). At the same time, the number of X.25 customers was subject to stagnation or very moderate growth: In 1992, the X.25 network had 100, in 1993 96 and in 1994 135 subscribers (*Baltic IT Review*, no. 1, 1996: 57).

Still in 1997 Eesti Telefon provided Internet access only through its Estpak subsidiary (Eesti Telefon, 1997d). In October 1997, however, Estpak merged with its mother company, Eesti Telefon. This was officially motivated by ongoing structural changes in Eesti Telekom in preparation for introducing the company on the stock exchange (*Äripäev*, 10 October 1997), but to an even greater extent it reflected the now acknowledged importance of the Internet access market. In December 1998, Eesti Telefon began to offer Internet access through the new brandname package ATLAS. From then on, the Internet became a main focus of Eesti Telefon and was regarded as a core activity of the company (as explicitly stated in Eesti Telefon, 1998a).

The emergence of commercial Internet access provision

The global Internet boom that started in the mid-1990s came at a suitable point in time from the perspective of Eesti Telefon. If this worldwide boom had occurred a few years earlier, the situation would have looked completely different, for the simple reason that the physical infrastructure in Estonia, as described above, was in a miserable state throughout the early 1990s. For a long time, for example, Internet connections between Tallinn and Tartu took place solely through an expensive satellite connection that routed the traffic through Internet exchanges in the Nordic countries, as no adequate direct connection existed between the two towns (see Chapter 5). The intense activities aimed at constructing a fiber-optic network throughout Estonia and building fast international connections took place in parallel with the emergence of the Internet access business, and this circumstance enabled other actors to take advantage of the situation. From a legal perspective, there was nothing that could prevent private firms from entering the Internet business. Not only was there a lack of exclusive rights in the field of data communications, but firms were not even required to apply for a license to provide such services. The resulting wave of new firm formation and market entry in this field is discussed below.

EENet, as mentioned in the preceding chapter, was one of the key actors in bringing the Internet to Estonia. It had its origins in the cooperation between the Institute for Chemical Physics and Biophysics (KBFI) in Tallinn and the Biocenter in Tartu. EENet remained focused on users in the educational, scientific and cultural spheres, and although a diversification of the customer base in the direction of commercial users took place at a later point, this appears to have happened too late for the organization to emerge as a serious competitor on the commercial market. However, many new entrants offering Internet access to private and corporate customers following the introduction of the first web browsers were, as a matter of fact, spin-offs from EENet (*Baltic IT Review*, no. 1, 1996: 55). EENet was thus important as a basis for new firm formation and the provision of a skilled labor force in the emerging Internet access subsector.

The typical new entrant in the emerging Estonian access market was wholly owned by Estonian private capital and had been active for some years in related areas such as computer assembly, customized software production and general support. Thus the most popular commercial Internet provider became MicroLink, a company that had started in computer assembly (see Chapter 5) and later emerged as one of the leading IT groups in the Baltics. Its dial-up Internet access service started to be offered in April 1995, using equipment from dominant US firms such as Cisco and Annex (ibid.).

A similar path was followed by Teleport, which emerged as the second most important dial-up provider in these early years. Teleport used a leased-line connection from EsData to provide Internet access, as did Pennu, a company which similarly to MicroLink had started out as a computer assembler (ibid.).

An access provider with a somewhat different strategy was Uninet, which started operations in Estonia at the end of 1994 as one of the first commercial providers. Its main focus was on corporate customers, who were offered Internet access through leased lines. A peculiarity of the company was its cooperation with the Finnish telecommunications sector, which led to the establishment of its own commercial communication channel to Finland and the Finnet network there. In 1996, Uninet was also still the only company in Estonia to provide *ISDN services*. All in all, it was thus an aggressive and innovative player on the emerging Estonian data communications market (ibid.: 55).

A more specialized Internet provider was Data Telecom, which started offering Internet access in May 1995 through leased lines. Its customers were mostly travel agencies, some of whom (such as Estravel) were also major shareholders. Data Telecom cooperated with the European EUNet, through whom it was directly connected to international networks (ibid.: 56).

The emergence of EsData as the first commercial Internet provider was discussed in Chapter 5. EsData and Uninet had in common their being used as bases for other Internet access providers: as mentioned above, Teleport and Pennu used leased-line connections to EsData's network to access the Internet. In the same way, Baltic Computer Systems (BCS) and Sünerkom emerged as Internet access providers by way of Uninet's network (information from EsData, 1996).

The evolving organizational landscape was accompanied by a steady growth in the number of Internet users. A breakthrough came in 1995 and 1996, and Estonia emerged as a surprisingly enthusiastic Internet country in European comparisons, with strong articulation of demand. After a period of exponential growth during the mid-1990s, by 2000 Estonia had more Internet connections per capita than countries such as Germany. On the other hand, there was still a gap between Estonia and the Nordic countries, where Sweden had twice as many Internet connections per capita and Finland four times as many (Eesti Telefon, 2000a: 20).

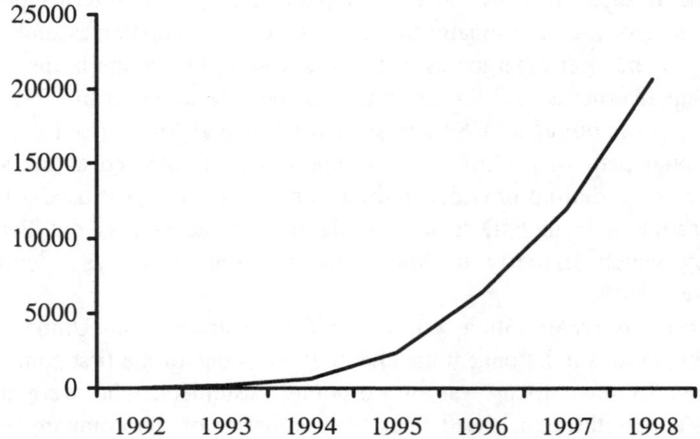

Source: CIESIN (2003).

Figure 6.6 Number of computers connected to the Internet in Estonia, 1992-1998 (mid-year)

5.2 The Development of Data Communications Applications

Once the Internet access business took off, vast opportunities emerged for totally new lines of business related to the Internet. For example, Sünerkom's main business, while also being an access provider, was the provision of media on the Internet. Thus Sünerkom hosted from 1995 onwards many of the first Internet-available Estonian newspapers, such as the big daily Eesti Päevaleht. With the immediate availability of search functions in the newspapers' archives, an important added value was demonstrated in terms of media on the Internet.

Teleport, for its part, was through ownership and collaboration tightly connected to a firm called Meediamaa, which emerged as a leader in the provision of web-design and related services – fields which until recently had been totally non-existent both in Estonia and worldwide. It was created in 1995 by the coordinator of the 'Tiger Tour', Linnar Viik (who had earlier been studying and working in Stockholm, Helsinki and the US) and Tarvi Martens, the director of ASO (the Data Communications Department of the

Institute of Cybernetics, which had started the governmental inter-agency data network – see section 5.3). Meediamaa, located in the former Soviet-era Pöögelmann electronics factory, also created an important overall Estonian portal, WWWärk, at the prominent web address http://www.ee. It was launched in 1994 with the goal to create a catalogue of all Estonian web pages, which at that time was less than 100 (*Äripäev*, 9 October 1997).

From an innovation perspective, however, even more interesting trends could be sensed in the fields of banking and public administration services. In particular the development of the banking sector in this period was later to prove decisive for the entire sectoral system of innovation, and it will therefore be discussed in detail in the following.

The evolution of electronic applications in the banking sector

Banking was a sector that differed in the Soviet Union from most Western countries. Commercial banks practically did not exist before the emerging transformation of the economy in the later 1980s, and the banking landscape that began to take shape in the early 1990s was therefore almost totally dominated by new start-up banks. These had to create, among other things, their information technology systems from scratch, a task that was approached in different ways in different banks. The young Estonian banking sector was in a somewhat peculiar position in this respect, because both the smallness of the country and the emergence of the new banks coincided with rapid developments in computer and communications technologies worldwide. For Western banks, which had been accumulating their IT systems in a relatively incremental way and along technological trajectories that typically stretched back several decades in time, it was in many ways natural to continue improvements along these trajectories, which were as a rule based on mainframe technologies. But in Estonia the situation was completely different, as there hardly existed any inherited banking systems at all.

The potential benefits of imitating existing foreign solutions were not at all clear from this perspective. In a sense, the start-up banks in Estonia were squeezed between two options, as formulated by one of the new banks' IT manager: either 'to build the business around decades-old technology and mature software, or to be on the leading edge, which is very often a bleeding edge' (Kaal, 1996: 38). In reality, the situation was not necessarily that black and white, but it was certainly clear that from the point of view of the young Estonian banks, there was considerable need for the ability to make difficult technological and entrepreneurial choices, both as to whether or not to imitate already existing technologies, and if not, then what technological strategy to follow, how much to develop in-house, etc.

It soon turned out that the commercial banks, as a rule, *were* able to make the important choices pointed out above. Moreover, their overall strategies in this respect crystallized at a very early stage, the breakthrough coming in 1993. Several banks then concluded that the best alternatives would have to include substantially new solutions – at least with respect to *process innovation* (if process innovation is here understood as the ways in which banking services are produced). On the *product* side, the aims hardly included anything new, i.e., the service ideas implemented in Estonia had long existed in Western countries if viewed from the customer's perspective. But on the process side, the situation was different. Moreover, several Estonian banks also found the field of information and communication technology so central to their strategy and future competitiveness that they typically preferred to develop their software solutions *in-house*. This was also related to the expected growth of the systems, starting from zero and creating a strategic need for 'secure timely software support for the extensive changes that were likely to occur in the course of the dynamic development' (Liik, 1996: 37). A common motivation to use domestic solutions was also the simple fact that the young, small Estonian banks could not afford the more expensive foreign systems (Sildmäe, int., 2003).

While hardware was always imported and the basic software platforms – such as operating systems and international database solutions – were also typically purchased from Western firms (usually Anglo-American producers), the applications themselves had little in common with available foreign solutions. For example, Hoiupank based its IT systems on small RISC computers because 'a mainframe-based system would have been beyond our means and a rapidly depreciating investment', and though its software solutions were based on the Progress system, the processes and products built on these platforms were developed independently (Liik, 1996). Similarly, Forekspank, which emerged as a first mover in electronic banking services in Estonia, selected a simple PC network consisting of OS/2 computers, complemented by 'Equation' software from the UK-based Kapiti. Starting from this, the bank developed its own in-house solutions (Kaal, 1996).

In Hansapank, which later emerged as the dominant bank of the Baltics, technological competencies accumulated initially through inter-organizational networking rather than in-house activities. The basis for this was the bank's close links to the small IT firm Crebit, which had an interesting history. The founders and managers of Crebit had backgrounds as mathematical modeling specialists, and the general manager Tõnis Sildmäe, with a PhD in this field from Moscow, had been working as the head of a computing center of an Estonian electronics R&D organization in Tallinn. Immediately preceding the birth of Crebit, in the period from 1989 to 1991, its founders had been working together in a private company in Finland, specializing in IT application for conference centers, including billing and related problems (see also Chapter 5). Cooperation with Hansapank began initially with minor consulting activities in 1991, but quickly grew as the Estonian banks' concern with telecommunications-related solutions increased (Sildmäe, int., 2003).

From a services perspective, IT systems were necessary in different fields. The first and most immediate concern was the need to link different branches of the new banks in an all-Estonian network, so that real-time operations between different branch offices could be launched. Secondly, there was the issue of creating a network of ATMs ('cash machines') linked to the banks, which was something that had never existed before in Estonia. Thirdly, there was the emerging trend in the advanced countries towards electronic transactions directly between a customer and the bank via telecommunications networks (at that time usually known as 'home-banking').

With regard to the *first* issue, i.e., interconnecting the branch offices, the banks represented a somewhat special category of organizations, due to their need for high-security real-time data communications in combination with their omnipresence not only in the bigger cities but also in most smaller rural towns. From the perspective of the banks, these often distant places were a nightmare in terms of the inherited Soviet telecommunications networks. The most natural idea – not least from the perspective of security – was for the banks to use the X.25 network of Estpak Data, i.e., the subsidiary of Eesti Telefon dealing with data communications provision (see section 5.1 above) and the only provider of access to data communications outside Tallinn during the first half of the 1990s. But Estpak's network had, as pointed out in Chapters 4 and 5, its historical roots in the Ministry of Communications' participation in the Soviet Akademset project, whereby the network had been designed to be used by the county-level public administration. This meant that Estpak's services were only available down to the level of county towns – not enough for the banks, which in the case of Hoiupank, could have 407 different branches (Liik, 1996). This peculiar situation motivated the banks to become active also at the *network layer* (cf. Chapter 3) of telecommunications services. Thus major banks with a strong presence in the smaller towns, such as Hoiupank and Hansapank, started to construct their own data communications networks in order to complement Estpak's X.25 system.

In addition to communications between different branches of the new commercial Estonian banks, there were considerable problems with regard to *inter-bank communications*. These had to be built from scratch. In 1991, the data needed for inter-bank payment transactions had initially to be transferred via regular mail, and the speed of the entire inter-bank payment process was thus determined by the efficiency of Eesti Post. In 1992, electronic data communications were enabled for the first time between the central bank and the commercial banks, and the clearing division of the former was given authority to accept electronic information transfers between banks. These and other activities are said to have been a precondition for the successful launch of monetary reform in June 1992, through which the Estonian kroon was introduced (Minka, 1996: 30). In late 1993, inter-bank communication was expanded through joining six Estonian banks – including the central bank – with the international inter-bank information exchange organization, SWIFT (ibid.: 32).

The relationships between the Estonian banks were also a main concern with regard to the *second* issue pointed to above, i.e., the creation of a network of ATMs linked to the banks. Magnetic cash, payment and credit cards started to be issued in 1993, and in order to support card transactions, an organization called Pankade Kaardikeskus (Card Center of the Banks) was created, which turned out to function well as an overall important hub for networking activities between several of the most important banks in the mid-1990s. Thus the Card Center provided the foundation for the cross-use of ATMs and point-of-sale (POS) terminals, a project initiated in 1996, following a Nordic tradition (Uba, 1996: 34).[18] The founders of the Card Center were Tallinna Pank and Hansapank, which were soon joined by Hoiupank, Forekspank and EVEA Pank. A similar organization, called Inpass Estonia, was created by another group of banks. While Inpass Estonia was a profit-oriented organization, the Card Center was non-profit-oriented, being inspired by a similar organization in Finland, Luottokunta, which according to Hansapank 'provided invaluable assistance, both in practical matters and in expert advice, for the development of the bank card in Estonia.' Luottokunta also functioned as a mediator during the first period when international card transactions started to be carried out (ibid.).

Not surprisingly, ATM *hardware* was always imported, but the *software* was often developed in-house. In-house development was here probably a result of the high costs of foreign systems rather than a result of strategic reasoning, as indicated by the fact that Forekspank successfully licensed-out its cheaper, in-house-developed ATM software to other Estonian banks. The Card Center also developed its own software, thus in a sense forming an interesting software development consortium shared by different banks (ibid.).

Concerning the *third* issue pointed to above, i.e., electronic services in the sense of information exchange and transactions between a customer and the bank via telecommunications networks, the breakthrough came in 1993. In the summer of that year, two commercial banks – Hansapank and

Forekspank – introduced electronic banking services for their clients for the first time. Hoiupank followed in 1994. These services were based on 'off-line' solutions, in which the communication was not interactive in a true sense; the customer – almost always a firm at this stage as the service was usually seen as too expensive for private customers – did not 'log in' to the bank's system, but sent all information needed for a transaction at a single point in time, to be processed later by the bank. This solution was based on the use of banking customers' personal computers, in which special software was installed. The reason for providing off-line rather than online services was directly related to the underdeveloped infrastructure at the physical network layer, i.e., the still mostly analogue and overly slow networks of Eesti Telefon, which according to the banks, 'ruled out all on-line systems' (Kaal, 1996: 39). Nevertheless, the importance of these off-line electronic banking services soon became clear, and in 1996 the share of electronically initiated payments in overall domestic retail payment transactions amounted to about one-third at Forekspank and nearly 40 per cent at Hansapank. The figures were somewhat lower for the rest of the banks (Minka, 1996: 32).

The off-line electronic banking services were based on in-house solutions as well as intra-Estonian collaboration and licensing activities. Hansapank's 'Telehansa' off-line solution was developed with the people from Crebit with their inherited Soviet-era competencies and Finnish experiences (see the discussion on Crebit above). Forekspank's 'Microbanker', which was later licensed to three other Estonian banks, was based on a program written in-house primarily by a 25-year-old Estonian computer expert who had earlier been working at MicroLink and at the Department of Radio and Communications Engineering at the Technical University (information from Madis Kaal, 2004).

A further step taken in these early years of electronic banking services was the decision to establish *telephone banking*. This segment of banking services was in the early 1990s perhaps the most quickly growing field of innovation in the banking sector globally. It had an obvious attraction both from the bank's and the client's perspective: it did not demand any complex technologies on the part of the client, except for digitally connected telephones, and the technological solutions rested basically on specialized telecom-related software and a system of pin-codes. These systems were growing rapidly in the West. For example, Nordbanken, the Swedish leader in telephone banking, had by mid-1995 some 130,000 telephone banking clients and a little more than a year later 230,000 (*Finanstidningen*, 6 June 1995; *Datateknik*, 26 September 1996).

In Estonia, activities in the field of telephone banking took off in 1994 and 1995. Similar to the case of off-line electronic banking, Estonian banks seeking to establish telephone banks found the available foreign solutions too expensive and therefore considered developing the required hardware and software in-house. However, as an alternative, it was soon discovered that closely related competencies already existed in Estonia, embedded in the Tallinn-based companies Univoice and Voicecom.

These two firms had an interesting background that stretched back at least as far as the early 1980s. At that time, an Estonian researcher from the Institute of Cybernetics, where he had been involved in the development of speech synthesis apparatus, emigrated to Finland and started to look for opportunities to commercialize the ideas developed at the Estonian R&D institute. Having worked at a number of Finnish firms, he founded Univoice OY in Finland in early 1990 and Univoice AS in Estonia towards the end of the same year (i.e., one year before the country's independence). A former colleague from the Institute's speech technology department became the director of Estonian Univoice, which focused on working out – in cooperation with its Finnish sister company – telephone service systems to be used in banking and related areas. In 1993, the firm was split into two different companies, Univoice and Voicecom. The former came to concentrate on the development and sales of small and medium-sized private telephone exchanges (PBXs), whereas the latter continued the focus on the emerging banking sector in Estonia. In 1994, Voicecom developed the first automated telephone banking system in Estonia for Tallinna Pank (information from Voicecom, 2003; *Äripäev*, 26 September 1996 and 12 November 1996).

Voicecom delivered similar systems to Hansapank (1995), Ühispank (1995), EVEA Pank (1998) and Hoiupank (1998), and thus became the dominant supplier of telephone banking systems in Estonia. The larger banks also became important and demanding customers of other telecommunications solutions developed by Voicecom, such as the digital recorder 'Direc', of which major customers were Tallinna Pank (1996), Hoiupank (1996), Ühispank (1996), Krediidipank (1998) and the Card Center of the Banks (1998) as well as a number of state and municipal organizations and even Eesti Telefon and EMT (with whom Voicecom would later enter collaboration in more radical terms – see further Chapter 7). Voicecom was also involved in developing application software for Tallinna Pank's ATMs (information from Voicecom, 2003).

In the mid-1990s, telephone banking was thus a dominant and rapidly growing trend both worldwide and in Estonia, and there were many ideas as to how to develop it further – for example, through 'fax banking'. In spring 1996, however, the Estonian bank Forekspank, which had so far not been involved at all in telephone banking, suggested a somewhat different approach to the further development of technology-intensive banking services: to provide *Internet banking solutions* for the first time in Estonia.

The Internet represented a competing technology to telephone banking, but was at this time hardly regarded as a serious competitor – because of the still very low number of Internet users (cf. Figure 6.6 above) and because of unresolved difficulties with security issues. There were thus bottlenecks both at the supply and the demand side.

However, while many banks – especially the large banks in Western countries, where telephone banking was developing much more rapidly than in Eastern Europe – hesitated and preferred to await both market and technological developments, some Estonian banks, despite their shortcomings, chose to enter the field of Internet banking almost immediately after the introduction of graphical web browsers and the first appearance of organizations' web pages.

The Internet services of Forekspank mentioned above – which were the first such services in all Eastern Europe – were initially more of an information than communications service, where clients could get information about recent transactions, although it was also possible to order a bank card and open a new account. In June 1996, however, the competitor Hoiupank – which similar to Forekspank had no heavy investments in telephone banking – introduced its 'PangaNet' product, which enabled users to carry out domestic bank *transfers* (*Baltic IT Review*, no. 2, 1996: 36). These were in-house developments that built upon competencies that had been accumulated from earlier computer-based banking services ('Microbanker' at Forekspank and 'Telepank' at Hoiupank).[19]

On the other side of the Baltic Sea – in Sweden – it was only later the same year that Internet banking was first introduced – and then only by two smaller niche banks. The larger Swedish banks did not enter the Internet banking market until around half a year after Forekspank's introduction of Internet services in Estonia. Nordbanken, which was the Swedish leader in telephone banking, had not even a homepage on the web until the fall 1996.

In Estonia, the combined effect of rapid growth of the banking sector in general, and of new information technology applications in particular, meant that both the absolute and the relative importance of IT systems increased, as the potential cost-cuts inherent in the use of IT became greater with a growing mass market. This implied enormous strains on the part of banking IT departments, which faced a seemingly never-ending steep expansion of technological applications that never seemed to come to a rest before they had to be renewed and expanded. As a matter of fact, the banks, which had started a few years earlier with a few employees responsible for IT activities, gradually grew to host the largest IT teams in the entire country. In addition, these IT departments frequently turned to other local IT firms for outsourcing of programming. The banks thus became extremely important hubs for the entire Estonian ICT sector.

Government applications

Apart from the banks, a major group of actors with far-reaching interests in data communications applications were government agencies. As discussed in Chapter 4, the Soviet-era Akademset project used interconnections between county-level government organizations in Estonia to send e-mails and files to each other over the X.25 network. However, this does not appear to have had any far-reaching impact on public administration before the dramatic events in the years around 1990. In the early 1990s, the question soon arose as to how government agencies could more fully take advantage of the new developments in the field of data communications and information technology. Totally new opportunities opened up as a result of Estonia's re-independence, but as discussed earlier in this chapter, the situation with respect to IT in public administration initially developed chaotically, without any real coordination.

With respect to data communications, important technological choices had to be made. In particular, it was not clear what type of networks should be built. It was not self-evident that the old X.25 network was really the best basis for further developments. Representatives of the Institute of Cybernetics, who, as pointed out earlier, were the most important advisors on issues related to ICT policy, had in the years around 1990 become more and more interested in the *Internet* as the main standard to be used in future data communications, and this strongly influenced current directions of development in Estonian public administration. However, it was not clear at this time that the Internet would dominate the future of data communications globally, and the resulting uncertainty demanded advanced knowledge of technologies.

In spring 1993, in connection with the creation of the Department of State Information Systems (RISO), the Institute of Cybernetics was awarded the responsibility for the creation of a new information and data communications system to be used by the State Border Guard Department. The network's purpose was to interconnect the different border control stations in the newly independent country and thus coordinate the flows across borders as well as the supervision of borders. Despite the uncertainties related to the future of the Internet and other data communications standards, the Institute of Cybernetics decided upon a system based on the TCP/IP protocol. Soon the Customs Board joined the project, as it was considered unnecessary to create two separate networks for these closely related but organizationally separate functions of border-related public agencies. The design of the network was completed by the beginning of 1994. Cooperation between the Institute of Cybernetics and the government agencies in this area was strengthened in December 1994 through the establishment of the Andmeside Osakond (Department of Data Communications, ASO) at the Institute. It had the character of a separate state-owned organization with its own accounting system and statutes, and its activities consisted of designing and constructing the governmental data communications network. Considerable expansion followed in 1994 and 1995, as an increasing number of government agencies joined the network (Kauber, 1996: 57f.; Martens, 1997a: 68f.). The quality of the network improved through the simultaneous modernization of data communications through Estpak Data and the general digitalization process driven by Eesti Telefon. By then, it was starting to become clear that the choice of the Internet protocol as the basis for the network was indeed suitable and fortunate. When the Internet started to take off in 1995 and 1996, the Estonian government sector was already surprisingly well prepared.

From an applications perspective, the government data communications networks were originally created in order to sustain the traditional activities of public agencies and to exploit synergies. In particular, the cross-use of databases was seen as necessary. For example, it would have been inconvenient for the Tax Board to create its own register of tax payers; the most effective way was, of course, to use the population and business registers. Much emphasis was thus put from the mid-1990s onwards on creating online connections between different government agencies (Agur, 1996b).

With the arrival of the world wide web and the rise of Internet use, these developments became intertwined with the development of new public services directed towards citizens. In May 1994, the FTP and WWW servers of the government and the State Chancellery were opened (Vallner, 1998). This marked the emergence of public agencies in Estonia as providers of Internet content services. However, data communications service provision in this sense developed slowly during the following years. In late 1998, Estonian information society advocates complained that

> most of the WWW pages [of public agencies] do not contain sufficient information, they are not regularly updated, and they do not include client-friendly interface features. Even if search engines exist, they are not very easy to use. The result is that the public does not see the Internet as an official channel of information dissemination. (Kalvet, 1998: 38f.)

It took two years before the first government databases were open for the public on the Internet, and it was only in 1998 that a relatively simple service such as providing public forms on the Internet could be launched. Moreover, this service, as well as the popular searchable database of legislative acts, was introduced through the activities of non-governmental actors, including both NGOs and private persons, and financed mostly externally (e.g., by PHARE) (ibid.). It thus appears that the public sector was much slower in creating new services as compared to the surprisingly early developments in the banking sector.

The Department of State Information Systems and the Informatics Center, which were the main state organizations responsible for this development, thus seemed more focused on the process of computerization in government agencies and on connecting the agencies with each other in a proper way, than on addressing the issue of providing new services with the help of the Internet. This focus was strengthened in 1997, when ASO (the Data Communications Department that had so far been a part of the Institute of Cybernetics) was transferred to the government's Informatics Center (*Baltic IT Review,* no. 11, 1998). In 1997, the government agencies also started to more actively use *intranet* technology. In addition, the first major *extranet* information system was realized: the tasks assigned to the ministers could now be observed, in addition to the Internet, in all ministries, from computers with a fixed IP-number (Vallner, 1998).

However, with respect both to government-internal developments and to new services development, attention gradually shifted to a somewhat different problem area: the lack of proper legislation. It started to become obvious that the development of new services was not merely a *technical* problem, but that a further development towards Internet applications also demanded substantial *institutional* changes. For example, once the idea emerged to provide forms on the Internet, it turned out that although the forms themselves could be made available, it was legally difficult to introduce the additional service of letting forms be *signed* and *returned* through the Internet.

Thus towards the end of the decade, a new focus in the process of sectoral evolution became legal issues related to Internet applications. These developments are discussed in Chapter 7.

5.3 The Middleware Layer

The increasing popularity of the Internet, and in particular the introduction of web-based online banking services directly related to the transfer of money, tended to raise for the first time a public debate about *security issues* in data communications applications. The issues around the extent to which Internet banking could be made secure enough to motivate its use were the main reasons many established banks globally were more retiscient in their approach to Internet banking than the Estonian banks.

In the Estonian R&D community, however, the importance of information security had been recognized much earlier, and considerable efforts had already been spent on knowledge-creation activities in this area. These efforts were a direct response to the new political situation in Estonia after regaining independence in late 1991, which suddenly created a need for information systems in strategic but completely new areas such as customs, border guards and national defense. In March 1992, the Informatics Council, whose members typically had a background associated with the Institute of Cybernetics, recommended the creation of a special *national information security program* for these sensitive state-controlled information systems. The task of elaborating a strategy and solutions in this new area was delegated to the Institute of Cybernetics. The Institute had hardly any experience in this particular field at the time, but the field of knowledge in question was obviously one that lay sufficiently close to the Institute's existing competencies for enabling a relatively rapid accumulation of new expertise. The Institute took the task seriously and began to deal with the field of information security in a thorough way, addressing not only the immediate needs of government bodies, but also entering basic research in cryptology (Oit and Praust, 1998).

In 1993, the strong drive for national information security projects even resulted in the creation of a whole new information security department within the Institute of Cybernetics. In line with the long-standing tradition of the Institute, basic research was here combined with the development of new products – now first and foremost in response to the government as an advanced customer. As described in the preceding section, the Institute in spring 1993 entered collaboration with the Customs and the Border Guards for creating a data communications network, which was based on the Internet protocol. In parallel to these activities at the network and connectivity layers, the Institute of Cybernetics launched the development of its own information security products. These were mostly developed fully in-house.

The first main product was 'Barrikaad', which was a firewall, i.e., a solution for protecting local area networks in their communication with the outside world. Firewalls were not customized products, and they could thus be imported as ready-made solutions, but the Institute of Cybernetics saw a potential market among smaller business users for alternative variants, especially in Estonia itself, since the imported firewalls were mostly overly expensive for smaller firms in Eastern Europe. If the average price of a US-made firewall amounted to some $16,000 in the mid-1990s, the Institute of Cybernetics had the vision to develop a firewall that was to cost some 30 per cent of that. The strategy followed by the Institute to achieve this goal was to use a maximum of available freeware, which in the end was shown to constitute no less than 90 per cent of the software in Barrikaad. To this was added a number of in-house developed software components, and a web-based user interface. A domestically attractive feature was that for the first time there now existed an Estonian-language firewall (Martens, 1997b: 29f.).

High expectations were held that Barrikaad would be successfully marketed *internationally*. However, the Estonian firewall remained relatively unknown in the West, and although it was recognized to have 'some interesting features', it was typically seen not to 'represent a viable company choice' (as evaluated in a test performed by *Network World*, 12 March 2001). In Estonia itself, on the other hand, where demand for Internet services was booming at a surprisingly early stage but where the price was inevitably a much more significant factor in the choice of product, it was usually seen to be 'more than enough'. On the domestic market it thus became a great success, with a share of almost 70 per cent of the Estonian firewall market (Martens, 1998: 48). The problem was that it would hardly be possible in the long term to develop information security products for the domestic market only. This was clearly recognized by the Institute of Cybernetics, where strategies during the ensuing years focused heavily on entering foreign markets. This, however, would turn out to be considerably more difficult than expected, as discussed further in Chapter 7.

With respect to cooperation in the field of information security between the Institute of Cybernetics and the government, the years following the Internet boom paradoxically saw a radically *decreased* devotion to further developments. After the elections in 1996, which led to the installation of a new government in Estonia, state expenditures on the information security program dropped dramatically. In the period 1993-1996, the budget for this had been annually more than EEK 1 million, but it dropped in 1997 to EEK 0.6 million, in 1998 to EEK 0.3 million and in 1999 to no more than EEK 0.14 million (Oit and Praust, 1998).

Innovation-related activities in information security, which remained the dominant segment in the middleware layer, were thus in a somewhat unclear situation towards the late 1990s. At the same time, there was also a gradual shift in focus. With the explosion of Internet users and applications in the last few years of the 1990s, it was realized more and more that information security could not be addressed solely in terms of advanced technical solutions, but had equally much to do with legal issues and establishing trustworthy procedures relating not only to bank transfers but also to the more general field of 'digital documents'. This will be discussed further in Chapter 7.

6. EDUCATION AND RESEARCH

The Late Restructuring of the Academic Landscape

As discussed in Chapter 5, the collapse of the Soviet Union and Estonia's re-independence had a dramatic impact on education and research activities in the country. In particular, it was noted that financial support for R&D activities in the universities and the academy of science was almost completely withdrawn due to the lost links to Soviet customers of research results, who had typically accounted for around 90 per cent of R&D financing. On the other hand, the new political situation opened completely new opportunities in terms of cooperation with the Western academic world, with respect to research cooperation and student exchange programs.

With an economic recession that continued until 1994-95 and a correspondingly problematic state budget, the educational system – which had so far been less hurt in terms of finance as compared to R&D – came to face enormous difficulties in the mid-1990s. This was especially true in the field of engineering education, which similar to R&D activities in general was often associated with the old Soviet society from which Estonia tried to distance itself in every possible way. 'Hi-tech' was typically imagined as something that was connected to the huge electronics factories belonging to Moscow and the military Soviet ministries (Rang, 1996). In February 1996, the Council of the Tallinn Technical University (TTÜ) turned with a public letter to the minister of education, protesting against the principles applied to the distribution of state funds to higher education (TTÜ Nõukogu, 1996). A similar protest was formulated a month later by 36 TTÜ professors – four from the Department of Radio and Communications Engineering – at which it was pointed out that the support of EEK 90 million (€6 million) in the year 1996 to engineering education at the Technical University corresponded to a mere EEK 16,000 per student, whereas the University of Tartu received EEK 137 million (€9 million), corresponding to 22,300 EEK per student. This could be compared to other countries, the majority of which did not discriminate with respect to the financial means per student in different universities and scientific areas – and if discrimination did take place, it was usually in the direction of providing *more* support to engineering education, not *less* as in Estonia. At the same time, with respect to research activities, TTÜ received only 10.6 per cent of the total available state budget funds. It was argued that despite parliament's approval of a law on the regulation of science, the financing of R&D within the framework of state programs had not been initiated (TTÜ professorid, 1996).

In parallel with the problems of *finance* at TTÜ and the Academy of Sciences, major *organizational* changes were carried out in the academic world. In 1995, the old diploma study structure in higher education was replaced by an Anglo-American-inspired three-stage system with bachelor's, master's and PhD degrees. The nominal time for bachelor studies was set to four years, while for master and doctoral studies two and four more years were added, respectively (Min, 2000a: 47).

Until 1996, all academic fields related in one or the other way to telecommunications sorted under the broad Faculty of Computer and Systems Engineering. The Faculty consisted of the Department of Computer Engineering, the Department of Automation and Systems Engineering, the Department of Electronics, the Department of Informatics, the Department of Radio and Communications Engineering, the Center for Biomedical Engineering and the Electronics Competence Center. The latter was opened in October 1995 within the framework of the TEMPUS project of the European PHARE program. The competence center was created 'as a new structural unit, which in cooperation with the Faculty of Computer and Systems Engineering offers modern electronics and information technology equipment and study courses.' The center consisted of a microelectronics design laboratory with 12 places and a UNIX classroom with 45 SUN work stations (TTÜ, 1995: 16).

In May 1996, it was decided upon request from the Faculty of Computer and Systems Engineering to *split* the Faculty into a Faculty of Information Technology and a Faculty of Systems Engineering (TTÜ, 1996: 13). To the former now belonged the Department of Computer Engineering, the Department of Informatics and the Department of Radio and Communications Engineering. The reasons for the split are unclear; some teachers argue that it had mainly to do with problems at the personal level (Rang, int., 2003). In practice, the two faculties continued to be closely related to each other through the study programs, all of which were organized under the broad 'Study Area of Information Technology'. Starting from 1996 and 1997, four different 'study directions' crystallized within this broad field: 'Informatics', 'Telecommunications', 'Computer and Systems Engineering' and 'Electronics and Biomedical Engineering'. A total of 220 students were accepted annually to these programs (Min, 2000a).

In addition to launching the reforms leading to the introduction of the Anglo-American-inspired structure of studies, the Ministry of Education also took the first concrete steps towards reforming the academic system, through a merger between universities and the old institutes of the Academy of Science. As discussed in the preceding chapters, the latter had under the Soviet system clearly constituted the most important R&D organizations, but after the collapse of the Soviet Union it was utterly unclear what their role would be in the new era. Two basic ideas were discussed: the first, seeing the institutes as part of the economy, foresaw a privatization of the institutes of the academy, in the same way as the state-owned enterprises were privatized. The second, seeing the institutes as academic organizations, imagined their merger with the universities.

The reform that finally crystallized foresaw a painful combination of these two ideas, and in November 1997 – six years after the actual collapse of the old system – a number of institutes were merged with the universities and, to a lesser extent, privatized (TTÜ, 1997: 18). The Institute of Cybernetics was split into two different organizations of about the same size: half of the institute joined Tallinn Technical University, whereas the other, more applications-oriented half was transformed into the state-owned R&D firm Cybernetica. From the perspective of the technical university, the merger with a number of old academic institutes meant that, from a formal point of view, its research budget almost doubled in 1998 (Küttner, 1998).[20]

International Collaboration

Cooperation between Estonian organizations of higher education and foreign universities had taken place on a minor scale in Soviet times (see Chapter 4) and these contacts had expanded considerably during the 'perestroika' years (see Chapter 5). Having achieved full political independence, however, completely new opportunities opened up in this respect. The Technical University started to organize international conferences, send students on foreign exchanges (mostly with the help of foreign financial support), establish partnership agreements with a variety of foreign universities, and not least participate in international research programs.

Among the conferences that took place at the TTÜ belonged the biannual 'Baltic Electronics Conference', which in 1994 was arranged for the fourth time, and coordinated by the Department of Electronics and the Electronics Competence Center. It saw participants from twelve (mainly Baltic Sea) countries as well as from local institutes belonging under the Academy of Science and the Tõravere Observatory close to Tartu, where an electronics research tradition was still alive since the days of Soviet military contracts (TTÜ, 1994: 11). Two years later, the same conference saw participants from 20 countries (TTÜ, 1996: 19). There was also the conference 'Ringhääling' (Broadcasting) in 1995 with mostly Estonian and Finnish participants, arranged by the Department of Radio and Communications Engineering. The Informatics Department arranged a conference on 'Hypermedia' in May 1996 (ibid.: 14), and the Department of Radio and Communications Engineering organized another conference in 1998, on 'Telecommunications', which, however, was purely domestic (TTÜ, 1998: 174).

Of greater significance was probably the emerging participation in international research projects. The dominant frameworks for this were the

EU's COPERNICUS, TEMPUS and PHARE projects – which were directed specifically towards trans-European cooperation beyond the EU itself – but the Technical University also took part in intra-EU projects. In telecommunications-related fields, the latter included the ESPRIT program on information technology, RACE on telecommunications as well as a program on Teleinformatic systems (TTÜ, 1995: 104). To these were later added Estonia's participation in the EU's Socrates/Erasmus and Leonardo da Vinci programs and the Nordic countries' NORDPLUS (TTÜ, 1997: 123).

On the bilateral side, the Technical University had by 1995 no less than 44 partnership agreements for cooperation with foreign universities. The most important countries here were Finland (13 partner universities), Germany (9), Sweden (3), Russia (3) and the US (3). With regard to scholarships awarded to students at the TTÜ, the Nordic countries provided the vast majority of financial support (Figure 6.7 below).

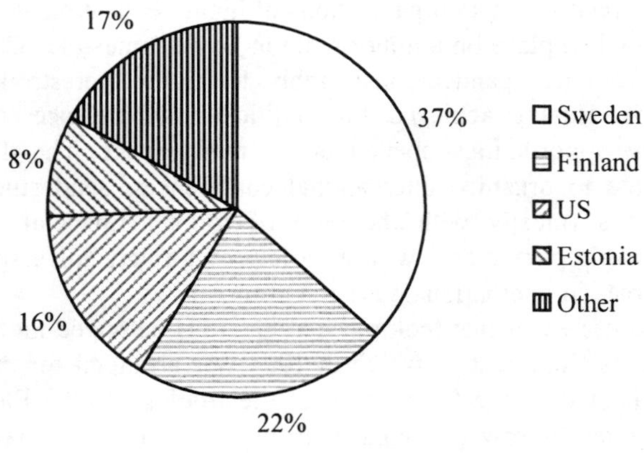

Source: Calculated from TTÜ (1994: 106).

Figure 6.7 Structure of scholarships awarded to students at TTÜ, 1995 (by country of origin)

An interesting early project was also 'ESTPID' – Electronics Cooperation between Stockholm and Tallinn for the Promotion of Industrial Development – which supported the development of an SME sector in electronics in Estonia. The project was planned for three years starting in 1993, and the goal was to make it possible for Estonian engineers and scientists to take part in training in Swedish companies and educational and research organizations, as well as to develop the Electronics Competence Center in Tallinn (Ubar, 1994: 4).

Creating Links between Academy and Industry

In the Soviet era, the universities and the academy of science had had very few direct links to production enterprises (see Chapter 4). Even more than in Western countries, the academic and the economic worlds constituted totally different cultures, with the academic realm being the more prestigious of the two. This contributed to the problems that arose after radical changes in the years around 1990 in terms of the lost Soviet markets for research results produced by Estonian academic organizations. As discussed in Chapter 5, these were now forced to look for new sources of finance, and it was a common expectation that Estonian firms would start to act as customers of R&D results, turning to the domestic universities and institutes to solve complex problems related to the development of new products and processes.

However, those firms that might have contributed to this were typically the very same electronics enterprises that collapsed during the first half of the 1990s. And to the extent that new firms were established, these turned as a rule to foreign ready-made technologies and solved their problems in cooperation with foreign partners rather than with the domestic R&D complex. From the perspective of the latter, this was sometimes seen as a result of a too liberal economic policy on the part of the Estonian government, which did not engage in any initiatives aimed at stimulating collaboration between foreign-owned Estonian firms and the local research organizations (Jaaksoo, int., 2003). But the question was also raised as to whether Estonian R&D would have been able to respond at all to the new demands of private business. The academic and economic sectors, having evolved for such a long time in relative isolation from each other, did not necessarily suit each other as collaborative partners in research, and the academic sector's negative attitude towards the world of economic production did not make it easier. At the Institute of Cybernetics, it was recognized in 1993 that

> we have inherited a truly Soviet attitude towards technology, seeing it, as before, as some kind of second-order thing, which does not deserve to be dealt with seriously in its own right. Indeed, it is more prestigious to solve scientific problems, to philosophize over themes of the information society and to brag of one's technical education. And unfortunately this begins with the engineering teaching in higher education. As a matter of fact, on paper we are able to create very complicated systems and appliances and even to produce working models of these – but our abilities are confined to this. From good and, what is essential, cheap *production* we are very far away. (Tani, 1993: 14ff., my translation)

However, there were exceptions to the general pattern. For example, in 1994 the Department of Electronics at TTÜ could report that it had successfully completed the modernization of the old Soviet surveillance signalization system 'Neva', in which software was developed that enabled data communication between surveillance centers via a modem connection (TTÜ, 1994: 68). In 1997, the TTÜ's annual budget from research contracts with business firms amounted to EEK 10.1 million (€0.6 million); if the institutes of the academy that joined the university in November 1997 were included, the amount grew to around double that (Küttner, 1998). This corresponded to as much as roughly a fourth of the university's total research budget, but in absolute numbers it was still very modest. In addition, the contribution of telecommunications-related research to industry-financed research at TTÜ was only around 5 per cent, whereas around 75 per cent of industry-financed research at TTÜ took place in the fields of mechanics, building and energetics; this probably reflected the strong inherited domestic academy-industry links from the Soviet era in those sectors (calculated from figures in TTÜ, 2000: 117; see also Chapter 7).

An interesting special case with respect to finance was the Division of Telecommunications at the Department of Radio and Communications Engineering (RSTI). As mentioned in Chapter 5, the Division was newly established in 1992, with its research and education responsibilities directly financed by Eesti Telekom (TTÜ, 1998: 121). A more common way for private business to support the TTÜ, however, was through co-sponsored gifts in terms of equipment to be used in teaching and research. For example, RSTI's cooperation with the Finnish-based firm ICL, with whom links had existed since almost a decade, resulted in 1995 in the Finnish firm supporting the Department by providing expensive software to be used in teaching (TTÜ, 1995: 13). In 1996, a microwave engineering laboratory was opened at the Department, this time sponsored by an Estonian professor. At the Department of Automation, a new laboratory would open in 1997 with the help of technology from Siemens and the Estonian computer firm Astrodata (TTÜ, 1997: 15). In 1996, support from the US firm Northwest Fourslide was recognized by the Department awarding it a 'gold plate' for its sponsorship of the Department of Radio and Communications Engineering (TTÜ, 1996: 17). A gold plate was awarded also to Eesti Telekom for its support of a modern laboratory designed for testing analogue apparatus connected to digital networks (ibid.: 19). In 1998, more gold plates were awarded to the Nordic telecom operators Telia and Sonera for their support of the university (TTÜ, 1998: 175).

RSTI's links to private business also took the form of joint seminars. For example, in 1995 a seminar was organized on the initiative of Ericsson on the theme 'Multiservice backbone networks' (TTÜ, 1995: 90). Similarly, in May

1996, Hewlett-Packard's Finnish subsidiary arranged a seminar on telecom and computer networks together with the Division of Telecommunications (TTÜ, 1996: 13).

Another important area of potential collaboration concerned the provision of places for student practitioners within private firms. In May 1997, TTÜ signed an agreement with Eesti Telefon, whereby the firm would provide scholarships and themes for Bachelor's and Master's theses in the fields of telecommunications and data communications as well as access to space and equipment required to carry out their projects, opportunities to gain work experience at the company as student practitioners and employment possibilities upon graduation. The perceived importance of the agreement was demonstrated by the fact that it was signed by the Rector of TTÜ and the CEO of Eesti Telefon (TTÜ, 1997: 11). Similarly, at the Department of Electronics, the director considered it 'very positive . . . that electronics firms have come to a clear understanding of the need to take part in the process of educating young specialists', initiatives which affected not only student practice but the firms' own training programs as well (Min, 2000b).

A variety of activities were thus taking place aimed at creating various types of links between academy and industry related to the Estonian telecommunications sector. However, it is not clear to what extent these emerging connections were significant enough to influence sectoral evolution. The general impression was rather one of continuing severe difficulties in academy-industry interactions. For example, with clear inspiration from Western developments, it was argued that TTÜ would have to take a much more active role in creating an environment more encouraging of SME linkages with academic departments and thereby transform the Mustamäe region (the location of TTÜ) in Tallinn into a 'science city' (Küttner, 1998). An interesting step was taken in February 1998 through the opening of the new 'Innovation Center' at TTÜ, with the goal 'to spur science- and technology-intensive business and to support innovation-oriented scientific work' (TTÜ, 1998: 172). It remained to be seen whether this marked the beginning of a new era for the Estonian university in the sectoral system of innovation.

7. SUMMARY AND CONCLUSION

The period dealt with in this chapter saw the Estonian economy overcome the disastrous recession following Soviet collapse and enter a period of high-level GDP growth. A natural end of the period can be marked by the Russian crisis in August 1998, after which both the Estonian economy as a whole and the telecommunications sectoral system of innovation developed according to a very different logic, as will be shown in Chapter 7.

Technologically, the period was characterized by a more or less rapid transition to Western technological trajectories. Key technological developments concerned the introduction of GSM and the take-off of Internet-related services in data communications – trends that largely coincided with worldwide developments. In the fixed networks, a time-consuming transition from Soviet to Western technologies began but had not been completed by the end of the period, and the same was true with respect to equipment production, where Estonian firms typically continued to use a *mixture* of Soviet and Western technologies. Apart from remaining Soviet and new Western technologies, there were a few examples of alternative developments, including the use of the Wireless Local Loop (WLL) in the fixed networks, the development of local firewall technology for data communications and original domestic developments with respect to IT systems, notably in some banks.

In general, telecommunications technologies were, in the period of investigation, in a highly fluid stage internationally, which created high demands for the ability to make technological choices in Estonia. An interesting aspect here was the surprisingly early choice of the Internet protocol as a basis for future data communications in both public and private sectors. While other countries typically hesitated, Estonia was an early mover in this field.

The structure of the sectoral system evolved in close relation to the build-up of new and destruction of old links to foreign systems of innovation. The reorientation that had been initiated in the years around 1990 continued, with eastward linkages playing increasingly unimportant roles. The only exceptions were a few electronics producers from the Soviet era, which did not succeed in attracting foreign investment and therefore continued to be oriented towards other ex-Soviet republics. In the West, it was the Nordic countries that continued to provide Estonia's most important partners in telecommunications, but some of the Nordic linkages, notably those of Eesti Telefon and EMT, were, as a matter of fact, somewhat weakened in favor of a greater freedom for these Estonian firms. Domestically, the organizational landscape evolved more in response to technological than institutional change. GSM and the Internet were thus associated with the emergence of a large number of new firms and inter-organizational linkages. After the great institutional transformations in the years around 1990, institutions were now in a process of stabilization, although towards the end of the period there were signs that institutional bottlenecks were emerging, particularly as a consequence of the rapid developments in the Internet subsector. Further elements of structural transformation will be pointed out in the discussion of the evolution of activities below.

Let me in the following discuss the different elements of the Estonian style of innovation, as it developed.

The *visions* that came to dominate this period differed from the preceding period in that it appeared less and less likely that any old industrial enterprises could possibly become self-standing players in the global electronics sector. The strategies that emerged were instead focused on exploiting the advantage of a low-cost environment, through which many Estonian equipment firms found a new role as subcontractors to large Western, mainly Nordic, multinationals. Innovation thereby played only a marginal role. On the service side, in contrast, Estonian firms and organizations developed a new self-confidence, as they discovered the advantage of acting in a country which had lost much of its old-established institutions and technologies. This enabled them to focus aggressively on early adoption of Internet and mobile communications services. In this connection the idea of a 'Test-Site Estonia' also emerged, which was seen as a possible way to quickly move into new and futuristic fields of technology. Estonia also identified itself more and more as a 'Nordic' country, perceiving itself as having a 'Scandinavian' mentality with the same particular interest in new ICT developments that was seen to be characteristic of the Nordic world. In contrast to the Nordic countries, however, Estonian actors had as a rule a stronger belief in the benefits of the neo-liberal economy. Consequently, the government sought to intervene as little as possible in matters relating to innovation and R&D, with its main task being understood as building an institutional environment hospitable to the development of private enterprise.

The *articulation of demand* in the period was, on the one hand, dominated by the overall need for investments in digital telecommunications technologies and information technology. Important customers were in particular the government, the emerging banks and the operators, but this demand was mostly for ready-made hardware and software rather than for the development of new products. Moreover, much of this demand was directed almost exclusively towards Western vendors. Among consumers, on the other hand, there was an unexpectedly high demand for mobile and Internet services, and this was to become a driving force and inspiration for service developers in these market segments. Research institutes and old Soviet-era production enterprises continued to suffer from the continuously vanishing demand in their fields, leading to further reductions in employment, whereas some firms, such as Elcoteq and Tarkon, managed to attract both investment and customers from the West (particularly from the Nordic countries).

Foreign demand thus came to play an important role in manufacturing, although this was only to a minor extent associated with innovation.

With respect to the *creation of new knowledge*, the emerging East-West connections – particularly in equipment production – tended to steer the Estonian sectoral system away from domestic involvement in R&D. A factor that contributed to this pattern was the neo-liberal institutional landscape, which did not stimulate the emergence of any links between industry and the domestic R&D complex. The focus was, instead, on imitation and adoption of Western developments. R&D institutes experienced severe difficulties and decreasing employment, but to some extent it was also a question of *reorganizing* knowledge production: some of the knowledge creation that had traditionally taken place within the Institute of Cybernetics continued in different forms in business firms that were either spin-offs from the Institute or founded by entrepreneurs that had earlier been working there. The character of new knowledge creation thereby tended to move in the direction of more business- and applications-oriented activities (as opposed to deeper R&D). On the service side, much of the activities were concentrated on learning from the West, by trying to master existing telecommunications technologies that were already in use there, such as digital fixed networks, mobile telephony and developments in the Internet subsector. In some cases, however, Estonian firms were very early followers, as reflected in, for example, the introduction of Internet banking, which took place earlier than in Sweden.

Competence-building developed in a rather hesitant way during the period, as a number of organizational changes and the introduction of the Western bachelor-master-doctoral system were shadowed by severe shortcomings in university funding. At the same time, however, the rise of mobile telephony and the Internet made the inherited Estonian strengths in radio engineering and computer and information technology increasingly relevant for the overall field of telecommunications. These inherited strengths were complemented by a new Division of Telecommunications (focusing on fixed telecommunications), financed by Eesti Telekom and created in response to the lost proximity to the Russian universities and institutes. Formal relations between education and business remained very weak, but this was partly compensated for in an interesting way by professors and teachers who, due to very low salaries at the university, worked part-time in business or government agencies. Electronics education experienced declining numbers of students, and when the electronics industry collapsed, some argued that electronics education should be removed completely from Estonia. With the emergence of a number of new or restructured electronics firms, however, the existence of electronics education seemed motivated anew. An important trend was also the increasing opportunities for foreign student exchange, particularly with Swedish and Finnish universities. Apart from the university level, an interesting project was the 'Tiger Leap', which addressed computer and Internet skills among students at secondary schools.

The wave of *new firm formation* that began around 1990 continued in this period, particularly in response to the emerging Internet business. Some small computer firms now diversified into Internet access provision, as did the telephony provider Eesti Telefon. The wave of entry in Internet services started around 1994. In addition, from 1996 a wave of entry into Internet services followed among local banks. Important was also the entry of no less than three GSM operators, all of which were largely based on Nordic capital. This created a competitive environment that was not matched by any other country in the world. On the equipment side, foreign electronics firms sought opportunities for subcontracting arrangements in Estonia, and both Elcoteq and Tarkon showed strong growth and thereby became inspiring examples for further Estonian entrants in this subsector. Many old electronics firms, on the other hand, were radically downsized and transformed, and seemed to be close to exit.

The *adaptation of organizations* was characterized above all by Estonian firms intensively learning the *modus operandi* of capitalism, while at the same time responding in various ways to the growth of new technologies in data and mobile communications. Mobile operators, Internet access providers, banks and other firms had to deal with exponentially growing consumer markets, which were associated with great efforts with respect to organizational adaptations. Elcoteq, too, grew through an increasing focus on mobile equipment. Many Soviet-era electronics firms, on the other hand, were busy with downsizing and responding to the chaotically lost demand. Many failed to adapt and to attract investments. The Institute of Cybernetics had to cope with spin-offs and researchers leaving to become entrepreneurs, and in 1997 the Institute itself was split in two parts – a university department and an R&D firm. With the spectacular growth of some service firms that were associated with new technologies, notably the mobile operators and Internet providers, organizational culture underwent an interesting evolution and divergence from that of, for example, Eesti Telefon. Mobile firms had also for the first time to learn to deal with a competitive situation, which had interesting strategic implications in terms of innovative profiles. Eesti Telefon, for its part, concentrated on increasing efficiency, by carrying out process improvements, outsourcing some non-core activities, while also taking an increased interest in Internet-related fields of business.

Networking remained at a somewhat uncertain stage, although some interesting trends could be sensed. For fixed and mobile services, the links between the Estonian firms and their Nordic investors were of special but decreasing importance, as the foreign partners, particularly the Swedish and

Finnish state-owned operators, became more passive shareholders. More important were user-producer relationships, both in services (between operators and equipment suppliers) and in manufacturing (between the Estonian subcontractors and their Western, primarily Nordic, customer firms). However, the innovative dynamics in these links were still generally weak, the Estonian firms having little to say in new product and process developments. With respect to university-industry relationships, new links between business firms and universities were built in a multitude of ways, but in the actual introduction of new products and processes into the economy, the universities were still almost totally absent. Hence, strong partnerships aimed explicitly at innovation remained very rare in this period – but at the same time, many embryos of potential future innovative linkages could be found. Thus the links between Eesti Telefon, EMT and Ericsson were considerably strengthened during the period, while at the same time the operators engaged in contacts with the banking sector at a minor scale. One of the most interesting trends was the development of links between, on the one hand, larger actors such as the operators, the banks and public administration, and, on the other hand, the dynamically developing local IT sector. One thus meets interesting software firms with respect to almost all activities that take place in the system: as emerging partners of operators, as partners and subcontractors of banks and government agencies, as the predecessors of Internet service providers, as supporters of the Tiger Leap project, etc. In addition, many software firms were linked to the Nordic countries through subcontracting or ownership. Later, the dynamic local IT industry and the close relations that emerged between the IT suppliers and large customers such as public administration bodies, banks and state-owned infrastructure companies, would turn out to be crucial for creative innovative activities and joint development of new products. Such relationships were exactly what had been so severely missing in the Soviet years.

The *provision of finance* was an activity in which there was a radical difference between Estonia and the West, due to the simple fact that Estonia remained extremely poor (the bottom mark for the whole economy was reached in 1994). Most organizations therefore sensed that they had no room for any far-reaching and uncertain innovative activities. The severe lack of funds influenced, among other things, ICT developments in business and public administration in the direction of highly cost-efficient solutions. The Technical University also remained severely underfinanced and felt discriminated against in relation to the non-technical University of Tartu. On the other hand, this period saw the emergence of contracts with international organizations such as the PHARE program. To the more well-financed organizations belonged EMT and the banks, whose rapid growth enabled the build-up of relatively strong in-house development departments.

With respect to *consultancy, advice and lobbying activities*, the Nordic advisors initially played important roles at Eesti Telefon and EMT, but the relevance of these advisory activities decreased surprisingly rapidly and had disappeared almost completely by the end of this period. On the side of the government, the Institute of Cybernetics and university researchers continued to play important roles as consultants for institutional change and policymaking. Important inputs came also from Estonians who had lived or still lived abroad, who came to play key roles in the initiation of projects such as the 'Tiger Leap'. In addition, several small firms (often created by Soviet-era researchers) acted as consultants for firms such as the banks. A consortium of the latter (the Card Center) was also advised by the Finnish Luottokunta. In equipment production, an interesting aspect was Elcoteq Tallinn's emergence as a key advisor for the Finnish-owned corporation's further expansion to Hungary and Russia.

Concerning the *creation and change of formal institutions*, Estonia was in general keen to follow Western standards in telecommunications-related fields, including the anticipated needs that would become important in connection to EU accession. In fixed telecommunications, however, the temporary monopoly remained in place, although it continued to be contested in the discussions. On the other hand, and more importantly, competition was enabled both formally and practically in mobile as well as Internet service provision. This stood in sharp contrast to the enormous problems experienced in neighboring Latvia at the time. The response to the emerging Internet revolution in terms of adapted legal structures, however, remained surprisingly hesitant, despite the eager developments in business in this field. Towards the end of the period, there was therefore a clear sense of institutional bottlenecks that had to be resolved for more far-reaching innovation to be enabled in Internet-related areas.

Public policymaking evolved through the formulation of a number of priorities within the broad field of ICT, but government expenditures remained at such a low level that it was doubtful whether the formulation of these priorities would influence real-world developments to any significant extent. The most important concrete project was probably the 'Tiger Leap' initiative, launched in 1996 and aiming at boosting general awareness and skills in computer technology and the Internet. No clear innovation policy for ICT emerged, probably as a result of the continuing neo-liberal political climate and, above all, the general lack of funds in this area. The positive side of this was above all the clarity with which competition was supported in services – first and foremost in mobile and data communications, but it also seemed clear that the temporary monopoly in fixed telephony would have no chance of getting prolonged after 2001. This inspired potential future entrants in that segment to prepare themselves for entry by first establishing themselves as service providers in other segments.

NOTES

1. The exchange rate of the Estonian kroon (EEK) was fixed at 8 EEK = 1 DEM.
2. In June 1994, the EC approved the report 'The European Information Society', stressing the importance of IT for employment and industrial competitiveness as well as for the improvement of social well-being and in general the quality of life. In Finland, the Ministry of Finance began in fall 1993 to prepare an 'information technology strategy up to the year 2010' ('TIKAS'). At a time of stagnation and high unemployment in Finland, information technology became a potential way of achieving economic revival, particularly increased export opportunities, while at the same time indicating the need for this development to be in balance with the general well-being of society. In March 1994, Sweden created an information technology commission, headed by the prime minister, with the main goals being to raise international competitiveness and to improve quality of life.
3. In Estonian: 'riigi tööstus- ja innovatsioonipoliitika'; note the early explicit use of the term '*innovation* policy'!
4. The Center was not the only government body that was involved in financing innovation projects; other actors were various ministries and the State Chancellery.
5. The corresponding amounts were $3.2 million in 1994, $6.6 million in 1995, $9.0 million in 1996 and $10.0 million in 1997.
6. In an investigation carried out by Eesti Telefon in 1997, customers' satisfaction with the company on a scale from 1 to 7 was found to lie in the interval 4.2-4.6 in major cities, but at only 3.7 in the countryside. See Eesti Telefon (1997a: 15).
7. Lippmaa was a famous researcher who had worked at the Institute of Cybernetics and the founder of KBFI, which had played an important role in connection to the first Estonian Internet connections.
8. Eesti Telefon won the prize of the 'best foreign investor' in 1995, a project arranged by the Estonian Investment Agency.
9. There were at that time six earlier RAS networks in operation world-wide.
10. A prejudice was seen to come from Sweden, where according to Grünberg two firms recently had licenses to operate NMT-450 networks (Telia and Comviq), though by now Comviq had practically liquidated its network. In reality, however, Comviq had never had any NMT licence, but had been operating an analogue network with a different standard.
11. There never appears to have been any problem for Radiolinja to initially acquire its GSM license in Estonia, which is interesting if compared to the problematic procedures in Latvia somewhat later, where the awarding of a second GSM license gave rise to big scandals. The Estonian licenses were awarded by the Ministry of Transport and Communications, where the tiny communications department consisted of only four people. The director was the former head of Tallinn's telephone network, which in the years around 1990 cooperated with the local Finnish networks. The latter also happened to be the main founders and owners of Radiolinja in Finland, a fact that may thus have motivated a certain respect in the relationships between Radiolinja and the Estonian ministry's representatives.
12. The network was thus launched on 31 January 1995. See EMT (1995).
13. That is, the service enabling a message to be sent to all mobile handsets located within the area of a certain base station.
14. Among the remaining 15 per cent was above all the Japanese NEC group, which supplied radio links and transmission equipment to EMT.
15. The same pattern was true with respect to firms such as HP, Fujitsu, IBM, etc., which similarly preferred to enter the new Baltic markets through their Finnish subsidiaries.

16. Different repair centers over the world were divided into five levels, depending on the repair capabilities. Elcoteq Tallinn's repair center gained the highest level and was one of 3-5 centers in the world on that level.
17. With respect to the steep decline in office machinery and computers in 1996, as shown in this figure, this is probably a result of the computer assemblers' diversification towards software and data communications service provision, thus shifting their classification.
18. As in Sweden, the cross-use of ATMs was enabled in different steps, first interconnecting various groups of banks and then interconnecting these groups to each other (cf. the relatively recently established Swedish cross-use of Föreningssparbanken's cash-machines with the other banks' ATMs).
19. See also Raivo Erik (Director of Finance, Forekspank), Madis Kaal (IT Support Department, Forekspank) and Tõnu Liik (Head of Information Technology Division, Hoiupank) in *Baltic IT Review*, no. 2, 1996: 36. Jaanus Soots at Tallinna Pank, i.e., the Estonian leader in telephone banking, did not consider Internet banking to have any bright future, referring mainly to the small volume of the Estonian market and the low number of Internet users.
20. The research budget for 1997 was EEK 44 million and would have been EEK 47.5 million in 1998, but through the merger with the institutes, the amount increased to EEK 81 million.

7. The New Dynamism (1998-2004)

'We have to do it, not tomorrow,
but yesterday – we are already late!'
(Peep Aaviksoo, CEO of EMT, int., 2003)

The telecommunications sector came to play an increasingly important role for the Estonian economy during the 1990s. By 2001, its share in Estonia's GDP had increased to more than 5 percent, which was a phenomenal figure even in international comparison. As we will see in this chapter, this turned out to be important from the perspective of innovation.

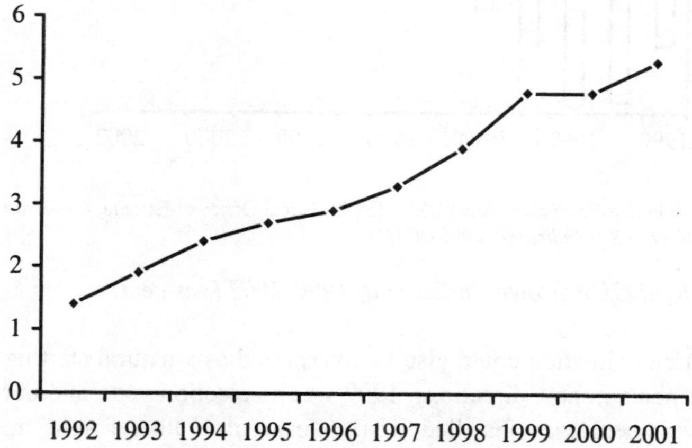

Source: Calculated from *National Accounts of Estonia* (NACE 64).

Figure 7.1 Contribution of telecommunications services to GDP in Estonia, 1992-2001 (percentages)

Much of the development that followed in the years around 2000 was closely related to developments abroad. To the all-European and global trends belonged important issues such as the process of liberalizing fixed telephony, the increasing outsourcing in equipment production and not least the global boom in mobile telecommunications and the Internet. From about 2001, these dynamic developments were followed by a worldwide crisis in several parts of the sector, the ultimate consequences of which were still far from evident by 2004.

At the same time, however, Estonia faced enormous challenges that were specific to the country and the former socialist bloc. In 1997, Estonia was invited to start negotiations for membership in the European Union, and this together with an enormous GDP growth in the same year seemed to confirm that Estonia was on the right path of development. But 1997 also marked the beginning of a more critical period. The national stock market faced a dramatic crash in the late fall, and in 1998 even darker clouds were approaching, as in August Russia was shaken by a giant financial crisis. The Russian crisis led to recession in Estonia in 1999, punctuating the gentle recovery of the economy that had taken place after the collapse of socialism.

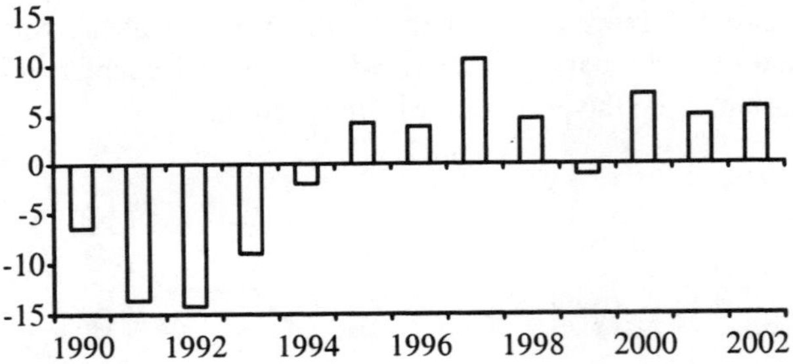

Source: *Transition Report Update*, April 2001: 15; Statistical Office of Estonia, *Lithuania, Latvia and Estonia – Key Indicators 2000-2002*.

Figure 7.2 Real GDP growth in Estonia, 1990-2002 (per cent)

But the difficult situation could also be interpreted as a natural starting point for developments in new directions. 1999 was an election year, and in March a new government was installed in the form of a liberal-right majority coalition. At the same time, the Russian crisis led to a radical down-turn in eastward trade and other linkages, and the second period of economic growth from 2000 onwards built upon a much stronger integration with the Western world and the EU. After a national referendum in September 2003, in which about two-thirds of the population voted in favor of accession to the European Union, Estonia joined the the latter as a full member in May 2004.

1. RESOLVING INSTITUTIONAL BOTTLENECKS

The Long Way Towards a New Regulatory Structure

Eesti Telefon's monopoly on fixed telephony had in the concession agreement of December 1992 been restricted to a period of eight years, with the possibility of extending this period to 16 years. With the acceptance of Estonia into the circle of EU candidate countries in 1997, it became clear once and for all that the option of extending the monopoly would not be realized, as that would violate EU directives.[1] Similarly, the already highly competitive markets in both mobile telephony and data communications, as discussed in the previous chapter, meant that there were many – and increasingly powerful – actors in Estonia who pressed for introducing a liberal regime in the fixed market. As a matter of fact, as will be further discussed below, several actors that entered as operators in the mobile and data market segments had their sights on a future liberalized fixed market. In other words, the expectations of liberalization became a driving force for increasing competition in the already liberalized market segments.

A strong private lobby against both EMT and Eesti Telefon appears to have developed steadily in the years after the introduction of GSM, and in particular after the appearance of the third mobile operator, Ritabell, in May 1997. Frequent complaints about what the new actors regarded as discrimination (for example, concerning the problem of interconnection between the private networks and those controlled by Eesti Telefon and EMT) served as a basis for trying to influence the sector's institutional landscape. The regulatory framework was frequently challenged and re-interpreted, sometimes through the courts. The adoption of the new telecommunications law, which would harmonize the Estonian legislation in accordance with EU and WTO demands, was considerably delayed. In September 1999, it was revealed that Levicom, the main company behind the mobile operator Ritabell and with strong interests in the Estonian telecommunications sector, had been working out an alternative telecommunications bill with a parliamentary faction (*BNS*, 20 September 1999). This indicated an obvious lack of consensus about the ways in which the experience and knowledge of telecommunications companies should and could be integrated into the important process of institutional change. In the early years around 1990, the involvement of Televerket and Tele in re-regulating the sector had been relatively unproblematic, but the sectoral structure had changed dramatically since then, and there were now a whole array of companies and other organizations with a direct interest in changing the rules of the game.

Thus a series of interesting but time-consuming debates characterized the working out of a new regulatory regime during the late 1990s. It was only in February 2000 – less than a year before the termination of Eesti Telefon's fixed-line monopoly – that a new Telecommunications Law could be adopted by parliament. The most important reform compared to the earlier version of the law was probably the explicit discrimination of firms controlling more than 25 per cent of a market segment (Telekommunikatsiooniseadus, 2000: §8). The aim of this paragraph was to encourage diversity and competition – although the law may also be interpreted as a result of lobbying against EMT and Eesti Telefon. The paragraph had considerable potential impact on innovative activities, as it obliged dominant service providers to notify the regulator some time in advance before introducing a new service. This was a problem especially for EMT, which was not only a firm with 'significant market power' (i.e., more than 25 per cent), but which also built its entire competitiveness heavily upon being a leader in technology and new services. Not surprisingly, therefore, there followed a series of new legal disputes after the introduction of the law (see, e.g., Sideamet, 2001b: 7f.).

As discussed in Chapters 5 and 6, the Telecommunications Inspectorate that had been founded in 1991 functioned for many years mainly as a *technical* regulator. Its main task was to coordinate the distribution of radio frequencies, and its first few years had also been spent intensively dealing with problems related to the remaining Soviet armed forces on Estonian territory. The Inspectorate had relatively little to do with Eesti Telefon, which had the monopoly on fixed telecommunications. Licensing procedures did not exist at all in Estonia in the field of data communications, whereas in mobile telephony this was an issue to be handled by the Ministry of Transport and Communications. When it became clear that liberalization of the fixed market was going to take place in January 2001, however, the situation changed. In correspondence with changes in the law in April 1998, the Inspectorate was transformed into the Communications Board (Sideamet), now responsible for tasks that were formulated much more in terms of *market regulation* in addition to the traditional field of technical regulation. The modified regulatory agency started its activities in August 1998 (Sideamet, 2001a: 36).

At the same time, the government was planning to sell a large part of its stake in Eesti Telefon and EMT. For that reason, the structure of the two companies was altered. Thus far, Eesti Telekom had been the state's holding company, which controlled 51 per cent of both EMT and Eesti Telefon. It was now decided that the Nordic investors' stakes (49 per cent) were to be transferred from Eesti Telefon and EMT to Eesti Telekom. In the new ownership structure Eesti Telekom was controlled 51 per cent by the state and 49 per cent by Nordic investors, and became sole owner of both Eesti

Telefon and EMT. The latter were thus more strongly integrated with each other from a formal point of view.

In January 1999, the Estonian state sold roughly half of its shares in Eesti Telekom (corresponding to 28 per cent of the company as a whole), and the company was listed on the Tallinn stock exchange. The move was inspired by developments in Finland, where Sonera (the former Tele) had listed its stock a year earlier. Eesti Telekom immediately became the largest company on the Tallinn stock exchange, representing almost half of its total value. The stock price quickly more than doubled, which led to complaints in which the government was accused of having sold its stake in Eesti Telekom much too cheaply (e.g., *BNS*, 11 February 1999).

The market opening in January 2001 was a unique event for a former Soviet republic and, indeed, for all former socialist countries in Central and Eastern Europe. But the formal alteration of the rules was not the only key issue. An equally major concern was whether competition would be enabled not only on paper, but also led to the entry of competing firms that offered fixed telephony services. It turned out that the regulatory changes *were* followed by a new competitive climate (as discussed further in section 2). Moreover, the competition was so fierce that major conflicts resulted. Again, the new formal institutions were interpreted differently by different actors, and new entrants complained that they were being discriminated against by the incumbent, which enjoyed a unique position through its control of almost all physical infrastructure. This led to legal disputes during spring 2001, resulting in the enforced lowering of interconnection rates charged by Eesti Telefon. In addition, the conflicts led to institutional adaptations in terms of changes in the telecommunications law itself, particularly in order to clarify the rules with respect to interconnection between operators, which through the court cases had turned out to be too diffuse and contradictory. Hence, as a result of the competitive relationships between actors, the process of institutional evolution continued for a long time after the adoption of the telecommunications law in 2000 (see Sideamet, 2001b: 9f.).

The Evolution of ICT Policy

In the broader field of ICT, institutional change and policymaking had during the mid-1990s been primarily oriented towards basic computerization and related areas, rather than towards the creation of Internet-related laws, other formal institutions or Internet services. Even the Tiger Leap project, launched in 1996, aimed initially at equipping schools with computers, although the project later changed its profile towards 'softer' activities, such as competence-building. On the applications side, the far-reaching developments in the private sector with respect to Internet banking and mobile services had so far only found a weak echo in the field of Internet-based public services, as noted in Chapter 6.

However, the outstanding popularity of Internet banking, which provided a powerful example of the huge potential for Internet applications, together with the exponential growth of the number of Internet users, inevitably inspired government agencies to engage in corresponding developmental activities. Once such activities started to be seriously discussed, it was soon realized that there were considerable institutional bottlenecks that hindered the launch of many Internet applications, particularly related to the field of 'e-government'. A key problem was seen to lie in the absence of a legal infrastructure that could regulate the electronic exchange of documents, public access to government databases over the Internet and related issues. In April 1997, a committee was established for the first time to define the required legal basis for the use of *digital signatures* and *digital documents* and to draft the corresponding legislation (*Baltic IT Review*, no. 8, 1998: 39). At about the same time, the Citizenship and Migration Board started to investigate the prospects for introducing an electronic ID card, which was seen as complementary to the expected use of digital signatures and documents (see further section 4.3).

Meanwhile, the Informatics Council also started to work out a set of 'principles of Estonian information policy', as the old documents originating mainly from 1994 (see Chapter 6) were seen to have become obsolete due to rapid technological development and quickly changing economic and political conditions in Estonia. However, the principles that were finally adopted by parliament in May 1998 were extremely broad and diffuse, without any particular priorities or new policy instruments identified. It limited itself to pointing to 'taxation' and 'legislation' (without specifying what this implied or giving examples) as the principal tools to be applied in building the 'information society' (Riigikantselei, 1998: §15). It also stated that private enterprises 'must be encouraged' to invest in R&D, but it neither pointed out any link between the information society and the innovation-based economy, nor indicated how it envisioned R&D activities to be strengthened through policy measures beyond taxation (ibid.: §22).

A more concrete path of change was set out after the shocking experience of Estonia's vulnerability in connection to the Russian crisis in August 1998, and in particular after the installation of a new government in March 1999. The new government was headed by Mart Laar, who started to take personal interest in the development of the information society in Estonia and who chose as his closest advisor in this respect Linnar Viik, a 34-year-old Internet entrepreneur and visionary who had also been working in New York, Stockholm and Helsinki. The new government, coming to power in the midst of the global IT boom, positioned itself as an active supporter of advanced developments in ICT. It showed an obvious dissatisfaction with the slow functioning of existing structures with respect to legislation and policymaking in the field of ICT, and far-reaching discussions aimed at changing these began in mid-1999.

The political will to focus on ICT was illustrated by the fact that IT costs as a share of the total state budget increased substantially in 1999 and 2000 as compared to earlier years (Ott, 2001: 23; Siil, 2000). Substantial new support from abroad was also acquired through the creation of the Baltic Sea IT Fund, which was part of the Swedish 'Baltic Billion' initiative for strengthening collaboration between Sweden and the countries on the Eastern shores of the Baltic Sea. The main projects that acquired finance through the new fund (altogether Estonia acquired SEK 18 million, corresponding to €2 million, for ICT projects) were carried out in 1999 and 2000 (see section 3 and 4 for a discussion of some specific projects).[2]

One of the first reforms initiated followed earlier experiences of the 'Tiger Leap', which focused on the computerization and internetization of secondary schools with related training for teachers. The field that was now addressed was higher education, resulting in the idea to create an 'information technology college' (see section 6). The slowly proceeding legislative work on digital signatures and documents was accelerated. The legislative process involved a broad array of actors, the driving organization being the Institute of Cybernetics, with its accumulated expertise in information security and previous experience in the creation of new ICT-related institutions in re-independent Estonia. The Institute evaluated a number of existing digital signature laws in other European countries, but found that many, such as the German law, had severe shortcomings and that most other countries had not realized what demands were posed on such legislation as a direct consequence of the nature of technologies. In particular, it seemed that few countries had been able to solve the problem of time-stamping and long-term validity of electronic documents. By early 2000, the Institute of Cybernetics was persuaded that it had found a viable solution, and in March that year the new law on digital signatures was adopted by parliament, to be valid starting December 2000 (Digitaalallkirja seadus, 2000). It was hoped that the new law would provide 'a good opportunity to develop various electronic services' (Siil, 2000).

The issue of digital signatures became part of a much larger reform program for information and communications technologies in Estonia, and shortly after the adoption of the new law an ambitious ICT strategy was presented by the government. ICT management structures were to be reorganized, and the Department of State Information Systems (RISO) moved to the Ministry of Transport and Communications (Ott, 2001: 24). This seemed to reflect the increasing awareness of the technological process of convergence between data communications and telecommunications in a traditional sense. (In this connection, the pan-Baltic English-language IT journal, the *Baltic IT Review*, was also renamed *Baltic IT&T Review*, thus explicitly including telecommunications.) At the same time, there were discussions to create the position of an 'IT minister', though this idea was never realized.

More important was probably the new emphasis on citizen-centered public IT services, which according to the prime minister's advisor were to replace the earlier agency-centered solutions (Linnar Viik in an interview with *BNS*, 23 March 2000). This related to the trend discussed in Chapter 6 about the creation of a single entry point (portal) for all governmental services. The provision of these services through the Internet was to be accelerated with the help of new legislation (in particular the digital signature law and the public information act), which was also to be seen in relation to the plans to issue electronic ID cards from December 2001 (see further section 4.3).

Also in 2000, the government introduced the 'E-citizen' project, which was supposed to make all (!) public services available online by 2004 (*Baltic IT&T Review*, no. 22, 2001: 23). Again, the knowledge-creation activities initiated in order to realize this vision soon led to the observation that the legislative situation hindered the new developments. The reason was the unclear situation with respect to the legal rights to access certain data. As a result, in late 2000 another law was adopted, the Public Information Act, which guaranteed citizens access to information that had been created as a result of the activities of public agencies. It was explicitly stated that such access *must* be provided also by way of the Internet (Meriste and Kalvet, 2001: 69). It thus also functioned as an interesting form of public technology procurement, where the government demanded from itself – by way of legislation – that new, still non-existent, public Internet services be introduced.

Intense activities were thus going on both with respect to the development of new services (as further discussed in section 4.2 below) and to institutional change and policymaking. The Estonian confidence in the country's capacity in ICT developments was strengthened by encouraging and surprising experiences with respect to innovation in the private sector, in particular mobile telecommunications and Internet banking (see further section 3 and 4.2). Abroad, too, Estonia tried to profile itself more and more as an advanced IT country, preferably comparing itself to the Nordic countries. In this connection an interesting channel of inspiration and interaction was the 'Northern e-Dimension', which was based on cooperation between the Council of the Baltic Sea States (CBSS) and the European Commission (*Baltic IT&T Review*, no. 22, 2001: 16-19).

With respect to innovation more generally, i.e., not only in the field of ICT, Finland especially seems to have served as a point of reference. For example, in order to strengthen developments with respect to innovation, it was suggested that an affiliation with the Bank of Estonia be created in the form of a development fund modeled after the pattern of Finland's Sitra. In the end, the Bank of Estonia did not agree, and the idea could not be realized. On the other hand, inspiration from Finland's Tekesi technology agency did result in the creation of the Estonian Technology Agency (ESTAG). It replaced the Estonian Innovation Fund, which in late 1996 had taken over the responsibility from the Informatics Fund for financing ICT-related innovative projects (cf. Chapter 6), and was integrated into the larger support organization Enterprise Estonia (Ettevõtluse Arendamise Sihtasutus, EAS) in January 2001. The new agency became responsible for a number of important projects relating to innovation policy in general (i.e., not specifically ICT-related innovation policy). Thus in 2001 it carried out the Community Innovation Survey (CIS III) for Estonia, and at about the same time produced an important five-year R&D strategy document for the years 2002-2006, which was approved by parliament in December 2001. In August 2001, Enterprise Estonia also concluded an important agreement with the City of Tallinn and Tallinn Technical University, foreseeing the creation of a Technology Park in the Estonian capital. There were also plans to create a number of 'competence centers', one of which was to focus on electronics, communications and information technologies. The contracts for these were finally signed in May 2004 (see section 7).[3]

The private sector also engaged in far-reaching support activities with respect to the further development of the Estonian ICT sector. In 2000, the major banks and telecommunications operators contributed heavily to financing the new IT college (as referred to above; see also section 7), and in February 2001 Hansapank launched the 'Look@World' project, which was aimed at further boosting the use of the Internet in Estonia. EMT, Eesti Telefon, Ühispank, MicroLink, Baltic Computer Systems and a few other private firms soon joined the project. Together these firms were to spend no less than EEK 250 million (€16 million) – an enormous amount for Estonian conditions – to reach the goal of the project, which was explicitly formulated in terms of overtaking (!) Finland with respect to the use of the Internet and Internet applications (Ehandi, 2001: 36ff.). The government, too, seemed to take inspiration from this ambitious goal, and in his speeches abroad during 2001, prime minister Mart Laar already talked about Estonia as aiming to become 'the no. 1 Internet country in the world' (*BNS*, 7 March 2001).

The relationships between these far-reaching institutional and policy developments and other activities in the telecommunications sectoral system are discussed in detail in the following sections.

2. THE FIXED NETWORKS: INTRODUCING COMPETITION

In parallel with the institutional reconfigurations that were being intensely debated in Estonia during the last years of the 1990s, Eesti Telefon continued the expansion and modernization of the fixed networks. Under severe pressure from private actors and the government, annual capital investments were further increased in order to boost basic telephonization as well as digitalization.

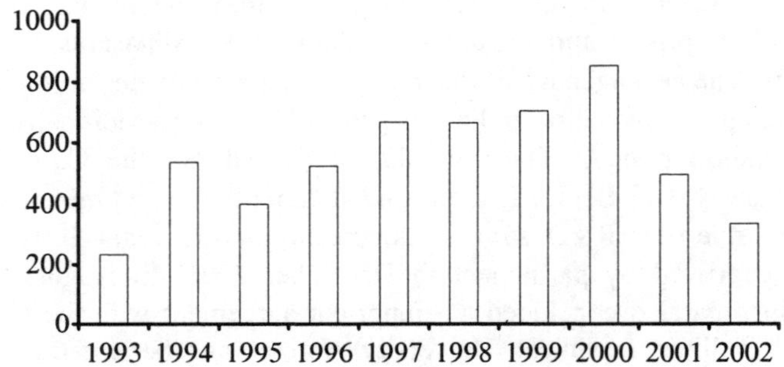

Source: Eesti Telefon (2002).

Figure 7.3 Eesti Telefon's annual capital investments, 1993-2002 (million EEK)

From January 2001, however, the introduction of competition in fixed telephony meant that Eesti Telefon lost both its exclusive rights and some of its special obligations with respect to modernization of the country's basic telecommunications networks. The concession agreement was formally ended through the signing of a new agreement between Eesti Telefon and the Ministry of Transport and Communications, whereby it was also stated that Eesti Telefon had met the (rather loose) terms and conditions in the concession agreement. The company remained responsible for providing universal service, and committed itself to providing a telephone connection to everyone currently on the waiting list by the end of February 2003 (Eesti Telefon, 2000a). But the earlier pressure for boosting the expensive process of digitalization was no longer mentioned.

The result was that capital investments in new networks and digitalization dropped dramatically starting from 2001, as shown in Figure 7.3. In 2002, investments amounted to only around a third of the capital expenditures of two years earlier. This had severe consequences for the process of digitalization, which had been dramatically boosted in the last years of operation under exclusive rights, but which now stagnated. Both Latvia and Lithuania, which had faced severe modernization difficulties during the 1990s, now overtook Estonia in terms of the level of digitalization (see Figure 7.4).

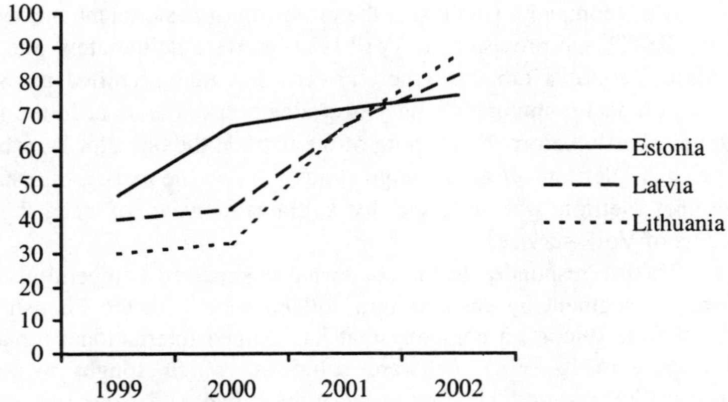

Source: *Enlargement Futures Report Series, Information and Communication Technologies Panel Report,* December 2001; *3rd Report on Monitoring of EU Candidate Countries (Telecommunication Services Sector)*, June 2003.

Figure 7.4 Level of network digitalization in the Baltic states, 1999-2002 (percentages)

One major group of actors seeking to push Eesti Telefon to fulfill its obligations during the last years of its exclusive rights were other operators planning to enter the market, i.e., as competitors to Eesti Telefon, in January 2001. These competitors had everything to gain from a high-quality basic infrastructure, to which they would then be able to connect their own networks. As discussed in further detail in section 4.1, the competitors were typically already active in other market segments – particularly in Internet access provision – where they sought to establish their brandnames and build up a customer base a few years before the actual liberalization of the fixed telephony market.

At the same time, while potential competitors were about to diversify from mobile and data services towards fixed telephony services, Eesti Telefon prepared itself for the increasing competition by defining itself increasingly not only as a *telephony company*, but above all as a *data communications provider*. The company's activities in this field are discussed in the section on the Internet below. However, there was also one field of services that tended to completely transcend the traditional division into telephony and data communications: Internet-based telephony.

In November 1999, more than a year before the liberalization, Eesti Telefon started on a trial basis to offer voice services over the Internet ('VoIP') (Eesti Telefon, 1999b). While seemingly contradicting the main activity of the company, which was the provision of basic telephone services over the PSTN, the provision of VoIP services were deliberately integrated with Eesti Telefon's other Internet services and thus signified the strong emphasis on data communications as a strategic business area. In the fall of 2000, i.e., shortly before the opening of the market, the operator launched its new product 'Netifon'. It garnered great attention on the market – to such an extent that Netifon was followed by a dramatic wave of entry by new providers of VoIP-services.

Eesti Telefon responded to the somewhat unexpected competition in this new market segment by entering into collaboration with the Finnish VoIP specialist firm Wicom, a company that had gained international reputation with respect to its VoIP software solutions, which sought to replace traditional PBX systems. The first result of the collaboration was presented in February 2001 and tried to attract corporate clients by offering the possibility to create a 'virtual' private branch system, while at the same time adding some new features to firm-internal telephone connections. Wicom seemed optimistic about the prospects for IP telephony in Estonia, arguing that in the more turbulent evolving market of the Baltics as compared to Western countries, it was 'possible to rapidly implement IP telephony services and to bypass certain stages of traditional telephone technology' (Eesti Telefon, 2001b). A pilot project for the new solution was announced in May 2001, when the Estonian-Finnish-Swedish ferry company Tallink became the first to launch an IP-based call handling system using Eesti Telefon's and Wicom's solution. The idea was to centralize the ferry booking system to a call center in Tallinn and to create teleworking opportunities (Eesti Telefon, 2001c).

At that time, the Estonian telecommunications market had already been fully liberalized. Not surprisingly, the opening up of the market in January 2001 – which at that time was a unique phenomenon in Central and Eastern Europe – led to a massive wave of entry and considerable turbulence in the whole sector. By June 2001, no less than 43 new service providers had entered the market, and half a year later yet another 29 providers (Sideamet, 2001b: 6). During the first two weeks of free competition, Eesti Telefon lost 8 per cent of the international calls market and 3 per cent of the domestic market (*Äripäev*, 8 February 2001). At the end of the year, the company had lost 9 per cent of the aggregate fixed voice services market. Revenues from international calls dropped by no less than 46 per cent as compared to the year before (Eesti Telefon, 2001a: 67).

Eesti Telefon's most serious competitors were the Finnish-owned Uninet and the Swedish-owned Tele2, both of which had earlier been active on the Internet and the mobile markets (i.e., if Uninet's alliance with Radiolinja, whose owners were basically the same, is taken into account – see section 4.1). These two actors pursued an aggressive strategy that, among other things, led to the amendments in the telecommunications law that were discussed in section 1. The general price level on fixed telephony services was considerably lowered. Tariffs fell particularly on international but also on domestic calls, whereas local calls became more expensive as a result of the removal of cross-subsidization (Sideamet, 2001b: 9).

3. MOBILE COMMUNICATIONS: TOWARDS CREATIVE INNOVATION

The Evolution of the Mobile Services Market

Mobile telecommunications was a field in which Estonia had taken a lead among Central and East European countries in the NMT era. Although GSM was introduced more or less simultaneously in most countries in Eastern Europe in 1995 – which also turned out to be the year of GSM take-off globally – Estonia, once again, emerged as the leading country, with a penetration rate of 18 per cent in 1998, double the figure of the second most advanced CEE country in terms of mobile phone use (see Table 7.1).

The entry of a third mobile operator in Estonia in May 1997 was, as pointed out in Chapter 6, globally almost a unique phenomenon, as there (at that time) existed only a very limited number of countries with three competing operators. This was the more astonishing if seen in relation to the size of the tiny republic. In terms of 'number of operators per capita', Estonia was many times ahead of all other countries in the world. The increasingly competitive market also resulted in more aggressive advertising campaigns and lower prices on mobile services as compared to other CEE countries, a fact that contributed heavily to boosting the number of users and the articulation of demand for mobile services. EMT, the leading operator, was surprised when in 1997 the aggressive advertising campaign of Ritabell in connection to the latter's market entry turned out to have a more dramatic effect on EMT's user growth than the company's own commercial campaigns had (*Äripäev*, 9 May 1997).

The market structure that crystallized in the late 1990s was one of EMT as the dominant player, with around half of all users, followed by Radiolinja and

Ritabell, both of whom took around 25 per cent. Later on, Ritabell advanced to challenge EMT's dominant position, with a market share of around 36 per cent in late 2003, while Radiolinja's market share tended to decrease somewhat (see Figure 7.5 below). In January 1999, Ritabell was fully acquired by the Swedish firm Netcom, which until then had owned 48 per cent of the shares. In 2001 Ritabell was renamed Tele2, which was already a well-known brandname used by Netcom-related operators in a number of other countries (Ritabell, 1999).[4]

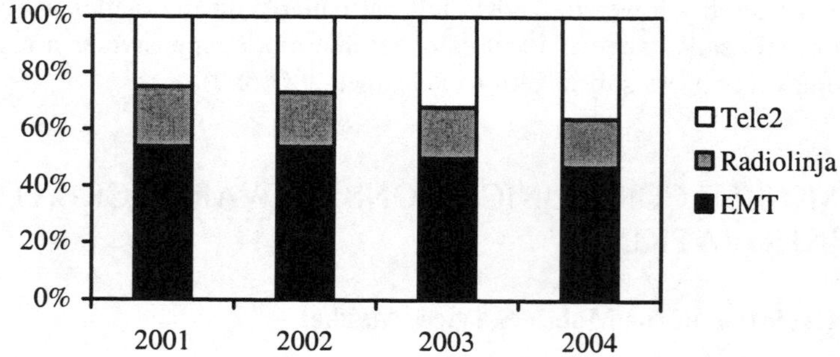

Note: '2004' = November 2003.

Source: *Äripäev*, 5 November 2003.

Figure 7.5 Market shares of mobile operators in Estonia, 2001-2004 (beginning of year)

In international comparison, other CEE countries caught up with and – in the case of Slovenia, the Czech Republic and Hungary – even overtook Estonia in terms of mobile phone penetration after 1998. This was presumably a result of Estonia's economic recession after the Russian financial crisis in August 1998, which hit the Baltic states much harder than the Central European countries. However, at the turn of the century, the very use of a mobile phone was no longer necessarily associated with any particular innovativeness; after all, GSM networks had by then already been in commercial operation for about a decade, if seen in global perspective. From the perspective of innovation, it may therefore be more relevant to discuss other aspects of the mobile communications subsector than pure market penetration.

Hence, it is important to point out that the three Estonian operators acquired somewhat different *qualitative* characteristics, as described in the following. Ritabell/Tele2, roughly speaking, emerged as the price leader on the mobile market. It succeeded in boosting the growth of the whole sector both through aggressive price competition and by launching, for the first time in Estonia, a prepaid service in February 1998. The importance of the introduction of prepaid services was clear from the fact that in 2003, no less than half of Tele2's customers were prepaid card users. The introduction of prepaid services was followed only later by the other two Estonian operators. However, Tele2's lead in this specific field was a rare exception from the general pattern with respect to the introduction of new services. The firm devoted only little resources to R&D and had a minimal development department as compared to Radiolinja and in particular EMT.[5] The strategy of Tele2 seems to have laid close to the overall characteristics of the Swedish-based global group even before the full take-over. However, there will be reason to return to Tele2 in the context of innovation further below.

Table 7.1 Mobile phone penetration in selected CEE countries, 1998-2002 (percentages)

	1998	1999	2000	2001	2002
Estonia	18	28	41	46	65
Latvia	7	11	16	28	39
Lithuania	7	9	14	28	47
Poland	5	10	17	26	36
Slovakia	9	12	21	40	54
Slovenia	10	31	55	74	84
Czech Rep.	9	19	41	68	85
Hungary	10	16	31	50	65

Source: *Statistical Yearbook of Estonia 2002*; ITU, *World Telecommunication Indicators*, 2002, 2003.

The Finnish-owned Radiolinja had been the first to launch GSM commercially in Estonia, and although it did not keep up with EMT in terms of number of subscribers and network coverage, it resembled the latter to the extent that it sought to establish itself as a technological leader. After all, it was the former R&D director of EMT who long acted as general director of Radiolinja – a fact that also tended to hinder any far-reaching cooperation between Radiolinja and EMT. In striving to reach the goal of technological leadership, the operator's close connection to its Finnish mother company across the Gulf played a very important role. Similarly, the traditional linkage that had already been built up in Finland between Radiolinja and *Nokia* became crucial for Radiolinja Eesti. In particular, Radiolinja Eesti was chosen by Nokia as a *test operator*, implying that many new technologies to be launched by the Finnish manufacturer were tried out in the Estonian network before being commercially launched globally. For example, in

March 2001 Radiolinja Eesti became the first operator in the world where Nokia's 'mCatch' positioning platform was tested (and later taken into use commercially – see also the discussion below on the evolution of innovation in positioning services). Similarly, in February 2002 Radiolinja Eesti became the first operator in Northern Europe to introduce Nokia's USSD service platform (information from Radiolinja, 2003; Viira, int., 2003; cf. also Kõstner, 1996).

This emergence of Radiolinja as a collaborative partner of Nokia seems to have been the result of a combination of different factors. On the one hand, Radiolinja's technological competencies were obviously recognized from the Finnish side, which together with the convenient geographical proximity – Tallinn being located just some 80 km away from Helsinki – made the idea attractive. The more conventional cooperation in terms of equipment supply had also functioned very smoothly all the way since the initial success in 1994-95, and considerable confidence and trust had thus had time to be built up between the two partners. On the other hand, Radiolinja Eesti also had the advantage, from the perspective of Nokia, that its mobile network was almost completely based on the Finnish manufacturer's technologies. By contrast, in Finland, where Radiolinja's network contained a considerable amount of Siemens equipment, it was more difficult to test new solutions from Nokia (Viira, int., 2003).

EMT, the jointly Estonian-Swedish-Finnish-owned operator and Eesti Telefon's sister company, overtook Radiolinja in terms of number of users and network coverage after the launch of GSM in 1995. In technology and the introduction of new services, EMT positioned itself as the most advanced player in the market, and used its reputation strategically in response to the aggressive appearance of Ritabell in 1997 and the increasingly tough competition in the Estonian mobile market. EMT's enormous profit margins of around 40 per cent (see Figure 7.6 below), which were extreme in global perspective, gave the firm considerable room for maneuvering. The firm was characterized by young and self-confident personnel with a general admiration for advanced and futuristic new developments, a self-image that also seems to have resulted in an increasing cultural separation from its sister company, the fixed incumbent Eesti Telefon. Having started as a minor venture within Eesti Telefon (though formally always a self-standing firm), EMT overgrew the fixed operator economically, in 1999 accounting for about 70 per cent of the value of the whole Eesti Telekom group (EMT, 1999).

Since 1994 the R&D department at EMT had been headed by a young telecommunications engineer, who had graduated from the Department of Radio and Communications Engineering at the Technical University, and had received his master's degree from Chalmers University of Technology in

Sweden. With technological opportunities in mobile service applications rapidly growing globally during the 1990s, EMT's R&D department under the new management team grew to become the largest among the three operators, and the central hub for innovation in the Estonian mobile telecommunications sector.[6] Its activities will therefore be discussed in depth below.

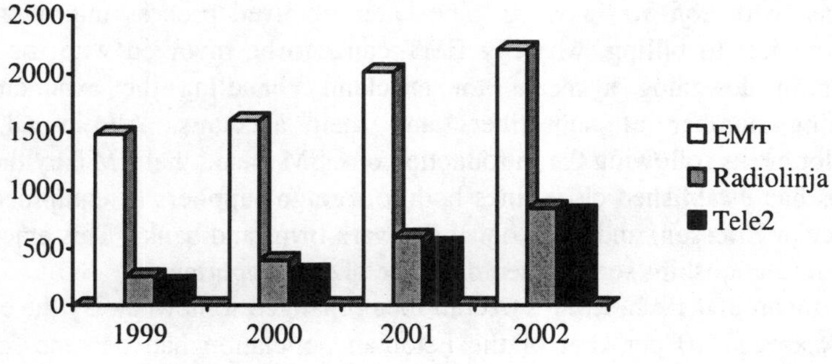

Source: *Äripäev*, 5 November 2003.

Figure 7.6 Turnover of mobile operators in Estonia, 1999-2002 (million EEK)

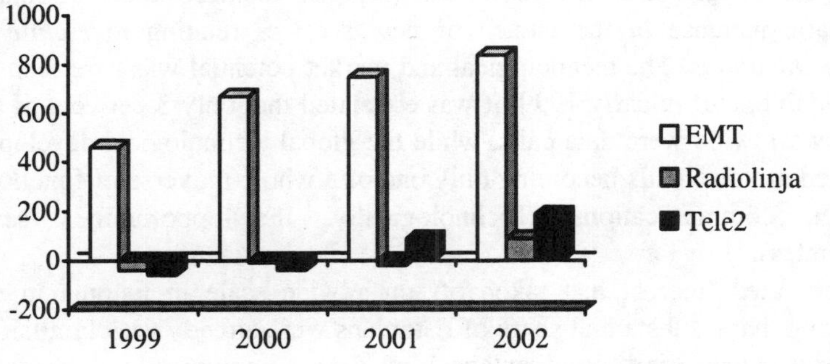

Source: *Äripäev*, 5 November 2003.

Figure 7.7 Profits of mobile operators in Estonia, 1999-2002 (million EEK)

Entering m-Commerce

As noted in Chapter 6, in the years immediately following the introduction of GSM, EMT heavily invested in network expansion, seeking to cover an ever larger territory in Estonia, while at the same time introducing a few basic data services such as SMS and mobile fax. There was also a focus on *process renewal* not least in terms of organizational change, as EMT grew from a small start-up firm, hosted in the facilities of Eesti Telefon, to what soon turned out to be one of the largest firms in the entire Estonian economy. While the growth of the network meant that the operator's business relationships with (foreign) equipment suppliers became increasingly frequent and tight, customer growth also meant a heavy stress on the expansion of *software systems*. The latter involved process improvements with respect to billing, whereby EMT came to be involved with the local banks in designing a means for efficiently handling the exponentially growing number of subscribers and their accounts. All in all, the developments following the introduction of GSM meant that EMT by the late 1990s had established close links both to foreign suppliers of equipment (in particular Ericsson) and to Estonian software firms and banks. This emerging web of relationships soon turned out to be of vast importance.

From around 1998, EMT's overall focus changed somewhat. By the end of 1997, around 10 per cent of the Estonian population had become mobile phone users, forming an enthusiastic and rapidly growing customer base that was unparalleled elsewhere in Central and Eastern Europe. Since May 1997, there were three operators in the market, where in particular the newcomer Ritabell sought to catch up by aggressive advertising campaigns and by its launch of Estonia's first prepaid card in February 1998 (Ritabell, 1998). On EMT's part, the year 1998 saw – in response to these market trends – a dramatic increase in the launch of new services relating to *mobile data communications*. The technological and market potential was here seen to be formidable; still in early 1999, it was estimated that only 3 per cent of GSM calls worldwide were data calls, while the global technological development pointed at voice calls becoming only one of a whole universe of functions in mobile communications. Technologically, the opportunities seemed bottomless.

The fixed Internet had taken off on a wide scale in Estonia in 1995, meaning that a substantial share of Estonians were already very familiar with the advantages (and disadvantages) of data communications services in general. This was important for EMT, which saw in the experienced fixed Internet users of Estonia a potential market for *mobile data services*, such as mobile e-mail and Internet connections using a lap-top and mobile phone. Such services were the most important new introductions by EMT in 1998. In the case of Internet access, which at this time was not yet available on the mobile display itself, EMT collaborated, though only on a minor scale, with the fixed Internet provider EUNet (which was the new brandname of Data Telecom in Estonia – cf. Chapter 6). At the same time, value-added SMS services started to find their way to the market: in 1998 EMT introduced the possibility of acquiring information on the mobile account balance through SMS, and in July 1999, in cooperation with Optiva Pank (a successor of Forekspank, which had been a first mover in electronic banking) and later Ühispank (which had hired Hoiupank's IT director after the latter's merger with Hansapank in 1998 – see section 4.2), a similar service was launched for account information from *bank* accounts – a service that was obviously inspired by the popularity in Estonia of Internet banking.

The interest in account balances and the cooperation with the banks were in themselves hardly of any strategic importance at this stage for EMT's competitiveness and future growth. However, they were closely related to another potential field of innovation that was globally deemed to have a bright future: mobile commerce. In Estonia, where an unusually large share of the population was already accustomed to using new communications technologies in banking (a sector that also hosted the largest IT development teams in the country – see Chapter 6), EMT did not see any reason to hesitate with entering this field – in particular not in a situation where two aggressive competitors did everything to break EMT's overall dominance on the mobile market. The enthusiastic interest in experimenting with m-commerce solutions at EMT's R&D department seemed to be revealed publicly for the first time in October 1999, when the company, in cooperation with the local Coca Cola representation and a small Estonian telematics firm, Oskando, set out to reprogram a soft-drink vending machine – initially only at EMT's own building – to let it be operated by a mobile phone; the cost for the drink was added to the mobile subscriber's normal bill (*Mobiil. EMT Kliendileht*, no. 5, 1999: 7).

Although this service was little more than a curiosity for a long time, it did point to an interesting direction for the future. In March 2000, an agreement was reached between EMT and the City of Tallinn for an unusual venture: EMT was to offer the possibility of *paying for parking a car* with the help of a mobile phone (ibid., no. 1, 2000: 5). The background was the highly problematic situation with respect to car parking in Estonia, which differed dramatically from the situation in, for example, the Nordic countries:

> The payment procedure was very bad. People wanted to pay, but it was so difficult. There were very few payment methods around; you went to the town, and you had to park, you had to go to a specific place to buy a "scratch card" . . . and then go back to where you wanted to park . . . So the city government was very interested in having a new scheme, a new technology. (Haljand, int., 2003)

The idea of the new service, which came to be called 'm-parking', originated in loose discussions taking place in an informal network of personal connections between EMT, Hansapank and local software firms, some of which had already had the city government as a client and partner. Hansapank's interest was in the payment procedure, but the architecture of the service that was proposed by Oskando (the small firm mentioned above) and accepted by EMT excluded the connections to the bank, as it was decided that the payment would be handled directly between EMT and the City of Tallinn (Kõuhkna, int., 2003; Haljand, int., 2003). In practice, this meant that the mobile operator for the first time emerged as a *competitor* to the banks.

In the actual development of the service, Oskando did not take part. Instead, the development took place in collaboration between EMT and Voicecom, with whom the mobile operator had been collaborating already in the NMT era. As discussed in Chapter 6, Voicecom was a company that had emerged out of the split of Univoice into two separate companies, and several of its employees had a background as researchers at the Institute of Cybernetics. The company, which employed no more than 10 to 15 people, had in the Estonian telecommunications system of innovation so far played its most important role in the field of telephone banking (see Chapter 6), but through the m-parking project the firm established itself in a decisive way in mobile innovation.

The 'm-parking' service was developed during the first half of 2000 and was launched commercially in July that year. It quickly acquired an unexpected popularity among car parkers, and within a short time almost half of all parking payments were collected through mobile phones. In a global perspective, the service was unique, and it also represented one of the world's first broadly utilized and profitable applications of m-commerce. At the global GSM Association's annual meeting in 2001, EMT's 'm-parking' was nominated as the 'most innovative service of the year', and it also won a special prize at a conference in Stockholm.

International interest in the innovation seemed to open prospects for *exports*. In early 2001, Norway showed an interest in licensing the solution from EMT. However, this turned out to be more difficult than expected. In Estonia, it was EMT itself that had offered the service in cooperation with the city government. The interested partners in Norway, however, were not mobile operators themselves, but third party entrepreneurs. This circumstance created problems with respect to billing procedures, ending up with the idea of charging customers by way of credit card payments (VISA). As a result, a practically new service architecture had to be designed. Thus a development consortium, mPay, was formed with EMT and the Norwegian entrepreneurs as shareholders. It was only in April 2002 that a pilot project could be launched in a limited area in Oslo. Moreover, further developments were hindered by legal uncertainties in Norway, and the project was again delayed. In addition, it was highly unclear whether the service would really acquire any popularity, as there – in contrast to Estonia – already existed a well built system of 'pay-and-display' machines for parking all over Oslo (Haljand, int., 2003; see also *Baltic IT&T Review*, no. 25, 2002: 25).

In Estonia, EMT licensed the new service to Radiolinja. Tele2, however, preferred to develop its own in-house solution and launch a similar service of their own. Similar trends could be seen in Latvia and Lithuania, where the service appeared in a number of different variants, often in-house developed by each operator. The export success that EMT had hoped for seemed to vanish.

Launching Mobile Positioning

Mobile parking was a service that was developed exclusively by Estonian firms, at least to the extent that EMT itself could be regarded as an Estonian firm. There are good reasons to see this as being the case. As a matter of fact, although EMT was owned 49 per cent by Telia and Sonera, the relationships to the Nordic operators beyond ownership per se were relatively weak, and the management had been exclusively Estonian since the mid-1990s. Only at the level of the supervisory board were Nordic representatives involved.

On the manufacturing side, the Estonian-Nordic links were stronger. Ericsson Eesti was essentially to remain a sales office for the global company's business in Estonia, and as such it had little opportunity to go its own way. The Estonian subsidiary was in no way meant to be a center for Ericsson's innovative activities; the only exception was in implementation-related R&D, where a number of Estonian engineers in the Tallinn office supplied radio network design services on a minor but worldwide scale (Sepp, int., 2001). However, this transnational integration did not prevent Ericsson Eesti from developing an increasingly local profile in the course of the 1990s, and in this sense it resembled the development of EMT. A main step was taken in this direction when the Finnish general director of Ericsson Eesti, Matti Lehtimäki, was replaced by a young, 32-year-old Estonian, Veiko Sepp, in late 1998. The latter had started his Ericsson career in 1991 in Helsinki, where Ericsson had offered the opportunity to write a telecommunications-related diploma thesis for the Tallinn Technical University (ibid.). Similarly to EMT, with whom Ericsson by the end of the 1990s could already look back upon a ten-year long experience of close cooperation, Ericsson Eesti from now on hosted a young, enthusiastic, and fully Estonian staff.

In 1999, Ericsson Eesti and EMT, both of which having recently installed new CEOs,[7] jointly decided to enter an interesting project: it aimed at launching a new service based on Ericsson's mobile positioning system (MPS) platform, which had been developed at Ericsson in Sweden during the preceding years. GSM-based positioning was at this time a technology that was still in its infancy and not yet commercially available. The satellite-based GPS positioning system (under an American lead) provided a much higher quality and accuracy of determining the location of a handset connected to the satellite communication (although it did not work inside houses). Internationally, an interesting debate was going on at this time; the field of positioning was already widely recognized as a promising future field of business, but it was unclear as to whether such services were to be provided through the satellite-based system or ordinary mobile communications networks. EMT and Ericsson headed for the latter variant, although MPS was not yet commercially available anywhere else in the world.

In November 1999, Tele2 in Sweden announced that it had launched the world's first GSM-based positioning service. It was aimed mainly at logistics firms as a cheaper way (compared to GPS) of keeping track of, for example, a vehicle (*Nyhetsbrevet Telekom*, 1 December 1999). However, this demanded special SIM-cards to be used in the mobile handsets. In contrast, Ericsson's system, MPS, built on technology whose intelligence was located in the mobile network itself rather than in the handsets, thus making any handset positionable without additional user software. This was attractive from the perspective of major attempts going on at this time with respect to making it possible for *rescue agencies* to locate emergency calls made from mobile handsets. EMT sought to develop such a service based on Ericsson's MPS platform.

Ericsson's headquarters in Sweden were somewhat skeptical of the idea of launching the world's first commercial MPS service in the former Soviet Republic of Estonia. An open issue was above all the availability of necessary complementary competencies in the tiny and economically poor country. In particular, MPS services were by necessity closely related to the field of geographical information systems (GIS), and if the new service was to be launched in Estonia, it would have to find local suppliers of such systems.

Surprisingly, it turned out that advanced GIS suppliers did not merely *exist*, but that Ericsson had even been involved in *collaboration* with one of them a few years earlier. The firm was called Regio and was known as Estonia's main producer of maps. As noted in Chapter 5, this was a young firm created in 1988 as a spin-off from the University of Tartu, and it had diversified into software development in connection with the digitalization of maps. There, it enjoyed an interesting advantage compared to many Western map producers: the latter had typically tried to launch digitized maps as an extension of analogue map production (meaning that their maps still contained some hand-drawn elements), whereas Regio had early headed for a complete break with the industry's past, in which it had never been involved. This had allowed the company to evolve swiftly along the new digital GIS trajectory. In 1997, Ericsson had collaborated with Regio in the development of a system for the Estonian police, where it was attempted to create a solution based on Ericsson's EDACS radio communications platform. The idea, which was successfully realized, was to display the positions of police cars on a map (Jagomägi, int., 2003).

Ericsson remembered the fruitful cooperation with Regio, and despite the hesitation connected with Estonia's being a former Soviet republic, the Swedish multinational agreed to let EMT, Regio and Ericsson Eesti enter the new positioning project. Also involved was, of course, the local Rescue Board. Similarly to m-parking, the project thus involved a government agency. At about the same time as Tele2 launched its new services in Sweden, i.e., in the fall 1999, EMT's R&D department announced that the MPS project was being launched. During the following months the project was developed, and in May 2000 it had arrived at a stage where EMT's CEO, the Rescue Board director and Ericsson's President Kurt Hellström could meet ceremonially in the medieval City Hall in Tallinn to sign the official memorandum preparing for the actual launch. It thereby also turned out that the project had managed to receive financial support from the Swedish Baltic Sea IT Fund (cf. section 1). From the beginning of June, it became possible to use the new service in Tallinn and the surrounding county, Harjumaa (*Mobiil. EMT Kliendileht*, no. 2, 2000: 4).[8] At the international positioning conference in Rome a few months later, EMT attracted surprised attention for being one of only two operators worldwide that at the time offered working positioning services in their networks.

Developing Mobile Telematic Services

EMT had developed its 'm-parking' service in collaboration with Voicecom and the City of Tallinn, and it developed mobile positioning together with Ericsson, the Rescue Board and Regio. These collaborative links for innovation were strengthened further during the following years, as was also EMT's collaboration with Oskando, the small telematics firm that had also initially been involved in the case of mobile parking.

Oskando was originally a firm doing electronics design, but it diversified into IT and underwent significant changes in 2000, when it was merged with a small software firm, Interplus, which had been supplying Internet solutions to the City of Tallinn. The manager and owner of Interplus, Tarmo Kõuhkna, was an experienced Estonian entrepreneur who had recently completed an MBA in Helsinki; his decision to invest in Oskando and merge it with Interplus was based on the fact that a Swedish car accessories wholesaler, K.G. Knutsson, which had recently been established in Estonia, had shown considerable interest in procuring *car alarm systems with mobile communication functions*, to be marketed in the Baltic states (where car theft posed a serious problem and the markets thus seemed promising). EMT had shown interest in implementing mobile phone-based car alarm services in its GSM network (based on a SIM-card in an alarm module to be mounted in the car, from where relevant messages could be sent over the GSM network in case of theft), and Knutsson had committed itself to procuring a substantial amount of alarms – if EMT could deliver the corresponding system. There thus already existed a guaranteed initial market for the future product.

However, it was clear that the project would be heavily based on electronics hardware, where EMT had no competencies. Yet, through its earlier cooperation with Oskando (in mobile purchases of soft drinks and mobile parking), the operator was well aware of the existence of the necessary hardware capabilities within its own network of partners. Based on a collaborative agreement with EMT and Knutsson, Oskando began developing the system. It decided to rely on technology from Ericsson, which had a domestic subsidiary aggressively promoting a special GSM module, GM-22, that was deemed appropriate for the project. The device was still in a relatively fluid stage of development at Ericsson and had been used only minimally so far. Through the early interest in the module in Estonia, Oskando grew to become Ericsson's most important customer worldwide for GM-22. Based on this module, Oskando developed its solution in-house. Manufacturing was outsourced, but final assembly and testing were taken care of by Oskando (Kõuhkna, int., 2003).

The service that built on this product, named the 'Car Controller', was launched by EMT in fall 2000. By then, it had also come to merge with developments of MPS (the positioning services discussed above). The MPS platform that had been used in the rescue service found a new application in the 'Car Controller', where the client – or a security company – was thus enabled to determine the geographical position of the car in case of theft. The system could also control various functions of the car from a distance, such as switching on the heat and blocking the fuel pump (in order to prevent theft).

Oskando managed to export its solution to a number of countries, primarily former Soviet republics (in particular Latvia and Lithuania, but also Russia and Belarus). By building on existing personal links to Finnish entrepreneurs, in particular the mobile positioning-based security firm 4 Technology Solutions in Lahti, an export market was found there. In general, however, Oskando viewed it unrealistic to aim at the Nordic or Western markets to any greater extent. One reason was the simple fact that car theft was a much smaller problem in the West. More important, however, seems to have been the skepticism of Nordic car accessories wholesalers at the thought of offering Nordic customers solutions developed by firms in the former Soviet Union. In Sweden, for example, the wholesalers preferred to rely on corresponding Swedish technology firms such as Satsafe (which offered solutions that competed directly with Oskando on the Lithuanian market). In Estonia, however, Oskando had more or less a monopoly (ibid.).

Further Developments

Through the launch of m-parking, mobile positioning and telematic services, all in 2000, this became the breakthrough year for EMT and its partners in terms of innovation. During the following years, the new services were further developed, and the network of partners that had begun to crystallize around EMT was strengthened. For Regio, which supplied the key middleware for EMT's positioning services (under the trademark 'Reach-U'), the partnership with EMT and Ericsson had interesting consequences. Regio acquired the status of a highest-level Ericsson partner, with the implication that its software products could be marketed all around the world through Ericsson's corporate network (Regio, 2001). The outlook seemed highly promising.

In order to enable and support further innovative activities, Regio decided to look for additional capital, which led to the merger between the Tartu-based firm and the Finnish IT company DONE. A formally separate company, Mgine Technologies, was created jointly by the two firms, to be responsible for further developments in mobile positioning. In October 2000, EMT decided to invest in Mgine and acquired 5 per cent of the shares. In January 2001, EMT took another, similar step by acquiring 26 per cent of the shares in Voicecom, with whom m-parking had been developed (EMT, 2000, 2001b). With both Regio and Voicecom, deepening and increasingly strategic partnerships thus seemed to take shape.

In December 2002, a first international success became reality, as Regio's mobile positioning software – marketed through Ericsson's global network – was chosen by Slovakia's largest mobile operator for the introduction of positioning services there. Negotiations were also going on with operators in other countries, whereby Regio's representatives traveled around the world with the Ericsson marketing team to present the Estonian technology (Jagomägi, int., 2003; see also Regio, 2003). At the same time, Regio continued to work together with EMT, but cooperation was launched also with Radiolinja, who, as mentioned above, had become the world's first operator to test Nokia's 'mCatch' positioning architecture in March 2001. On this basis, which was strengthened further through Nokia's USSD service platform starting in spring 2002, Radiolinja developed a number of new products, such as the interesting 'Leia Lähim' ('Find the closest') service for getting information about nearby cash-machines, public Internet access points, petrol stations, etc. This was launched in November 2002 (Radiolinja, 2002). At this time, Regio had got into trouble due to the difficulties of its

Finnish investor DONE in connection to the IT crisis, and Radiolinja therefore hesitated to enter into cooperative agreements with Regio. However, when DONE filed for bankruptcy in 2003, Regio bought its shares back from the Finnish investor and thus became a wholly Estonian-owned firm again. In 2004, Radiolinja re-initiated its collaboration with Regio, which was selected for the renewal of the older positioning software (information from Regio, 2004).

Oskando, for its part, remained independent, but its close collaboration with EMT continued, and in 2002 Oskando, similarly to Regio but on a secondary level, became a selected partner of Ericsson. In reality, however, Oskando's relationship with Ericsson was developing in troublesome ways, as Ericsson soon had problems delivering the new versions of the GM-22 module. Oskando was dissatisfied with delays in delivery as well as with the price and quality of the module. This motivated the firm to look for partnership with other large equipment suppliers, and in spring 2002, after a few months of negotiations with Siemens' Swedish subsidiary in Stockholm, Oskando started to purchase the German firm's corresponding product as a component in future developments of the car controller.

Oskando also developed a 'home controller' (enabling the remote control of, for example, heating systems in a summer house), a 'gate keeper' (enabling the remote control of gates, usually for cars) and related products. In fall 2003, Oskando and another Finnish partner, the GPS-specialized firm Fastrax, jointly introduced a new and peculiar mobile phone: it had only one (!) button, and was intended for elderly people, who in case of emergency could push the button, thereby making an automatic phone call to the Rescue Board, which could determine the needs of the person (*Space Daily*, 25 June 2003).

Moreover, the breakthrough in the year 2000 of mobile data communications innovation in the various fields of applications discussed above, gave rise to vast public attention, and it was hardly surprising in this connection that a number of new firms were created, inspired by the recent Estonian success stories and aimed at further exploiting the business potential in value-added services. Perhaps the most interesting case of such new firms was Mobi Solutions, which was established in October 2000 by five very young students (not researchers) at the University of Tartu. The new firm initially focused on mobile advertising, but a new turn followed in mid-2001, when the prime minister's ICT advisor, Internet visionary and entrepreneur Linnar Viik joined the company and a new service was launched: 'SMS-voting'. The service provided responses to questions via a mobile phone and showed the answers immediately on a screen. The imagined usages included opinion polls, quizzes, evaluation of speakers and arranging mobile elections. Starting in fall 2001, SMS-voting started to be used at a number of events and conferences in Estonia (*Äripäev*, 16 November 2001). The service seemed simple, but it turned out to be the beginning of a whole array of similar value-added mobile service developments, which were typically pure software solutions that could be rapidly developed without any deeper cooperation with operators or equipment suppliers. In this sense, Mobi's developments also pointed in the direction of the vast universe of innovations in data communications applications that used the world wide web without having any stronger relationship to firms in the other layers of data communications. Mobi and other firms saw their future in a corresponding expected growth of mobile Internet (see, e.g., Viik, 2002).

Mobi and Regio, both of which were located in Tartu and had started out as spin-offs from the university there, also began to cooperate. This took place in connection with a public tender for a mobile phone-based ticket-vending system to be introduced in Tallinn's public transport in the summer of 2002. Mobi and Regio formed a consortium that competed with Regio's partner in the positioning field, EMT (which also submitted a tender in a consortium formed with Hansapank). Mobi and Regio won the tender and developed the system, which was introduced in July 2002. Soon after this, in February 2003, Mobi was chosen as an 'associated partner' of Ericsson within the latter's global network 'Ericsson Mobility World'. This was regarded by Mobi as key to the global market – the means to export its innovations. A first taste of this was sampled as Mobi's SMS-voting service was used at the international Ericsson Mobility World conference in Stockholm in March 2003.

Oskando also diversified from its focus on telematic services towards SMS-based services and m-commerce. In January 2003, EMT and Eesti Loto launched the new service 'm-lotto', which was based on development work by Oskando and which was seen to be unique not only in Estonia but in the whole Baltic Sea region, at least. Customers using the service were able to purchase, via mobile phone, tickets for two popular lotto games ('Keno' and 'Viking Loto'). By calling a special short phone number, the customer received a ticket with a random computer-generated selection of numbers, confirmation of inclusion in the draw, as well as an SMS message in case of winning. The service sounded simple, but in Oskando's view it was based on a key vision of m-commerce as being most profitable with respect to 'impulse purchases' (Oskando, 2002; see also Teemant, 2003).

Tele2's Estonian subsidiary was also inspired by the new Estonian developments, although it did not take part in any of the innovative projects referred to above. The Swedish-owned operator had a highly interesting software department, which in 2000 spun off to form a free-standing firm, Trigger Software, which remained wholly-owned by Tele2. In 1999, it developed Tele2's Internet portal Everyday.com for all the countries in which

Tele2 was active worldwide, and with this experience it was not surprising that Tele2 became the first Estonian mobile operator to launch WAP services, i.e., the emerging field that was hoped to become a mobile variant of the world wide web. Trigger became involved with a large number of projects for Tele2 Sweden, relating to such diverse fields as WAP and SMS applications as well as (future) UMTS services (information from Trigger Software, 2003).

Trigger's position in the larger Tele2 international group reflected the most usual way of organizing innovation in the Swedish-based operator. As mentioned above, Tele2 had a dramatically smaller R&D department – including Trigger – as compared to EMT and Radiolinja. This reflected the price rather than technology focus of Tele2 in the competitive landscape, but it also reflected Tele2's international character. In contrast to EMT and Radiolinja, Tele2 in Estonia could directly use substantial accumulated competencies from a whole array of other Tele2 subsidiaries abroad, and this contributed strongly to the possibility of reducing the number of R&D staff. Just as Trigger's software solutions worked not only for Estonia, Tele2 Estonia could use developments originating elsewhere in Tele2's corporate network.

This internationalization of R&D in mobile telecommunications was an important trend, and it was followed by other Estonian operators. Radiolinja had long since had strong links to its Finnish mother company, but in 2003 it also became a partner in the global Vodafone's strategic network, opening up completely new perspectives in terms of mobile service developments. Similarly, after the merger between EMT's Nordic investors, Telia and Sonera, in 2002, EMT seemed to move in the direction of increasing integration with what was becoming a 'TeliaSonera sphere' in the whole Baltic Sea region. As a matter of fact, the three Baltic countries had in a sense provided an experimental field for cooperation between Telia and Sonera prior to their merger, and the importance of this Baltic experience was explicitly referred to by the operators in connection to the merger (Telia, 2002).

A sub-group within this sphere was formed by the three Baltic operators in which TeliaSonera was heavily involved: EMT in Estonia, LMT in Latvia and Omnitel in Lithuania. Cooperation between these took off after the Telia-Sonera merger through the formation of the 'Baltic Sea Alliance'. The idea behind the alliance was illustrated by a number of developments that gradually tended to blur the very borders between the three Baltic countries: customers of the three operators traveling to another Baltic country were exempted from paying for incoming calls from abroad, a development that was extended in November 2003 through the introduction of a joint tariff for

GPRS independent of which Baltic country a customer was located in at the moment (EMT, 2003b).

At that time, Estonia had also taken a first small step towards 'third generation' mobile telephony, through the first successful UMTS call in EMT's trial network (EMT, 2003a). By the time of Estonia's EU accession, however, 3G developments lay still in the shadow of further innovation in GSM. It remained to be seen whether and when the new standard would have a real impact on the sectoral system of innovation.

4. THE INTERNET: FROM E-BANKING TO E-GOVERNMENT

In parallel with far-reaching developments in the field of mobile telecommunications in Estonia and the liberalization of the fixed market in 2001, the country also saw a radical transformation and further development of data communications and in particular the Internet. However, whereas the initial focus with respect to the Internet had been largely concentrated on the provision of Internet access and a few application trends such as Internet banking, the years preceding Estonia's joining the EU saw a shift in the direction of an increasing emphasis on applications in terms of both private and public new and improved services. The center of gravity in this chapter therefore lies in the applications and middleware service layers, rather than in Internet access.

4.1 Provision of Internet Access

The Internet access market in Estonia saw a gradual wave of entry during the mid-1990s, as discussed in Chapter 6. Towards the end of the 1990s, this wave was followed by strong consolidation trends, which also involved a new role for foreign capital.

Thus in summer 1997, the wholly Estonian-owned Internet provider Uninet, which had been active since 1994 and already had far-reaching cooperation experience with Finnish partners, was taken over by Finnet International, which bought 62 per cent of the shares (*Äripäev*, 9 September 1997). At first sight, this seemed surprising in view of the fact that Finnet International was an operator offering mainly international telephone calls. However, the ongoing institutional developments in the Estonian telecommunications sector meant that telephony could no longer be seen separately from data communications. In the words of Uninet's general director:

> When Finnet International came as an investor in 1997, it was not offering any

> Internet services for end customers. So the main reason why Uninet shares were bought was because it was already known that the [fixed telephony] market would be liberalized in 2001 . . . It helped a lot, of course, to have existing customers, existing know-how about the Estonian market, when the year 2001 came. (Nisula, int., 2003)

Entering the Internet access market was thus seen as a strategy to prepare for market liberalization in *fixed voice services*, and attracting Internet clients on a broad scale was seen as a way to build a client base in (future) voice services. In other words, the expected institutional change appears to have created strong incentives to boost the number of Internet customers.

The investment by Finnet International also meant that strong links were created between Uninet and the mobile operator Radiolinja Eesti. The reason was the tight collaboration between Finnet International and Radiolinja in Finland, which both had origins in local Finnish telephone organizations such as HTF (Helsingfors telefonförening) in Helsinki.

A similar strategy to that of Uninet was followed by the Estonian communications company Levicom, which held the majority in the mobile operator Ritabell. In October 1997, Levicom acquired all the shares of Meediamaa. As noted in Chapter 6, Meediamaa had been a first mover in Internet access (under the name Teleport) as well as content. Through the acquisition, Levicom, well aware of the to-be-opened fixed voice services market, had now a client base both in mobile and Internet services, which was to form a strong point of departure for acting on the (future) voice market. Two months later, Levicom also acquired all the shares in Netexpress, through which the firm had until then been providing Internet access services together with Baltic Computer Systems. Through the full acquisition, Levicom sought to strengthen its Internet access provision by merging Netexpress with Teleport (*Äripäev*, 15 December 1997). In January 1999, Swedish Netcom acquired 90 per cent of the shares in Levicom, which – together with Netcom – was already the owner of the mobile operator Ritabell. Netcom's experience (as Tele2) on the international voice market clearly pointed further in the direction of an Estonian provider of the full range of services in fixed, mobile and data communications.

At the same time, Eesti Telefon, too, sought to integrate voice and data communications provision. As already noted in Chapter 6, Eesti Telefon's subsidiary Estpak Data merged with its mother company in October 1997, thus attempting to strengthen Eesti Telefon's position on the Internet market. Eesti Telefon also operated a leading Internet portal in Estonia, Neti.ee, and in January 1999, the company declared that its future priority – after accomplishing the build-up of the basic national telecommunications infrastructure – would be to further develop the field of data communications rather than voice services (*BNS*, 14 January 1999; see also Eesti Telefon,

1999a: 3). One year later the company declared its intention to start developing Internet *content*, probably through a subsidiary (*BNS*, 8 March 2000). These plans, however, were delayed in connection with the global Internet crisis.

MicroLink, for its part, had during the mid-1990s emerged as the most popular dial-up Internet access provider. In January 1998, MicroLink strengthened this position through the takeover of Hansapank's advanced inter-office data communications network, which had been built a few years earlier by Crebit. From Hansapank's perspective, the idea from the beginning was that the data network should be sold when the market was ready, as it was hardly economically justifiable to retain an all-Estonian network for internal use only (Sildmäe, int., 2003). In June 1998, furthermore, consolidation proceeded through MicroLink's merger with Astrodata, which similar to MicroLink had started out as a computer assembly firm, but which now also offered Internet services. Astrodata's founder and CEO, Jaak Anton (a former researcher at the Soviet-era Institute of Astrophysics and Atmospheric Physics in Tartu), for a time kept working in the new MicroLink group, which through the merger had become the largest IT group in the Baltics. Somewhat later, however, he moved to become the director of Voicecom, one of the mobile operator EMT's key partners (cf. section 3) (Anton, int., 2003).

In December 1997, Data Telecom also sold 51 per cent of its shares to EUNet of Holland, with which the Estonian firm had already been cooperating for a few years. EUNet itself was acquired by Qwest in March 1998 (*Äripäev*, 12 December 1997).

By 1999, the Internet access industry had thus been radically transformed through consolidation and considerable foreign investment. At the same time, the number of Internet users was growing exponentially. In September 1999, the increasing competition for new users led to a dramatic price war in Estonia. It began with Tele2 starting to offer – and above all aggressively advertise – 'free' dial-up Internet access. MicroLink then announced that it actually had been offering a similarly free service since April, although it had not been advertising this in any large campaigns. A few weeks later, Eesti Telefon also responded to Tele2's move by starting to offer 'free' dial-up connections (see, e.g., *Kõnetraat*, no. 7, 1999: 5f.).

Uninet, which by this time had become Finnish-owned, was more focused on business clients and therefore did not take part in the price war. Instead, the company set out to introduce Internet technologies that were not yet available in Estonia. A few years earlier, Uninet had been the first to introduce ISDN services in Estonia, and in April 1999, Uninet also became the first in Estonia to launch DSL connections (Nisula, int., 2003). The move was followed somewhat later by Eesti Telefon, and the competition that followed in this field primarily between Uninet and Eesti Telefon seems to have contributed substantially to Estonia's clear lead in DSL penetration among the former socialist countries.[9]

In terms of the number of Internet users in Estonia, the country had by the turn of the century emerged as an outstanding leader among the Central and East European countries (see Table 7.2), and it was well ahead of a number of Western countries. For example, Estonian Internet penetration was greater than that of Great Britain, Italy and France (in the latter case, the Estonian figure was twice as high).

Table 7.2 Internet users in CEE countries, 2000 (per 10,000 population)

Country	Internet users
Estonia	2721
Slovenia	1508
Croatia	669
Slovak Rep.	939
Hungary	715
Czech	973
Latvia	619
Poland	725
Romania	357
Bulgaria	527
Lithuania	609
Russia	197

Source: ITU, *Internet indicators: Hosts, Users and Number of PCs 2000.*

From the perspective of innovation, however, similar to the situation in mobile communications, the very *use* of the Internet was as such no longer necessarily related to innovative activities. Rather, the focus was shifting towards innovation in Internet *applications* and related technological and institutional tools (the middleware layer). The activities in these fields will be be discussed in depth below.

4.2 The Development of New Data Communications Applications

The evolution of electronic banking applications

Similar to the Internet access industry, the Estonian banking sector also faced dramatic consolidation trends towards the end of the 1990s – and just as in the case of the Internet access providers, foreign capital came to play an important role. Up to around 1997, the banking landscape in Estonia had been quite diversified, although some cross-ownership did occur among the different banks. Just a few years later, in contrast, 90 per cent of the banking market was controlled by only two banks.

Hansapank concluded a series of mergers and acquisitions in June 1998, when the bank merged with Hoiupank, which was an important actor in terms of electronic services (see Chapter 6). Soon thereafter, the position of the bank was strengthened through a stock emission which was fully covered by the Swedish Föreningssparbanken, which entered for the first time as an investor. In December 1999, the Swedish owner increased its ownership to over 50 per cent.

Ühispank, which had earlier played a rather minor role, similarly grew through a series of acquisitions, in particular the merger with Tallinna Pank in mid-1998, to become the second largest bank in the country. Moreover, similar to Hansapank, a foreign investor entered as a major shareholder, in this case the Swedish SEB group. By October 1999, the latter had acquired the majority in Ühispank.

In parallel to these trends, the two Estonian banks also expanded to Latvia and Lithuania. In Latvia, Hansapank acquired the technologically leading Deutsch-Lettische Bank, which was renamed Hansabank-Latvia in October 1996. In Lithuania, Hansapank later opened its own subsidiary. Ühispank, for its part, expanded through cooperation with two existing Latvian and Lithuanian banks, the majority of the shares in which were later acquired by SEB. This development also marked a dramatic overall strengthening of the role of Swedish capital in Baltic business from about 1998 onwards. The increased power of the Estonian banks was symbolized by the erection of new buildings in the center of Tallinn, in particular Ühispank's skyscraper.

From the perspective of innovation in telecommunications-related services, however, the internationalization of the Estonian banking landscape hardly influenced the introduction of new products and processes in any decisive way – at least not during the first few years. As discussed in Chapter 6, Internet banking had been introduced in Estonia even before it arrived in Sweden, and the IT systems and electronic solutions offered by Hansapank and Ühispank were at the time of the arrival of the Swedish investors at a *higher* rather than *lower* level than in Föreningssparbanken and SEB in Sweden. For this reason, Swedish ownership hardly opened up any radically new prospects for technology transfer across the Baltic Sea or any acceleration of technological advancements in Estonia. In addition, the systems that had been built separately in Sweden and Estonia typically differed too much in their very logic for it to be reasonable to borrow solutions from the one to the other side of the Baltic sea (Tamm, J., int., 2003). This was so at least up to about 2003, when, similarly to the trend in mobile communications, integration between branches on the different sides of the Baltic Sea seemed to enter a new stage of development.

If the new links to the Swedish banks were mostly a matter of ownership and hardly involved any organic integration initially across the Baltic Sea, there were other new links being built up within Estonia itself. For example, as pointed out in section 3, the mobile operators had started to cooperate with the banks in some respects, such as providing bank account information through the mobile handset. From the perspective of the banks, the mobile operators (in particular EMT) also emerged as interesting *pilot customers*. This concerned the development of services such as 'direct debit' and 'e-billing', where EMT, for example, functioned as the most important pilot client for Ühispank. Somewhat more peculiar in global comparison was the emergence of the banks as *authorizing centers* for other actors, including the mobile operators but also government agencies such as the Tax Board (see further below for the Tax Board's Internet activities).

The experience with Internet banking had made the banks the main centers of expertise in Estonia with respect to electronic identification applications, and at least up to the arrival of the new digital ID card (see section 1 and 4.3), they had more or less a monopoly on providing 'hard' identification services, i.e., involving money transfers and similarly important information exchange. Instead of creating their own identification systems, large firms such as the telecom operators and government agencies typically chose to use the banks' systems, implying that a customer had to log into his or her Internet bank, from where, after having been electronically identified by the bank, s/he could proceed to services offered by other firms and agencies. In this way a mobile customer could, for example, change the parameters in a subscription contract, or submit an income tax declaration at the Tax Board.

With respect to larger and more far-reaching developments, the banks' external management of innovation also involved strong collaborative links to *competitors*. In practice, this was identical to cooperation between Ühispank and Hansapank, which together controlled no less than around 90 per cent of the market after the consolidation in mid-1998. This collaboration was usually organized through the existing Card Center of the Banks, with origins linked to the similar Finnish organization Luottokunta (as outlined in Chapter 6). Ühispank, originally a member of the competing Inpass Estonia, remained with this group for only two years in the mid-1990s, thereafter joining the Card Center, which in practice thus became a national center (ibid.). From the perspective of innovation, the Card Center was in an interesting way well suited for developmental projects, both because as a non-profit-oriented firm it had no interest in short-term profitability, and because it was jointly owned by the banks themselves, enabling valuable competencies and resources to be pooled in the center for important infrastructural projects. In addition to functioning as a processor of card transactions, the Card Center thereby evolved to become an organization that resembled an innovation consortium.

A notable case of joint innovation in this sense was the development of an infrastructure for mobile commerce. As discussed in section 3, the banks had been excluded from the first applications of mobile commerce, in particular with respect to the success of 'm-parking'. The mobile operators emerged as competitors to the banks by starting to administer also the very commercial transactions. The operators soon started to offer so-called 'm-accounts', implying that the mobile subscriber transferred a sum of money to the mobile operator, to be used in different mobile commerce services (see, e.g., *Baltic IT&T Review*, no. 22, 2001: 31ff.). The banks responded to this challenge by developing a joint infrastructure for m-commerce, seeking to link mobile transactions to ordinary bank accounts. It was therefore based on the same logic as with standard card transactions, but instead of sending the bank card number to the Card Center for authorization, the mobile phone number was sent as an identificator. The merchant was identified through a code that was sent together with the amount by way of a phone call to the IVR of the Card Center, which returned a confirmation to the customer and the merchant that the transaction had been carried out (Tamm, J., int., 2003).

Jointly developed by Ühispank and Hansapank, this technologically simple logic emerged as the national standard in Estonia, ready to be commercially launched in spring 2003 (although it was expected to take some time before it would spread). M-commerce in this sense was not intended to compete with ordinary card transactions, but rather aimed at situations where card transactions were not available. In 2004, a further development took place through the introduction of a person-to-person mobile money transfer system. This, it was hoped, would become a large-scale take-off in the field of m-commerce, and at the time of Estonia's EU accession it seemed to be a success (information from Jaan Tamm, 2004).

Towards e-government

With respect to the public sector, Internet-related activities had during the mid-1990s been mostly focused on computerization and establishing new data communications networks. ASO, the IP-based data communications network created in 1993 and used thereafter on a minor basis by the Institute of Cybernetics and the State Border Guard Department, entered a new phase of development in 1997, when it was transferred to the Informatics Center. With the Internet boom and wide prospects for new data communications applications in public administration, the network was extended, and in fall reopened under the new label 'EEBone'. It was seen as

> a major step towards the Internet related cooperation between the governmental

agencies. The Internet connection of governmental agencies has become cheaper, of higher quality and more organized. (Vallner, 1998)

In 1998 a *common access point* of Estonian governmental agencies was also opened on the Internet. This was said to follow recommendations recently decided upon by a G7 committee and International Council for IT in Government Administration (ICA). It was to ensure

> public access to information of constitutional institutions of the state and PAAs, from one address, http://www.riik.ee/, reflecting in a well-balanced way the functions of all state institutions, making the state more transparent to the citizen, by Internet service, directly addressed to him, e.g. the archives of blanks of PAAs, common e-mail system, universal documents archive, etc. (ibid.)[10]

This project, referred to as the 'State Web Center' and organized under the Department of State Information Systems (RISO), seems to have been a reflection at the applications level of long-standing joint activities occurring at the connectivity level (EEBone). The opening up of the new common state portal motivated an investigation and inventory of the existing web pages of government agencies. RISO was 'pleasantly surprised' by the high level of Internet applications in county governments. Most other government agencies, however, turned out to have organized their Internet applications (mostly merely informative web pages) in a somewhat random way, and typically differed widely in style between different agencies.

At this stage an interesting addition to furthering the development of governmental IT systems was a project launched with support from the Baltic Sea IT Fund (cf. section 1). Its goal was to provide IT management training for Estonian public sector IT specialists, and for this purpose the government concluded an agreement with Baltic Computer Systems (BCS) and the Swedish Dataföreningen (whose chairman was Tony Hagström, the key person in Televerket's original cooperation with the Ministry of Communications – cf. Chapter 5). BCS, as mentioned in Chapter 5, had started out as a Swedish-Estonian joint venture and still had a Swedish Estonian as an owner. Its program manager became an Estonian professor who had worked at Chalmers University of Technology in Sweden (and who, before that, had studied in Tallinn together with the CEO of BCS). He had early on been one of the key persons in the 'Tiger Leap' project (cf. Chapter 6) and was an important supporter of the renewal of IT education, and thus the prospects for smooth Estonian-Swedish collaboration seemed to be excellent. In 1999 it was agreed that Dataföreningen was to provide training for Estonian trainers, who in turn were to educate public sector IT specialists. The Estonian trainers, who were trained by the Swedes, came from BCS, which itself had much experience in the design and delivery of local IT courses. During 1999 and 2000, around 200 Estonian public sector IT specialists attended the Swedish-Estonian IT management course that was developed (Sild, int., 2003).

A particularly interesting government organization significant to the further development of Internet-based public services was the Tax Board. The Tax Board had participated from an early stage in ASO, the data communications network, and towards the end of the 1990s the Board was about to complete electronic interconnections between different branch tax offices throughout the country. At the same time, as discussed in Chapter 6, Internet banking was gaining popularity in Estonia. The question then arose as to why it was possible to handle all financial transactions except taxation issues via the Internet. The Tax Board sensed a clear demand for Internet services enabling the submission of income tax declarations directly over the web, and at about the same time as the internal interconnection of branch offices was completed in 1999, the Board started to think about the actual development of the new Internet services (Lapõnin, int., 2003).

The planned services did not exist anywhere else in the world at that time, but there was considerable expertise available in Estonia with regard to the closely related field of Internet banking solutions. The main inspiration thus came from the local Estonian banks, rather than any existing applications of Tax Boards in neighboring or other foreign countries. The banks and many local software firms had gained experience in the field of Internet banking applications since its introduction in 1996. Towards the end of 1999, the Tax Board's IT department started to design a general scheme for what came to be called the 'e-Tax Board', which was thus to include not merely informative services on the web, but interactive services enabling the submission of tax declarations online. For developing security issues, the Board made good use of existing informal relationships with the banks, with which it had a great deal of IT interests in common. The services developed included the interesting feature of requiring a person to log into the e-Tax Board by entering his or her Internet bank. The Tax Board also offered the possibility of acquiring a special identity with password, etc., directly from the Tax Board itself, but in practice entry through the banks emerged as the dominant method among users (ibid.).

On the software side, the Tax Board worked closely with the local firm Abobase Systems, a firm whose early history was discussed in Chapter 5. Abobase Systems had early experience with the Nordic market and in the course of the 1990s emerged as a leader among Estonian software firms in general and database-related software in particular. The company's chairman, Kaido Saarma, was originally a graduate from the Department of Radio Engineering at TTÜ, and not surprisingly the firm thus also counted a number of telecom operators among its clients (for example, in software related to billing systems and later the development of the mobile portal

'Airport' for EMT). From Finland, Nokia had also showed an interest in the company, looking for outsourcing partners in software for mobile devices, but Abobase had rejected this offer, as the firm was specialized in open languages rather than microcode. The company now emerged with Hansapank (in which Föreningssparbanken was just about to get hold of the majority of shares) as the Tax Board's main partner in developing Internet-based tax declarations (Saarma, K., int., 2001).

With these joint efforts, a pilot phase of the new project was completed in early 2000, when as a first step private persons were offered the possibility of filing their income tax declarations directly over the Internet. Around 12,000 people used the new service at this early stage, making the project a clear success. The project was then extended so that people could view their social tax notices over the Internet, and in a second main phase, launched in fall 2000, the service was also opened to firms (Estonian Tax Board, 2000: 24f.). By 2003, around 50 per cent (!) of all tax declarations were filed electronically in Estonia (estimation by Lapõnin, int., 2003). In Sweden, a similar kind of service was launched for the first time two years later.[11]

While Internet banking became the main source of inspiration for the innovative activities of the Tax Board, the popularity of the 'e-Tax Board' in turn inspired the development of further public Internet services. The basis for this was the above-mentioned 'state web center'. Thus in 2000, the Department of State Information Systems (RISO) launched the 'e-citizen' project, with the goal of making all public services available online by 2004 (*Baltic IT&T Review*, no. 22, 2001: 23). After the creation of new formal institutions that formed a legal basis for the far-reaching development of such services (see section 1), concrete innovative activities started to be initiated – mainly by RISO, which now fell under the Ministry of Transport and Communications – with respect to specific new Internet services. These activities were also stimulated by new legal acts which explicitly stated that public services *must* also be provided by way of the Internet (Meriste and Kalvet, 2001: 69).

One of the first new services created was the 'Today I Make Decisions' program, initiated by RISO. This service addressed the ability of local residents to take part in the state's decision-making processes. People were enabled to submit ideas and comments about draft legislation, and portal users could consistently monitor the progress of law-making. Comments that were posted were freely available to everyone. The service received a European Commission prize at an e-government conference in Brussels in November 2001 (Vallner, 2002: 40f.).

The big challenge, however, was to make real the idea of offering all public services from a single Internet page (http://www.riik.ee). In 2002, RISO complained that

> many [government agencies] have been designing their own e-services, and they think that residents will be willing to search all over the Internet to find the necessary services. (ibid.: 42)

The idea was instead to create a truly service- and citizen-oriented Internet environment, from where there would be direct links to specific public *services*, rather than to specific *agencies*. However, RISO complained that even the very recent legislation in the field, such as the Public Information Act, was not in line with this ideology:

> The [public information] act is centered on [agencies]. People need access to services without having to know which department handles them, but the act assumes that residents will look at a list of [agencies] rather than a list of services. (ibid.: 39)

The interdependencies between the innovating organizations, the legal structures, and the technological solutions could thus be seen to pose obstacles to the further evolution of services.

Other obstacles were found to lie in the underdevelopment of the *middleware layer* of services, in particular with respect to the interconnection of public databases, the certification and time stamping of digital documents and the provision of other tools needed for creating actual services using digital signatures and documents over the Internet. A breakthrough came in January 2002, when the first certification authority had been acknowledged, an infrastructure for interconnecting databases had been created, and the first electronic ID cards were issued (the development of these are discussed in section 4.3 below). By using the ID card (connected to a computer with the help of a card reader) and entering the state web center portal in the Internet, it was then possible, for example, to check out data about oneself as recorded in various public databases. Many public services could, as a matter of fact, also be used by the long-established digital signature offered by the banks in connection to Internet banking, but some services could only be used with the new electronic ID card (Vallner, int., 2003). However, it was still not possible to file different public forms directly over the Internet to the relevant government agency, 'because of the ongoing absence of secure means for authenticating digital documents' (Vallner, 2002: 41f.).

More importantly, it was obvious that although the technical solutions developed by RISO and its partners were evolving in an aggressive way and often aimed at world leadership, it would require many years before the public sector would be able to fully exploit the new Internet services to the same extent as the banks were already doing. At the time of Estonia's accession to the EU in 2004, it appeared possible that, as envisaged a few years earlier, if not 'all', then at least a large share of public services would be available over the Internet by the end of that year. For the services to be *used*, however, the prospects were much more unclear.

4.3 The Middleware Layer

Towards a Public Key Infrastructure (PKI)

In section 1, the institutional evolution with respect to *digital signatures and documents* was discussed. In section 4.2 it was then pointed out that a complementary component to the law was the creation of electronic ID cards, i.e., chip-based smart cards that could store information about its holder in various respects. The idea was far from new internationally, but its diffusion in Western societies had been very slow throughout the mid-1990s. The reasons for this were to be found in a complex web of technological, legal, demand and other factors. In Estonia, the idea of electronic ID cards was enlivened in connection to working out the digital signatures legislation, whereby promising prospects for the new ID card were seen in the fact that the first Estonian passports that had been issued after re-independence in fall 1991 would expire in fall 2001 (after ten years). This seemed a good occasion to introduce a revolutionary technology.

The state agency that first raised the idea was the Citizenship and Migration Board (which was responsible for the passports), but very soon other government agencies also became involved. Not surprisingly, the agencies turned to the state's old collaboration partner, Cybernetica (the privatized part of the old Institute of Cybernetics), for elaborating the idea of new ID cards. What was at stake in a broader sense was the development of a 'Public Key Infrastructure' (PKI), a field much in focus globally at this time and which envisaged the creation of a technological architecture that would pave the way for future innovation in Internet services in virtually all fields of society. Supported by the Informatics Center, a pilot project was launched in 1998, but its failure put a halt to development. Actors recognized that the existing technology was far from sufficient, but they also complained that there was an institutional vacuum in the field of digital signatures and documents. The issue appeared unusually complex not least due to the delicate *interdependencies* between technological and institutional (particularly legal) changes. In addition, no one really knew what the specific areas of actual *use* were going to be (Reimo, int., 2003).

While the idea of the new ID card was thus evolving very slowly, innovative activities at Cybernetica did not show any signs of stagnation. After the completion of the 'Barrikaad' firewall (see Chapter 6), the privatized R&D organization had set out to develop a VPN (Virtual Private Network) solution based on in-house competencies, which similar to 'Barrikaad' found good response in the domestic market, though less so with respect to exports. In early 1998, it was described as Cybernetica's flagship product (Martens, 1998: 46). A more far-reaching project was the development of 'TrueSign', a product that was directly linked to the issue of digital signatures and documents. The purpose was to provide long-term proof-value of digitally signed documents, and it was seen as a solution to major problems with respect to the time dimension in the context of digital documents. Again, great hopes were raised with respect to export potential (Reimo, int., 2003).

In order to support the marketing of Cybernetica's information security products, the R&D company decided to spin off this part of the value chain in March 2000 through the formation of the small firm Privador (which had earlier been the brandname of the VPN solution). This followed the general trend towards spin-offs in the booming global IT industry at this time, and Privador succeeded in attracting additional financing from a local venture capital firm. Cybernetica sought to focus on its core competencies, i.e., deeper R&D, as it in practice outsourced all marketing and product support to Privador. The latter was initially headed by Tarvi Martens, the 31-year-old former head of ASO (the governmental data communications network that had originally been an administrative unit of the Institute of Cybernetics), who later left and was replaced by a young entrepreneur whose computer firm had subcontracted to the Institute of Cybernetics (ibid.).

Privador was created at about the same time as the new digital signature act was adopted by parliament, which marked a new wave of activity with respect to information security and, again, the issue of a PKI. Only a week after the adoption of the new law, Privador signed a memorandum of understanding with the two main telecommunications operators, Eesti Telefon and EMT, for establishing their version of a PKI, which was now described by Privador's CEO as the very 'cornerstone of the information society'. EMT and Eesti Telefon were motivated to participate in the project by the need for secure and reliable Internet-based services that they were to introduce in the future (Eesti Telefon, 2000b).

A few months later, the new consortium was joined by the two large Swedish-owned Estonian banks, Hansapank and Ühispank. In February 2001, cooperation was strengthened through the creation of the jointly-owned firm Sertifitseerimiskeskus (Certification Center) (Certification Center, 2001). The firm was a response to the government's decision to regulate the use of digital signatures by relying upon private certification authorities rather than providing certification and time stamping services itself (cf. section 1). Organizationally, the new firm appears to have been inspired by the long well-functioning cooperation between the banks within the framework of the Finnish-inspired Card Center of the Banks (see Chapter 6 and section 4.2). Similar to developments of the Card Center, the partners in the Certification

Center, who controlled the lion's share of the telecommunications and banking markets, aimed at using the new Center to establish a *de facto* national standard in Estonia. The Center argued that the considerable problems PKIs faced in many other countries were due to PKIs not being universal, but rather limited to certain parts of society and the economy. The Estonian solution was argued to be unique (Kase, 2003: 42).

Similar to the Card Center, the Certification Center also became involved in far-reaching innovative activities. The whole architecture around the certification and time stamping services had to be invented, which also involved substantial software development. Somewhat surprisingly, Cybernetica's products played little role, although its spin-off company Privador acted as a consultant and the informal links were strong between the Certification Center and the R&D organization. The Center used the information security products of a US firm, Baltimore Technologies, which was also a partner of Privador. Domestically, the Center cooperated with small software firms such as IT Grupp and the Swedish-owned SEB IT Partner Estonia (which had been created in 2000 to support SEB's IT projects in the whole Baltic Sea region). The main result was in the end an Internet portal called 'DigiDoc', through which digital signatures could be created, handled and verified using the electronic ID Card (connected to a computer via a special reader), without the installation of any special software from the side of the user (ibid.).

However, developments towards the wide use of digital signatures were slower than expected. The Certification Center was approved as a certification authority by the Communications Board only in late 2001, and the ID cards themselves started to be issued in January 2002. The first official digital signature that was actually produced on the basis of the DigiDoc portal occurred in October 2002. By the end of 2003, the situation with respect to the diffusion of the electronic ID card looked satisfactory, with over 350,000 cards having been issued. However, at the same time only a few thousand ID card *readers* had been sold, which in practice meant that the new cards were not used at all for the available Internet services (ibid.; Reimo, int., 2003).

Nevertheless, the Certification Center remained optimistic about the future, and it was hoped that the requirement posed in the legislation that all public agencies *must* accept digitally signed documents would act as an engine for further growth. Once the public sector started to provide services involving digital signatures and the ID card, it was hoped that the private sector would follow. The Certification Center thus continued to expand its innovative activities. A highly interesting trend was the *international interoperability* of digital signatures. In June 2003, the Center signed an agreement with its Finnish counterpart, Väestörekisterikeskus (Population

Register Center), to 'harmonize the concepts and practices between the two countries regarding digital signature and document format and exchange'. The goal was to 'make legally binding digital documents a reality within and between Finland and Estonia'. The agreement was seen as a natural continuation of earlier cooperation that had taken place between the two centers in connection with the creation of electronic ID cards, whereby the Certification Center acknowledged the assistance it received from its Finnish counterpart in working out the Estonian solution (Certification Center, 2003).

The development of 'X-Road'

An important requirement of information technology in the Tax Board was the capacity to connect to a number of very *different* databases, organized under completely different government agencies. The Tax Board needed smooth access to and interconnection with databases such as the population register, the business register, the customs board, etc. When the Informatics Council elaborated the state IT budget in June 2000, it was observed by some experts – notably from the Institute of Cybernetics – that the cross-use of databases in government agencies was a quite common phenomenon, and that many agencies were about to develop services based on such interconnections. For example, it was planned to develop a new visa register, which would need data from sources such as the population register, the business register, the vehicles register, the police register, etc. Around 20 such planned projects could be identified, each of which was to cost around one million EEK. The idea then gained ground that it could be effective to develop a systematic means of interconnecting different databases, so that the specific projects would involve only minor developmental work (Kalja, int., 2003).

Despite potential legal problems with respect to such an explicitly systematic interconnection of various databases – a development which might lead to the interlinking of sensitive information in controversial ways – the Department of State Information Systems (RISO) started to develop a corresponding project, named 'X-Road'. The idea was to create an infrastructure for interconnecting various state databases, but it also sought to combine the project with developing new services for citizens via the Internet. Interesting prospects seemed to open up not least due to the recent adoption of the law on digital signatures (in March 2000) and the related plans to issue electronic ID cards. RISO hired a project manager in January 2001 (a professor from the Technical University, who had earlier been head of the software department at the Institute of Cybernetics), and during winter and spring 2001 a series of seminars was organized, with a total of around one hundred participants from various branches of academia, business and the public sector. Experiences from other countries were also taken into account, but it turned out that similar systems did as a rule not exist elsewhere, with the exception of Iceland, where the interconnection of databases had been solved in such a way that the databases had simply been placed in one large server, thus eliminating the need for interconnection via data communications networks. It was decided that this would not be a viable solution for Estonia at all. What was demanded was instead to use the Internet to interconnect the databases. This idea had also long existed in other countries, but legal uncertainties appeared to have hindered the actual development of systems elsewhere in Europe corresponding to 'X-Road'.

RISO collaborated with the Ministry of Internal Affairs, which produced background material defining the most popular type of queries that were made to Estonian databases, and with Cybernetica (the privatized part of the Institute of Cybernetics), which performed a thorough analysis of data security issues (a field that advanced considerably in the past few years – cf. Chapter 6). Using these results, a public tender was then issued, to which five Estonian firms and one foreign firm replied. The foreign firm, IBM, sent representatives from the United States to bid on the project, which they offered to develop for an amount of EEK 26 million. Technologically, this was considered the best offer, but from an economic point of view, RISO considered it impossible to accept. Instead, the project went to Assert, an Estonian firm that had originally been created as a Swedish-Estonian joint venture (see Chapter 5). Headed by former researchers from the Institute of Cybernetics, Assert had grown to become one of the dominant IT firms in Estonia, and it had acquired a position also as a software subcontractor to Nordic firms such as Ericsson. Assert's proposed architecture for X-Road was considered slightly inferior to that of IBM, but the Estonian firm offered to develop the project for a mere EEK 5 million (Kalja, int., 2003).[12]

The first version of 'X-Road' was finalized in December 2001. When the first new electronic ID cards were launched in January 2002, the system found its first use among citizens, as discussed in the preceding section. Moreover, for the various government agencies, X-Road became a popular tool in the interconnection of databases. However, although the technology was already in place, the legal situation remained uncertain. It turned out that different government agencies interpreted the legal status of X-Road in radically different ways: for example, the population register refused to let its data be run on X-Road, viewing this as illegal, given the existing legislation. Other registers, however, interpreted the legislation differently, and most databases could thus in practice be run using X-Road. In October 2002, the legislation was changed, whereupon the population register also started to use X-Road (ibid.).

When Estonia became a member of the EU in May 2004, the country had advanced surprisingly far with respect to Internet-related solutions in public agencies. Although future prospects were unclear, the country played a major role in terms of creative innovation in this field, and in the context of 'e-government' the former Soviet republic was recognized as one of the world leaders in international evaluations (as shown in Table 7.3 below). This is all the more interesting from the perspective that the country was still far from Western levels of GDP and general economic development, which was still very much based on the exploitation of the low-cost economic environment. In this way, Estonia seemed an encouraging example for many other less developed countries in the world, by seemingly showing that it was indeed possible to arrive at a highly advanced e-government system without the enormous resources available in richer countries.

Table 7.3 Ranking of countries in 2001 with respect to achievements in 'e-government'

Country	Score	Rank
Singapore	5.43	1
Finland	5.40	2
Iceland	5.35	3
Sweden	5.10	4
Estonia	***4.95***	***5***
Canada	4.93	6
Hong Kong	4.90	7
Taiwan	4.90	7
Denmark	4.88	9
United States	4.88	9

Source: GITR (2002).

In 2002, the strong and internationally recognized e-government performance of Estonia led to the creation of a unique international 'E-Governance Academy' in Tallinn. It was funded by the United Nations (UNDP), the Open Society Institute (OSI) and the Estonian government, and was aimed at spreading the Estonian experiences through special training in Estonia to other countries in Central and Eastern Europe, the former Soviet Union and Asia. Not surprisingly, the main Estonian actor in providing the international training was Baltic Computer Systems, a firm that evolved in the direction of ICT training services provision. The existing and smoothly functioning cooperation with Sweden was thereby also continued, further supported by organizations such as SIDA (the main Swedish development aid organization). One pattern for the training was that a group of public IT heads from the above-mentioned countries spent a week in Stockholm, where the advanced and high-cost solutions of the Swedish government were studied. Thereafter, the group spent another week in Estonia, where the more realistic low-cost but, as emphasized by the Estonians, highly efficient e-government solutions of a less developed country were studied (Sild, int., 2003).

5. EQUIPMENT SUPPLY: GROWTH, CRISIS – AND CREATIVE RESPONSE?

In parallel with the dynamic evolution of mobile and data communications provision, the field of telecommunications equipment production underwent a dramatic development in the years around 2000. After the Russian crisis in August 1998, many old Estonian electronics firms declined further, as these firms continued to be heavily dependent upon ex-Soviet markets, while the lack of foreign investment seemed to make it impossible for them to penetrate Western markets. New firms such as Elcoteq and historical but transformed producers such as Tarkon, on the other hand, were hardly influenced at all by the Russian crisis, since their markets as well as their sub-suppliers were located almost totally in Western and Asian countries. They could now ride the global boom in the mobile communications industry.

Elcoteq, the Finland-based corporation, was clearly the most drastic illustration of this trend: its facilities on the outskirts of Tallinn had been expanded several times during the mid-1990s, and having started from a staff of 10-20 in 1993, employment grew to 1,900 in 1997 and to a maximum of around 3,500 by fall 2000 (Petersen, int., 2003). At that time, the company was also constructing a new plant (located half a kilometer from the first one) that was to be dedicated to the manufacture of infrastructure products for mobile telecommunications, including electronics for radio base stations. The already existing plant that produced 'box-built' mobile phones, starting with the Ericsson GA-628 model (cf. Chapter 6), had grown to become a great success, having the majority of employees and conducting the most sales (Elcoteq, 2000). Elcoteq Network's focus on telecommunications products grew to 71 per cent in 1998, and in terms of customers, Ericsson's and Nokia's share of sales was no less than 76 per cent by 1999 (*Äripäev*, 7 August 1998).

In Tartu, the historical firm Tarkon had been acquired by Swedish investors in 1996, and although its production was more diversified than that of Elcoteq, the share of telecommunications followed the same increasing trend. After a drastic reduction in personnel by more than 90 per cent from 2,500 around 1990 to around 225 in 1996, the factory started to grow again to 327 in 1998 and 600 in the fall of 2000. In 1996, around 90 per cent of production went to Nordic customers, and in the same year the company also became profitable for the first time. In addition to its early established customers Ericsson and Allgon, the company started to produce radio bases station antennas for the Swedish firm Smarteq in 2000. At that time, the share of telecommunications equipment in Tarkon's sales amounted to around 50 per cent. The company's sales increased from EEK 65 million (€4.2 million) in 1998 to EEK 150 million (€9.6 million) in 2000, and profits rose from EEK 3 million to EEK 15 million in the same period (*Äripäev*, 16 October 2000).

With respect to subcontracting, an overwhelming majority of agreements within the Estonian electronics industry were signed with firms from Nordic countries. Thus by 1995 Sweden accounted for 55 per cent and Finland for 34 per cent of all subcontracting agreements (i.e., together they represented 89 per cent) (Berg-Andersson, 1998: 24). Specifically in telecommunications, this Nordic dominance turned out to be self-sustaining: the strong presence of important suppliers to Ericsson, Nokia and other Nordic telecom producers made it attractive for other firms in the same supply networks to follow. Thus an entire array of supplier firms such as Nolato, Merab, Traction, Nefab and PMJ – all of which had Ericsson and/or Nokia as their most important customers – established production in Estonia, forming a local cluster within the global networks of the Nordic multinationals. This trend was particularly evident in 1999 and 2000.[13]

Most subcontractors in Estonia were in reality *second- or third-tier suppliers*. Elcoteq Tallinn was a striking exception through its direct links to Nokia and Ericsson, whereas most other firms, including Tarkon, counted their customers among firms that were suppliers to Ericsson and Nokia (such as Smarteq and Allgon). An interesting trend was the emergence of Elcoteq Tallinn as a customer of Tarkon's products: this was a result of Allgon's decision in 2001 to outsource some of its production to Elcoteq in Tallinn. Tarkon, which had earlier supplied Allgon directly, thus became a supplier to Elcoteq Tallinn (Carlsson, int., 2003).[14] However, this was an exception from the general pattern: both Elcoteq Tallinn and Tarkon were to a great extent mere islands in Estonian industry without any dynamic linkages to other Estonian firms. Despite Elcoteq Tallinn's size and importance for the Estonian economy, its vast demand for parts and components was almost totally directed towards Western and Asian suppliers, and had therefore hardly any effect on the evolution of other domestic electronics firms. Although a number of firms, as mentioned above, formed a cluster in Estonia, they were hardly linked to each other in terms of supply networks. Their presence was rather the result of Estonia's increasingly favorable reputation as a location for labor-intensive and low-cost subcontracting in telecommunications. This was indeed very much a result of Elcoteq Tallinn's success, but it did not stimulate the emergence of domestic inter-firm linkages among the various subcontractors. Inter-firm linkages remained, instead, dominated by linkages across the Baltic Sea.

But the evolution of equipment production involved much more than the reconfiguration of inter-firm linkages and the rapid growth of sales and employment. There were also important qualitative changes with respect to activities carried out by the firms. Subcontracting activities directed towards Nordic customers started in the early 1990s in a very simple way, focusing on goods that had long been in production in other countries and confining activities to the most labor-intensive parts of the value-chain. The products manufactured were typically intermediate goods, whereby final assembly, testing and marketing were carried out by the Nordic customers. Similarly, all components were delivered directly from the customers to the Estonian firms, and the customers also handled all issues related to logistics. Entire product lines were sometimes moved from Nordic factories (which were shut down) and rebuilt in Estonia, so that the local firms did not have to engage in process development (Petersen, int., 2003).

At Elcoteq Tallinn, these characteristics changed in interesting ways starting in the late 1990s. The 1997 Ericsson contract for the 'box-built' manufacturing of a new mobile phone from start to finish marked an important transition towards increasing involvement of Estonian actors. A further step was taken in the year 2000 through the creation of an engineering services division which belonged to the firm and which later became an independent business unit. It was later renamed the 'New Product Introduction' (NPI) center. It meant that Estonia had become gradually more involved in process innovation, to the extent that Elcoteq Tallinn's Estonian management was given responsibility for the creation of new production lines in Tallinn. The creation of the NPI center was the logical result of continuous developments over the course of the past years, through which manufacturing experience had gradually built up. The Nordic customers now had sufficient confidence in Estonian competencies that they left equipment specifications up to the Estonians, who could now choose their own equipment, procure components as needed or develop parts of the equipment themselves. In response to the increasingly favorable reputation of Elcoteq and growing pressure for streamlined production and logistics in the global telecom industry, Elcoteq Tallinn began to ship its completed products not to Ericsson and Nokia, but directly to *their* customers. In 2003, the NPI center in Tallinn employed 64 people (ibid.).

However, from early 2001 Elcoteq Network faced a dramatic crisis, which was linked to the general downturn in the global telecom business. Poor financial performance forced Ericsson to close down all its own mobile phone manufacturing plants in high-cost European countries. These plants were then acquired by one of Elcoteq's competitors, Flextronics. In doing so, Flextronics took over the entire manufacture of Ericsson's mobile phones which Ericsson had already agreed should be contracted out to Elcoteq's plants in Hungary and Tallinn. This was a major blow for Elcoteq Network, and the company's shares lost half their value in one day. Also in 2001, the markets altered course, the customer base was further weakened, both the president and the CEO of the company resigned, and 'adjustment measures' were initiated (information from Elcoteq, 2003).

Interestingly, however, the new CEO of the Finland-based corporation adopted a strategy that responded in a surprisingly creative and innovative way to the crisis. The focus on core competencies (mostly related to mobile products) became stronger, but at the same time it was decided to *broaden* activities from manufacturing, material services and logistics towards more intensive R&D. Design, engineering and after-sales services would thus be added and strengthened. The first steps towards increasing Elcoteq's design expertise were the above-mentioned establishment of NPI (New Product Introduction) centers. In April 2001, it was declared that Elcoteq Network would also set up a Technology Center in Sweden, 'in order to facilitate Elcoteq's early involvement in the development and manufacture of Ericsson's communications products'. This center would also provide services to other Elcoteq customers in Sweden. The new center was set up in the Stockholm area, and the manager was recruited from an Ericsson R&D department (Elcoteq, 2001).

The major step, however, was taken in July 2002, when Elcoteq acquired the R&D team of Benefon, a Finnish mobile phone and telematics company located in Salo, Finland. From then on, Elcoteq (in Finland) got directly involved in *product innovation*, starting at the stage that Ericsson, Nokia and other customers delivered prototypes, a point where Elcoteq could influence further developments related to the establishment of mass production. Around 90 'highly skilled design and engineering experts' were transferred from Benefon to the new design center of Elcoteq (Petersen, int., 2003; Elcoteq, 2002a).

In Estonia, the impact of the global telecom crisis and the corporate rearrangements that followed were ambiguous. Mobile phone manufacturing for Ericsson was discontinued in Tallinn in 2001, leading to massive reductions in personnel. The brand new plant that had just been completed was closed down, at least for the time being. By spring 2002, the number of employees had been almost halved, and at this stage another 270 people lost their jobs (Elcoteq, 2002b). At the same time, however, the global telecom crisis, which became associated with an increasing emphasis on cost-efficiency, tended to reinforce the overall global trend towards outsourcing manufacturing, a tendency that fitted well with Elcoteq Network's future prospects – in particular at the corporation's Estonian branch, where labor and other costs were many times lower than in Western countries.

After a year or two of dramatic decline, sales and employment in the Estonian equipment industry turned upward again. In May 2002, for example, Elcoteq Network signed an agreement with the Swedish company Wireless House to manufacture the 'GiSMO Communicator' at the plant in Tallinn. Building on earlier experience, Elcoteq started to manufacture the product box-built, so that it was ready for distribution to end-users. In addition to labor-intensive manufacturing, Elcoteq Tallinn provided Wireless House with NPI services including development of assembly and test lines (Elcoteq, 2002c). Similarly, in November 2002, Elcoteq Network and the Swedish firm LGP Telecom (which merged with Allgon in March 2003) signed an agreement concerning manufacturing of radio base station products for LGP in Elcoteq's plants in Tallinn (and in a plant in China). In the beginning of 2003, the new plant that had been shut down almost two years earlier was put into operation again.

In November 2002, Elcoteq Tallinn was also chosen by the Estonian Ministry for Economic Affairs as the winner of the first Estonian Quality Award contest. The assessment team was 'particularly impressed by Elcoteq's devotion to continuous improvement and Total Quality Management'. The award was handed out by the President of Estonia, Arnold Rüütel (Elcoteq, 2002d). But with respect to innovation – in terms of new products and processes rather than TQM – the Estonian branch of Elcoteq Network still remained somewhat in the shadow of developments in Finland. The Estonian development in the direction of increasing initiative in terms of process innovation continued, but R&D at the level of product design was still carried out only in Finland.

Tarkon also faced a crisis in connection to the global telecom crisis, although its broader product portfolio and reliance upon a greater number of clients made the company somewhat more resistant to the turmoil. The share of telecommunications in Tarkon's sales had by 2003 decreased to around a fourth, while employment remained relatively stable at 540 people. The decreasing importance of telecommunications products did not exclude Tarkon from riding the global trend towards increased outsourcing. Thus in 1999, for example, Ericsson decided to outsource some products to Tarkon that had earlier been produced in-house in Sweden (Carlsson, int., 2003). At the same time, Tarkon's Swedish owners, who were obviously more than satisfied with their Estonian investments, increased their shareholding from 60 per cent to 85 per cent (information from Tarkon, 2003). The trend towards increasing Estonian participation in innovation, however, was slower than at Elcoteq. A minor design team of around 4-5 people, some of whom had been working at their positions since Soviet times, was seen as enough to take care of process developments, such as the design of tools and fixtures (Carlsson, int., 2003). Also, Tarkon never entered 'box-built' production, but remained focused on the labor-intensive part of the production process, and the products were intermediary goods.

The overall importance of the telecommunications equipment industry for the Estonian economy, and the dramatic growth and decline in the years around 2000 is illustrated in Table 7.4 below. As can be seen, before the global equipment crisis, the share of telecommunications products in Estonia's total exports had already grown to almost 30 per cent. By 2002, however, the share had been halved, standing for 'only' 15 per cent of total exports. The turmoil in telecommunications is emphasized also by the much smoother developments in related sectors.

Table 7.4 Estonia's telecom-related exports, 1998-2002 (percentages of total exports)

	1998	1999	2000	2001	2002
Office machinery and computers	0.9	0.6	0.3	0.3	0.3
Electrical machinery and equipment	2.8	3.7	4.2	4.5	5.7
Radio, television and communications equipment	12.6	13.4	29.9	25.2	15.5
Medical, precision and optical instruments	1.8	1.6	1.9	1.6	1.8

Source: Statistical Office of Estonia, *Foreign Trade* (selected years).

There were also discussions about the potential establishment of a manufacturing-related subsidiary of Nokia in Estonia. As Nokia did not yet have any formal representation in Estonia, it attracted much interest when the Finnish telecom giant revealed that it had decided to investigate the opportunities for establishing a production facility or even an engineering center in Estonia. However, the results of Nokia's internal investigation of the suitability of Estonia for such a facility were depressing. According to the Finnish company, there was a serious undersupply of engineers and, in general, people with a technical education. Estonia therefore did not meet one of Nokia's most important prerequisites for the potential establishment of a subsidiary. Nokia also stated that the Estonian government did not seem to prioritize equipment-related hi-tech areas in any acceptable way for the company to be interested in entering the country (Rang, 2000: 10; Penjam, 2000b: 87f.).

Nokia therefore launched its engineering center in Hungary instead of in Estonia, while other foreign investors continued to base their interest in Estonia on the exploitation of the country's low-cost environment, which remained a good location for labor-intensive production.

At the same time, however, an interesting tendency in the Estonian electronics sector was the creation of small firms that were directly focused on advanced innovative activities. These firms – two examples were ARSmikro and Analoogdisain – specialized in *electronics design* (Rang, 2001: 110f.). This gained support from the Estonian Innovation Fund, which granted EEK 1.3 million (€83,000) explicitly for the revival of a more innovation-oriented microelectronics industry. The Department of Electronics of TTÜ participated in the project. In 2000, electronics R&D actually seemed to have made a comeback 'from below' (in small rather than in large firms), and more microelectronics were being designed domestically now than in the 1980s (Min, 2000b: 211f.). The new firms were established by Finnish partners but were based on microelectronics competencies from TTÜ and the old Soviet-era Tondi Elektroonika (i.e., Püügelmann) plant. The firms produced documentation to be used in electronics production in Finnish firms, which used the small Estonian firms as subcontractors, though in totally different ways from the case of the large telecommunications subcontractors (Tani, 1999).

There were also other small firms, such as Elvior, which did subcontracting R&D for large Nordic telecommunications firms, primarily Ericsson and Nokia. Elvior was founded in 1992 by a researcher who had just come back to Estonia after a few years of employment at the Finnish VTT research institute in Oulu. Through the informal contacts created there, the company was launched with the business idea of developing embedded software for telecommunications equipment on a subcontracting basis. However, the activities were later expanded in an interesting way, as the Estonian firm started to develop a tool that could simulate the telecommunications hardware environment. The purpose was initially to use this tool for internal use instead of buying expensive testing equipment, but the tool's high performance motivated Elvior to start marketing it externally. Business thus diversified (Kull, int., 2003).

At the time of Estonia's entry into the European Union in May 2004, the future of the Estonian electronics and in particular telecommunications equipment industry was still far from clear. The crisis that had started in 2000 now seemed to have been overcome, and most firms were generally optimistic about their prospects for survival and growth. The whole electronics sector was estimated to employ at least 5,000 people. Out of these, around 3,300 were employed at Elcoteq or Tarkon, which were the clearly largest Estonian electronics firms (*Äripäev*, 26 May 2004). But labor-intensive manufacturing was still clearly the most dominant segment, and while both Elcoteq and a number of small electronics firms showed interesting signs of developing in more creative directions, the sustainability of this movement towards increasingly innovative activities remained uncertain, and 'own-brand manufacture' (OBM) seemed a distant dream.

6. EDUCATION AND RESEARCH

The Evolution of Teaching at Tallinn Technical University

At the Technical University, telecommunications-related subjects had been taught since 1996 within two separate faculties, as the Faculty of Computer and Systems Engineering had been split into a Faculty of Information Technology and a Faculty of Systems Engineering. The two faculties were headed by deans who were well-known scientists both domestically and internationally. The Dean of the Faculty of Information Technology had graduated from TTÜ in the 1960s in the specialization of automation and telemechanics (cf. Chapter 4) and had continued to work there as a scientist. He had also been a guest researcher at Linköping University in Sweden as well as in one French and two German universities. In 1997, he left TTÜ to become head of the IT department at Ühispank (cf. section 4.2 above), but returned as a professor in 1999. The Dean of the Faculty of Systems Engineering had also graduated in the field of automation and telemechanics, but had thereafter been a researcher at the Institute of Cybernetics, where he had been one of the driving personalities behind EKTA, the 'Special Design Bureau' (see Chapters 4 and 5), before becoming a professor at TTÜ in 1992 (information from TTÜ with regard to the CVs of the deans).

In the rector election at TTÜ in February 2000, the two telecommunications-related deans were the main rector candidates, finally resulting in the election of the Dean of the Faculty of Information Technology – to which also the Department of Radio and Communications Engineering belonged – as the new Rector of the Technical University (TTÜ, 2000: 43). In 2001, the Dean of the Faculty of Systems Engineering left TTÜ, and the two related faculties merged again to form the 'new' Faculty of Information Technology.

In the 'old' Faculty of Information Technology, the number of graduates started to increase in 1997 after a previous decline. For example, at the Department of Radio and Communications Engineering, there were only nine graduates in 1997, but by 1999 this number was already three times as high. The number of graduates was somewhat more stable at the Department of

Computer Engineering and the Department of Informatics (see Table 7.5 below).

Table 7.5 Number of graduates from the Faculty of Information Technology of Tallinn Technical University, 1996-1999

	1996	1997	1998	1999
Department of Computer Engineering	19	15	30	28
Department of Informatics	21	24	30	32
Department of Radio and Communications Engineering	29	9	16	29
TOTAL	69	48	76	89

Source: TTÜ (2000: 32).

In the fall 1999, a total of 1,078 students were studying at the Faculty of Information Technology. Of those 234 were specializing in telecommunications, 479 in informatics and 495 in computer engineering. In terms of academic staff, however, the Department of Radio and Communications Engineering had the largest number of professors (four), although this was not balanced by a corresponding lead in docents, lecturers and mid-level senior researchers (Rein Jürgenson in TTÜ, 2000: 32f.).

With respect to the quality of studies, a report published in 2000 concluded that

> the curricula are already comparable to Western universities. However, whether teaching always takes place in accordance with the course titles, or rather in accordance with what the available teachers are accustomed to having taught for many years, and what the quality of teaching is, remains a question in itself. (Penjam, 2000b: 84f., my translation)

However, an external evaluation carried out at about the same time at the Faculty of Information Technology proved very satisfied with the quality of teaching and the implementation of modern curricula. Thus the bachelor's and master's curricula were 'unconditionally accredited', while only with respect to the PhD curricula was the situation somewhat unclear, due to the fact that the number of PhDs was still negligible (TTÜ, 2000: 33). It seemed that the *quality* of teaching was a much less serious problem than the *low numbers* of students and graduates, which were clearly not enough to satisfy the needs of telecommunications-related private business and public administration. As pointed out in section 5, this was a major argument for

Nokia to establish their new engineering center not in Estonia, but in Hungary.

Low numbers of people with high-tech skills were also a problem with respect to *research*. The majority of R&D groups in universities and firms were small, and critics pointed out that in order to participate more effectively in cooperative programs with the EU and the US, bigger research groups would have to be formed among researchers who were active in related fields. Only then would a critical mass of scientific strength have a chance to be formed (Penjam, 2000a: 13). On the other hand, there were also a number of fields in which Estonian telecommunications-related research was seen to have good future prospects. These areas included, for example, geographical information systems, component-based software technology, automatic projection systems in microelectronics and data security in computer networks (ibid.: 14).

With respect to the further evolution of teaching activities, there was also a strong tendency towards *international cooperation*, which involved discussions on the potential convergence of curricula. In an increasingly clear way, TTÜ identified its main foreign partner universities to be the Royal Institute of Technology (KTH) in Sweden and Helsinki University of Technology in Finland (TTÜ, 2002: 191). With Helsinki, the activities moved in the direction of creating a 'twin university', which was a part of the broader idea of building a 'scientific twin city' spanning across the Finnish Gulf (TTÜ, 2002: 53). In the Swedish case, close cooperation with KTH – where the new Rector, Anders Flodström, was a frequent guest in Tallinn – was complemented by an inter-university consortium called Baltech. It was created in 1997 by a number of technical universities or faculties in the three Baltic states (Tallinn, Riga, Kaunas and Vilnius) and in Sweden (KTH, Linköping and Lund). The long-term goal was to create a virtual 'Baltic Sea University of Science and Technology', whereby discussions also referred to the idea of double-diplomas: one from a Swedish and one from a Baltic university. Related to this, KTH and TTÜ in 2002 discussed the integration and convergence between curricula at the two technical universities (Information from Baltech, 2002; TTÜ, 2002: 50).

The Evolution of University-Industry Relationships

While the internationalization activities were seen to contribute to a 'Westernization' – and in particular a 'Nordification' – of the Technical University as well as increasing teaching quality, a bigger problem was seen to be student retention, i.e., students were not staying long enough at the university to acquire a degree:

> Of the students who enter bachelor's and master's studies, many are lured into the

> labor market [before graduation]. They are efficient young people, who are presented with flattering job offers that have far more definite future prospects than continued studies. (Rein Jürgenson in TTÜ, 2000: 33, my translation)

The same could be said of the situation with respect to academic staff:

> Information technology is one of those fields in which the workforce streams away from the university, because in software firms, industrial enterprises and banks there is a great demand for qualified ICT specialists. The university is simply unable to compete with these organizations in terms of wages, and therefore the majority of ordinary teaching staff and scientists also work outside of the university, serving as consultants, team leaders or even top level managers in firms. (Penjam, 2000b: 84, my translation)

However, apart from the problems inherent in this situation, an interesting side-effect was the strengthening of (person-embodied) academy-industry linkages, through scientists who pursued a double career: in business (or government) and at the university. This phenomenon had seldom anything in common with the trend in Western countries towards university researchers entering business to commercialize their own research results; in Estonia, this was only a secondary consideration. Instead, scientists working in private firms in Estonia did so primarily in order to simply secure their personal income, and their private jobs did not necessarily have anything to do with applying their own research results to a business context. A similar pattern could be seen with respect to PhD students, who could seldom get their PhD studies financed to any adequate extent. In any case, although such a double career often led to stressful situations for the persons involved, there is no doubt about the *positive* effect this circumstance had on both research and teaching, in terms of professors and others who were surprisingly capable of linking their university teaching directly to their personal daily experiences from up-to-date involvements in ICT-related private business or in public administration (e.g., Kalja, int., 2003).

The very size of the tiny republic, and the fact that nearly all higher technical education in the country remained concentrated in one single university, seemed to strengthen the informal links (which later often evolved to become formal – see below) between the universities and business. In the words of the director of the Department of Electronics,

> Estonia is so small that practically all the leaders of the companies who are working in the field of electronics and communications have come from our department . . . I have been in the university for more than 30 years . . . and since 1981 have been in Tallinn, a permanent member of this department. Nearly all the students come through us . . . And also at the other departments, it's the same situation . . . so in practice we know *all* the leading staff and technology managers in the sector in Estonia – as far as they are not imported from abroad . . . We can

call them any moment and ask them directly: 'Are you interested in doing this?' (Rang, int., 2003)

But university-industry collaboration was also strengthened in more *formal* ways. One interesting trend was the creation and strengthening of links between EMT, Ericsson Eesti and TTÜ. In January 1999, TTÜ and Ericsson Eesti signed a cooperation agreement for five years, according to which Ericsson was to follow the example of Eesti Telefon (see Chapter 6) by offering students themes for graduate theses in the field of telecommunications (TTÜ, 1999: 175). In November 1999, EMT, the mobile operator, signed an agreement with TTÜ giving birth to a scholarship program (TTÜ, 1999: 174).[15] This example was followed by Ericsson, again, in March 2000 (TTÜ, 2000: 45).

Having created opportunities for cooperation in terms of scholarships and graduate theses related to EMT's and Ericsson's activities, the two firms proceeded in April 2000 by developing a framework agreement with the Department of Radio and Communications Engineering, which aimed at the creation of a *testing and development center for mobile communications services* (TTÜ, 2000: 46). This was part of an already established worldwide Ericsson project, the 'Mobile Applications Initiative' (MAI), which now had an Estonian node. The project, amounting to EEK 9.4 million (€0.6 million), was supported almost 50 per cent by the Swedish Baltic Sea IT Fund. It enabled the modeling of GPRS and UMTS networks, and it was thought to be used both for study purposes and by private firms engaged in the development of mobile applications. However, the laboratory that was formally opened at TTÜ in May 2001 (although by then it had been in actual operation for almost a year) found its use during the following years almost exclusively for academic purposes. During the first three years, the center thus turned out to be a helpful tool for carrying out around ten master's theses. Only in rare cases, such as in connection to EMT's development of mobile positioning services, did the MAI center find industry-related use (Ots, int., 2003; Department of Radio and Communications Engineering, 2001).

On the equipment side, too, the links between the Technical University and private business developed in interesting ways in the years preceding Estonia's EU accession. The popularity of electronics engineering as a field of study increased among students after a low-level mark earlier in the 1990s, and in 1999 the number of applicants to the 'electronics and biomedical engineering program' (in itself an interesting heading) was for the first time greater than that for the informatics program (Rang, 2000).[16] This seems also to have motivated new efforts to improve the curricula and quality of teaching in electronics, which now came to include new university-industry partnerships. The university entered a cooperative agreement with Elcoteq, whose 40-year-old Estonian director was himself a graduate of the Department of Electronics. The agreement concerned 'system construction and design', which in addition to traditional course work enabled the students to spend 20 hours in an electrical circuit drawing course and to become familiar with the fundamentals of quality systems at Elcoteq. Cooperation was also developed with a number of electronics firms in connection to graduate theses (ibid.). However, the Department of Electronics criticized Elcoteq for remaining too isolated from Estonian society (Min, 2000a: 50f.).

The basis for the cooperative links that emerged were the already existing informal networks at personal levels, as pointed out above. The Technical University thereby seems to have been the main driving force in developing the relationships, by seeking cooperation with the domestic industry as a way to improve not only teaching, but above all its own research funding. The joint projects were initially bilateral, but sometime around 2000, with Elcoteq and other telecommunications subcontractors riding the worldwide mobile boom and a number of new small electronics firms appearing in the country (see the preceding section), the idea developed to create a more network-like form of cooperation and thereby seek financing from additional sources. Specifically, there were plans at the Faculty of Systems Engineering to create a 'competence center' – inspired by centers in Austria, Sweden and other countries – for R&D in electronics and communications technologies in Estonia. After the reunification of the Faculty of Systems Engineering and the Faculty of Information Technology, the idea of the competence center expanded to include information technology (after which the center was referred to under the cryptonym of ELIKO). The efforts to create ELIKO were strongly supported by a number of private firms, most of which were already existing collaborating partners with strong links to Finland, such as Elcoteq (the large telecommunications subcontractor), ARSmikro and AnaloogDisain (both of which were subcontractors in electronics design to Finnish firms), and Polar Electronics (a Finnish electronics firm, which was to act as an external partner).

TTÜ turned to the government with the suggestion of formalizing a national program for a whole array of competence centers in various research areas, and to acquire financial support. The initial response from government was very cool. However, it was seen that TTÜ's idea would soon have a good chance of being financed once Estonia entered the European Union, where programs focusing on university-industry collaboration were prioritized and organized as consortia-like organizations with their own structure and statutes. This was exactly the idea raised at TTÜ, and when it became increasingly clear in 2002 that Estonia would soon join the EU, the government became more active in supporting the idea of competence centers. Rules were developed and formal applications filed. In May 2004, shortly after Estonia's EU accession, the formal agreement to establish ELIKO was signed, and included nine partners. The first year budget amounted to approximately €1 million, of which about 60 per cent came from the state (information from ELIKO, 2004).

That sum was not to be underestimated in view of the difficult financial situation that R&D in telecommunications-related research faced at TTÜ. In 2000, the total amount of resources acquired for R&D from private firms was around €1.5 million for TTÜ as a whole. The most well-developed university-industry links in this sense were clearly in the fields of mechanics, building and energetics (together representing more than 75 per cent of total Estonian R&D funds from business), probably reflecting the strong inherited domestic academy-industry links from the Soviet era in those sectors. In the two telecommunications-related faculties, in contrast, the corresponding links remained extremely weak. The Faculty of Systems Engineering was responsible for acquiring a mere 1.5 per cent of TTÜ's total R&D funds from business sources, while the corresponding figure for the Faculty of Information Technology (including the Department of Radio and Communications Engineering) was 3.4 per cent (calculated from figures in TTÜ, 2000: 117). In this perspective, the establishment of the new competence center appeared as a revolutionary improvement.

Apart from the competence center, the realization of Tallinn Technology Park also seemed to come closer. As mentioned in section 1, an agreement foreseeing the creation of the Park was concluded in August 2001. In February 2002, a brainstorming meeting took place at TTÜ for completing the first business plan for the Park, with participating experts coming from Poland, Latvia, Finland, Germany and the United States (TTÜ, 2002: 43). At the same time, there was an intense search for investors and sponsors, as Estonia's conditions were seen to demand enormous investments. In March 2003, the Tallinn Technology Park Development Foundation was formed, with the goal to 'develop a Tallinn Technology Park that can compete successfully with Nordic analogies', as it was formulated by the Minister of Transport and Communications. The location was to be close to the House of Cybernetics (where the two successor organizations of the Institute of Cybernetics and a number of other related firms and organizations were located), not far from the main building of TTÜ, and ICT developments were to play a particularly important role in the activities of the Park (*Äripäev*, 13 March 2003). In 2004, it was still unclear how the creation and further development of the Technology Park would affect the Estonian system of innovation.

The Establishment of the IT College

At the height of the global IT boom in 1999, it had become increasingly clear in Estonia – as well as in many other European countries – that the education system could not meet the exponentially growing demand for information and communication technology experts in private business. The situation was even worse in public administration, where agencies, given the meager state budget, could not offer well-educated ICT specialists attractive positions with competitive salaries. The obvious need for ICT competencies, in particular in applications-related areas, gave rise in Estonia to a broad discussion about the prospects of establishing additional educational facilities related specifically to information and communications technologies, though at a more practical and applications-oriented level and with a shorter time to graduation than provided by traditional university curricula.

The idea was discussed at a meeting at TTÜ in May 1999, resulting in a loose suggestion to create an 'information technology college' to be hosted at the Technical University (TTÜ, 1999: 180). The idea seems to have been inspired by an agreement signed one month earlier on the largest Estonian island, Saaremaa (Ösel), which foresaw the creation of a new minor college to be administered by TTÜ and located in the island's capital, Kuressaare (TTÜ, 1999: 178). At the Kuressaare college, located in an area that traditionally lacked all higher education facilities, it became possible to study tourism and restaurant management as well as electronics. The motivation for the latter study direction was related to the establishment of a subcontracting plant on Saaremaa by the Finnish electronics firm InCap Electronics,[17] but the idea of the new college also seemed to be inspired by KTH's – i.e. TTÜ's main foreign partner's – initiative to launch a somewhat similar college in Visby on the remote Swedish island of Gotland a few years earlier.

In any case, the discussions aimed at establishing an IT college were echoed a week later at a conference organized by Sonera, the Finnish operator, in Tallinn, which brought together the ICT elite of Estonia. Support for the idea was impressive and broad, including such luminaries as Jaak Aaviksoo, the former Minister of Education and Chairman of the Informatics Council (who had now become Rector of the University of Tartu); Tanel Tammet, a TTÜ professor who had returned to Estonia after a few years at Chalmers University of Technology in Sweden; a number of current and former researchers from the Institute of Cybernetics; and the heads of a number of key telecommunications-related firms. To the latter belonged above all EMT and Eesti Telefon, as well as the Association of Computer Firms (which was headed by the CEO of Abobase Systems, the key partner of the 'e-Tax Board' – cf. section 4.2) (Tammemäe, int., 2003).

The broad consensus about the immediate need for creating the IT college seems to have contributed to launching the project at amazing speed. By September 1999, shortly after a conference in the city of Pärnu at which the idea of an IT college had been further discussed, a decree was issued by the Minister of Education on behalf of the Estonian government, deciding formally that the new college was to be created. In March 2000, the Estonian Information Technology Foundation (Eesti Infotehnoloogia Sihtasutus, EITSA) was formally created by the Ministry of Education, Tallinn Technical University, the University of Tartu, Eesti Telekom and the Computer Firms Association. EITSA in turn became the founder of the IT College. In contrast to the earlier idea, however, the college was *not* created as an affiliate of TTÜ, but emerged as a free-standing educational organization. The former minister of communications (who had once signed the concession agreement in 1992 and who had now retired from his position as CEO of Eesti Telekom) became the Chairman of EITSA, and a former researcher from the Institute of Cybernetics became Rector of the IT college. The supervisory board was headed by the CEO of Voicecom (EMT's innovative partner firm – see section 3), who was also a former researcher from the Soviet-era Academy of Sciences (information from IT college and interviews with the mentioned persons, 2003).

Teaching started in the fall of 2000, that is, slightly more than a year after the first discussions raised its potential creation. This extremely quick formation process appears to have been crucial. A year later, the situation with respect to funding opportunities for a new educational facility of this type looked completely different, as the IT boom had already turned to the dotcom death, stock markets had collapsed, and firms became much more careful in their financial support activities.

The college offered three applications-oriented study areas: IT systems administration, IT systems development and IT systems analysis. Studies lasted for three years, and in contrast to the traditional universities, there was a tuition fee. Beginning in 2003, however, the college was accredited by the government and thereby received public funding for 75 student places, a number that was increased to 90 in 2004. At the same time, the college also received powerful financial support from different sources, including large ICT-related firms such as Eesti Telefon as well as foreign support. Thus the Swedish NUTEK agency contributed EEK 5 million, and later Telia provided EEK 1 million (*Eesti Päevaleht*, 2 April 2001; Tammemäe, int., 2003).

In circumstances related to the global crises in the ICT sector, the number of students accepted decreased from 146 in the first year to 80 in 2001 and 75 in 2002, but increased to 160 in 2003, when the college offered 75 government-funded places. In spring 2003, the first students graduated from the IT college (Tammemäe, int., 2003).

7. SUMMARY AND CONCLUSION

This chapter covered a period that stretches roughly from the Russian crisis in 1998 to Estonia's accession to the European Union in 2004. In this period, a new dynamism seemed to characterize the Estonian telecommunications sector – a dynamism that was strongly influenced by both technological and institutional change. The global boom in the Internet and mobile telecommunications, which the Estonian system responded to in highly interesting ways, was about to change the structure of the sector completely. The boundaries between different technological fields tended to become increasingly diffuse, and much of the activities in the Estonian system of innovation were driven by the exploitation of the vast technological and market opportunities that resulted from this convergence of fields.

Structural evolution in the period was characterized by high turbulence with respect to both organizations and institutions as well as their relationships. A variety of new organizations and new partnerships were formed in order to exploit the increasing technological and market opportunities in the Internet and mobile subsectors (particularly with respect to innovation in services and applications), as well as in equipment production (where Estonian actors were mainly subcontractors to foreign firms). The structural tensions that could be sensed in the beginning of the period between organizations and institutions were at least partly resolved through institutional change, which enabled firms and other actors to pursue their innovation-related activities in new fields. The interaction between organizations and institutions was unusually intense, as firms sought and achieved adaptations in the new institutions (such as in connection to the new telecom law), or were even themselves established with the purpose of sustaining the new institutional paths (such as in the case of the creation of the Certification Centre, which was closely related to the new law on digital signatures). Further elements of structural change are included in the discussion below.

Let me in the following discuss the elements of the Estonian style of innovation in telecommunications as it developed during this period.

The *formulation of visions* was an important and interesting activity in this dynamic period. The idea of transforming Estonia into the 'no. 1 Internet country in the world' and to 'overtake Finland' in ICT developments signaled a new self-confidence and a new level of ambition, in which actors became increasingly aware of the advantages of a small and 'new' country, in an era where many other European economies found it difficult to abandon their inherited technologies and carry out rapid and necessary changes in the system of innovation as a whole. Similarly, actors in the mobile communications segment emerged as visionaries on the offensive, strongly believing they could actually become world leaders in certain segments of the rapidly developing sectoral system. In equipment production, Elcoteq continued its strategy to capture a greater share of the value-added in the globalized production of telecommunications equipment, with the belief that Estonian manufacturing firms had the potential to engage in more far-reaching processes of innovation and perhaps even product development. Related to that, there was the dream of an 'Estonian Nokia', i.e., of a large firm that could penetrate global markets with its own, Estonian brand and which could become a driving nucleus in the domestic system of innovation.

The articulation of demand was important in this last period both at the level of consumers and customer organizations. After the Russian crisis in 1998, consumer demand decelerated somewhat, presumably in response to the overall economic downturn, but at the same time consumers seemed to become increasingly interested in trying out new services. Demand thus became an important source of innovative activities in services. By 2001, domestic demand for telecommunications services, while small in absolute terms, corresponded to nearly 6 per cent of Estonia's total GDP, which in international comparison was a phenomenal figure. The government and the larger telecom-related firms – including the banks – also acted as demanding customers of software solutions and R&D from local firms. Also, firms such as EMT played a role as lead users of new Internet-banking services. In the manufacturing subsector, demand continued to come primarily from outside Estonia and grew rapidly until the global equipment crisis, after which it decreased in a disastrous way. At the time of Estonia's EU entry, the situation had improved somewhat, and the global trend towards further outsourcing also seemed to indicate increasing opportunities for Estonia to sell not only manufacturing services, but also knowledge-based *electronics design services* to foreign, mainly Finnish firms. A more problematic demand category was the demand for university-based R&D; this demand continued to be very weakly developed.

The creation of new knowledge involved considerably more creative activities than in the earlier periods. Estonian actors did not merely adopt worldwide developments, but started to initiate their own, globally unique developments in a few segments in mobile and data communications. These activities were performed by a variety of firms and other organizations that had been formed in the earlier periods. However, R&D in general was still very weak, with few firms seeing any financial possibilities to invest in more far-reaching knowledge creation activities. The focus was therefore more on applications and middleware than on any deeper, science-based activities. The notable exception was Cybernetica, the privatized research institute that continued to carry out impressive in-house R&D. In manufacturing, knowledge creation concentrated on processes and logistics, thus widening the range of competencies in the Estonian (subcontracting) firms, though the dream of 'own-brand manufacture' seemed still distant. A new situation was possibly emerging with respect to new knowledge creation in both business and academia, as a result of the establishment of ELIKO (the new competence center) in 2004, and the Technology Park in Tallinn.

Competence building continued to suffer from under-financing and, related to that, low numbers of university students. The corresponding lack of competent specialists in the country remained a serious problem for the whole system of innovation. Particularly during the IT boom, it appeared unattractive for students to stay at the Technical University up to graduation. On the other hand, the evidence points at a high quality of teaching despite the difficult financial situation. The new bachelor-master-doctor structure, which was an adaptation to Western standards and the Bologna process, began to crystallize more clearly after its introduction from the late 1990s. Tallinn Technical University also developed its cooperation with Nordic universities in interesting ways, one goal being to form a sister university with Helsinki University of Technology, and to develop double-diplomas in cooperation with other Baltic and Swedish universities. However, engineering education was also characterized by the specific Estonian circumstance that there existed only one technical university in the country, a fact that influenced university-industry relationships. The high demand for IT specialists led to the creation of a new 'IT College' in an amazingly rapid way – obviously enabled in a crucial way by small-country flexibility – from which the first students graduated in 2003. An interesting new trend was also that Estonia, through the 'E-Governance Academy', engaged in competence building for *other*, primarily post-socialist, countries in Europe and Asia.

With respect to the *formation of new firms and other organizations,* this period was generally turbulent. It began with a consolidation trend in the Internet access industry, with a wave of mergers and acquisitions, but at the same time the IT boom saw a wave of new firm formation in more applications-oriented Internet services. In connection to the liberalization of the fixed telephony market, there was also a wave of entry in this subsector, beginning early 2001. In mobile services, the phenomenal growth of subscribers and the innovative achievements of some Estonian firms inspired the formation of new small firms focusing on further development of value-added services. On the equipment side, Estonia saw a virtual boom in the establishment of new foreign-owned subcontracting firms in the years around 2000, but more interesting was perhaps the trend towards formation of new small design-oriented electronics firms.

Organizations also adapted in interesting ways to accommodate innovation. Electronics firms – in particular Elcoteq – concentrated on handling the phenomenal production growth during the boom years around

2000, and then on readjusting to the difficult conditions related to the global equipment crisis, whereby around half of the employees had to leave. Elcoteq also adapted to a more innovative style by trying to engage in a wider range of value-added activities and process developments. For many other firms, however, the need for a more innovative focus was not immediate, which led them to concentrate on growth based on already existing capabilities, rather than on innovation. This was to a certain extent also true for the local IT sector, where many firms were busy enough with relatively non-innovative subcontracting to larger private firms and government agencies. Among service firms, efforts continued to be spent on adapting organizational structures to accommodate further market growth. A difficult issue was the problems of finding competent specialists. Moreover, many firms carried out changes in their management boards in the turbulent years around 2000, resulting in something like a generation shift among business leaders. An interesting trend could be observed in the activities of foreign markets – in particular the other Baltic states – in which a number of Estonian firms established and developed important subsidiaries (particularly the banks). There were also attempts to facilitate the export of Internet and mobile services, but the results of these attempts remained meager.

With respect to *networking*, the period saw a vivid trend towards innovation-related partnerships and networks. In the Internet era, firms and other organizations discovered that they had a lot to gain from collaboration with firms whose activities until recently had seemed too far away from their own fields, but which through the ongoing convergence between different subsectors often became absolutely key. At the same time, competitive relationships became increasingly tough, and in general this competition turned out to be a healthy force in the Estonian system of innovation. Firms frequently responded to competition not only by adjusting price levels, but also by engaging in further knowledge-creation and, in the case of firms such as EMT, emphasizing exactly the innovative spirit of the company as a way to respond to competition. In addition, firms adjusted to competition by engaging in the redefinition of formal institutions and by building alliances, as well as by growing through acquisitions. Mobile operators, banks and government agencies built links to smaller domestic and sometimes also foreign (in particularly Nordic) firms for innovation. The Nordic firms were very happy to use the small neighbor as a 'test market' for new telecommunications products. While large foreign equipment producers, together with domestic operators and smaller software and Internet-related firms, thus played an important role in the Estonian networks, service providers remained almost totally isolated from domestic manufacturing firms. With regard to university-industry links, these were still weak but showed signs of improvement. It was networks of personal relationships that were the more well-developed here rather than any formal arrangements, but the formalization of relationships seemed to be clearly on the way through the establishment of a joint university-industry competence center at the end of the period.

The provision of finance was to some extent improving initially in the period through the IT boom. While the very limited government expenditures increased only slightly, a number of larger Estonian actors – notably EMT, Eesti Telefon, Hansapank and Ühispank – contributed heavily to projects of importance for the system of innovation, such as Look@World and not least the new IT College. Financing of telecommunications-related education thus took place with the close participation of dominant Estonian companies. In general, however, Estonian firms continued to spend extremely little on R&D, and this influenced the whole profile of the sectoral system, in terms of a concentration on cost-efficient solutions and applications and middleware innovations. With respect to concrete projects, there was a substantial contribution from abroad, for example, through the Swedish Baltic Sea IT Fund. There were nearly no funds at all from business sources for university research in telecom-related fields (in contrast to sectors such as mechanics and energetics), but a possible improvement could be sensed here through the establishment of the new competence center ELIKO.

With respect to *consultancy, advice and lobbying activities*, there was considerable lobbying in connection with the working out of new telecommunications legislation and other documents. These activities continued to play an interesting role after the liberalization of the fixed market in 2001, contributing to changes in legislation. On the consultancy side, many university professors and researchers, as well as business managers, worked part-time as advisors in various government agencies – both with respect to concrete government-led innovative projects and in connection with the working out of new legislation and policies, thus following up the tradition established in the early 1990s through the participation of the Institute of Cybernetics in these matters. In business itself, the earlier consulting from the Nordic countries in firms such as Eesti Telefon and EMT had disappeared almost totally, although there was also a wave of new interest from the side of the Nordic investors to increase their influence on their Estonian partners.

Creating, changing and abolishing formal institutions was an important activity in the period of investigation here. Considerable efforts were spent on resolving difficult institutional bottlenecks, which were seen to hinder further innovation, particularly in Internet-related services. Examples included new laws regulating digital signatures and the cross-use of databases. These activities succeeded rather smoothly and were pushed further by the unusually well-developed consumer demand for such services.

Thus important new institutions were created and implemented, including the liberalization of the fixed services market and a number of data communications laws such as the digital signature act. Much of this work on new institutions involved a variety of actors both as expert advisors (e.g., from Cybernetica and Tallinn Technical University) and lobbyists. The strong interests of the three competing mobile operators, which all planned to enter the fixed telephony market, made it unlikely that the temporary monopoly in that sector would be prolonged, and thus Estonia became the first country in Central and Eastern Europe to fully liberalize the sector in 2001. Many of the new institutions were also later amended after hot debates and conflicts among those actors who were influenced by the new laws, or in response to rapid technological change that tended to make them obsolete. Where institutions were seen to be insufficient, large firms – notably the operators and the banks – formed special consortia, with an important aim being to create *de facto* national standards as a basis for further Internet-related service innovation.

The *formulation of public policy* also involved the participation of many actors from outside the government itself, and was thus closely related to consultancy activities and lobbying, as discussed above. Activities related to telecommunications- and in particular Internet-related policies experienced a clear upturn after the installation of a new government with new advisors in 1999. A main emphasis turned out to be the field of e-government, where policymaking activities were directly related to Estonia's phenomenal development to become a world leader. In the background also lay informal industry-government relations, whereby especially the large banks became sources of inspiration for policymaking and public Internet services. Other important drivers behind many policy activities seem to have been the Technical University and developments in the Baltic Sea region. The whole period was characterized by an increasing awareness of the importance of innovation, a trend that went far beyond telecommunications-related policymaking.

NOTES

1. The telecommunications council of the European Union decided in 1994 that telecom markets were to be liberalized in all member states by January 1998. However, by the end of 1996, there were only three countries that had actually implemented the liberalized market: Great Britain, Sweden and Finland.
2. For an overview of the fund, see, e.g., *Baltic IT Review*, no. 4, 1999.
3. For more on these developments, see Knowledge-Based Estonia (2002). Results from CIS III are found in Kurik et al. (2002). Cf. also Min (2000b: 212f.) and *Äripäev*, 2 August 2001.
4. NetCom already owned 48 per cent before this, and it now acquired 90 per cent of the majority owner Levicom, thus raising its total share in Ritabell to 95 per cent.

5. This fact was recognized by interviewees from all three operators.
6. In 2003, EMT's R&D department employed around 40 people, around half of whom came from the technical side and half from product management. These two sides earlier represented two different departments, which merged in September 2003.
7. EMT's new CEO was the brother of the earlier minister of education, a physics professor who had also been head of the Informatics Council and who in 1998 had become Rector of the University of Tartu. The new CEO had earlier been the head of Norma, a car security company that was closely related to the Swedish Autoliv.
8. The Baltic Sea IT Fund stood for half of the development costs, which totalled EEK 12 million (€0.9 million).
9. See, e.g., *3rd Report on Monitoring of EU Candidate Countries (Telecommunications Services Sector)*, June 2003: 14. In particular, Estonia had an xDSL penetration rate of 2.24 per cent, many times ahead of Latvia and Lithuania, where the corresponding penetration rates were 0.40 per cent and 0.28 per cent, respectively.
10. The Estonian word 'riik' means 'state'.
11. That is, it was made possible in Sweden for the first time in 2002. See *Ny Teknik*, 30 April 2003.
12. The software architecture was based on the Unified Modeling Language (UML), which had originally been developed by Ivar Jacobsson from Ericsson, and which was now about to emerge as a world standard.
13. See, e.g., BNS's interview with Nolato's Lars G. Persson (*BNS*, 18 May 1999), who explicitly referred to the already strong presence of Elcoteq as an important reason for opening a Nolato factory in Tallinn.
14. In March 2003, Allgon merged with LGP Telecom, a somewhat similar firm which since November 2002 also used Elcoteq as a subcontractor.
15. For some reason, the Swedish and Finnish ambassadors to Estonia took part in the ceremonial meeting where the agreement was signed.
16. A potential explanatory factor for this is obviously also the IT boom, which stood at its maximum around 1998-2000, probably drawing potential students away from the Informatics program towards private IT businesses.
17. In 2004, there were three electronics subcontractors active on Saaremaa: Incap Electronics Estonia, Enerpoint Saare and Volex Estonia. They employed a total of 320 employees, while planning for further expansion. See *Äripäev*, 26 May 2004.

8. The Making of the Estonian Telecommunications System of Innovation: Analysis and Conclusion

This chapter builds on the empirical material presented in Chapters 4 to 7 in order to summarize the most important characteristics of Estonia's transformation and reorientation in telecommunications, and sort out the main findings with the help of the analytical framework developed in Chapter 2. The chapter is organized as follows. Sections 1 and 2 discuss the performance of the Estonian telecommunications system of innovation, and its ability to generate business and appropriate returns from innovation. Section 3 analyzes the structural evolution of the system over the four historical periods of investigation, with particular emphasis on the evolution of organizations and networks. Section 4 turns to the transformation of the Estonian style of innovation in telecommunications, in terms of the gradual post-Soviet crystallization of new ways of carrying out various activities in the system. Sections 5 and 6 then turn to the more specific issues of the role of history and the influence of the foreign environment, respectively.

1. SYSTEM PERFORMANCE: CREATIVE VS. IMITATIVE INNOVATION

Innovative Performance in Manufacturing

On the manufacturing side, the Estonian electronics industry was of considerable importance in the Soviet period, with a total work-force of about 25,000-30,000 people. The focus of the industry had almost nothing to do with telecommunications in a traditional sense, i.e., with fixed telecommunications equipment, but a major specialization was radio electronics, a field that would later (with the rise of mobile telephony) converge with the telecommunications sector. Production enterprises in the Soviet-era Estonian electronics sector were strongly dependent upon R&D carried out elsewhere in the Soviet Union, although large enterprises (such as

Punane RET, the Pöögelmann works, etc.) usually involved the establishment of applications-oriented R&D organizations in connection to factories located in Estonia. In particular, the electronics industry also led to a strengthening of electronics-related *education*, including the Department of Radio Engineering at the Polytechnical Institute (aimed at satisfying the need for engineers primarily at Punane RET). In the years around 1990, however, the Estonian electronics industry collapsed almost completely, resulting in a 90 per cent reduction in employment, as the previously secure Eastern markets were lost.

In the mid-1990s, this – in Western eyes almost unimaginable – collapse was followed by a wave of Western investment. This led to a revival of the industry, though not at all to its previous level. Western investors were not interested in Estonia's innovative capabilities, but saw the country as a conveniently located, low-cost and liberal economic environment with a reasonably skilled labor force, and above all wages that were only about 10-20 per cent of those in the West. Hence, the Estonian electronics firms – in which telecommunications equipment became the new principal specialization – emerged as subcontractors to large Western producers, focusing on the most simple and labor-intensive parts of the value chain and typically dealing much more with mature rather than new products, thus eliminating their involvement in creative innovation to a very great extent. The economic success of these subcontracting arrangements – which became the powerful basis for Estonia's post-socialist export performance – encouraged a further strengthening of the focus on the exploitation of low-wage labor and imitative adaptations rather than on any emerging creative generation of innovation. Estonian equipment production became in this sense locked into a largely non-creative style of innovation.

This trend went hand-in-hand with increasingly fierce global competition in the 1990s and the related global trend towards outsourcing electronics manufacturing to low-cost countries. The Estonian showpiece and success story, the Finnish-based Elcoteq, gave Estonia a good reputation in this field and served as a good advertisement for attracting a whole array of additional Nordic telecommunications-related manufacturers, which thus invested in a series of new plants in Estonia, especially in the years around 1999. At the time of Estonia's EU accession, practically all these firms – similar to almost the entire manufacturing sector of the country – were still extremely focused on labor-intensive production, with the Estonian involvement in creative innovation being more or less negligible.

There were two exceptions from this pattern: first, Elcoteq, clearly the largest of the telecommunications-related manufacturers, appeared to be moving in the direction of creative innovation, particularly process innovation and 'design for manufacturability'. The company became a typical example of firm-level evolution towards higher R&D intensity as observed in the Tiger economies of East Asia in the 1980s, which climbed the ladder towards 'own-design manufacture' (ODM) and 'own-brand manufacture' (OBM) – although Elcoteq at the time of Estonia's accession to the European Union still appeared to have a long way to go to the latter stage. Similarly, marketing and distribution were still functions to be handled by Elcoteq's customers.

Elcoteq Tallinn had in 2004 a relatively large – by Estonian standards – engineering department involved in process development, component procurement, advanced testing, complex logistics processes, etc. The tougher, more globally competitive climate in telecommunications manufacturing together with increasing technological complexity tended to drive the large Western telecom producers towards an external organization of innovation, which made their relationships with subcontractors more important. This gave Estonian suppliers an opportunity to get more creatively involved in the innovative process, and Elcoteq Network's strengthening of its R&D base *in response to* the global telecom crisis has to be seen in this light. However, it remains uncertain what role the *Estonian* part of Elcoteq Network will play with respect to the corporation's R&D activities.

For good or bad, the evolution of the Estonian equipment sector was totally dependent upon deeply integrated connections with global networks, especially with Ericsson and Nokia and their suppliers (the Estonian firms typically acting as second- and third-tier suppliers). This way of integrating with the Western world – in contrast to the idea raised around 1990 to establish a large domestic, independent telecom producer – has proved a viable strategy in a medium-term economic perspective. However, it appears that the Estonian government could have done much more to support a movement towards increasing Estonian involvement not only in *production*, but also in (creative) *innovation*. There appears never to have been any strong political will to actively strengthen the long-term evolution of the electronics sector in this sense.

Most probably, this attitude reflects the memory of scale-intensive electronics production as a typical Soviet-era contrivance, belonging to an 'old economy' in a much more emotionally charged sense than this term has in the West. Thus the field of electronics education, in particular, seems to have been seriously neglected and remained totally in the shadow of the strong political emphasis on telecommunications-related *services*. Some foreign firms did show an interest in establishing R&D and knowledge-intensive engineering subsidiaries in Estonia, but since the number of graduates in relevant fields from TTÜ could almost be counted on one's fingers, this remained unrealistic. From this perspective, public policy's neglect of electronics education was an obstacle to realizing the potential of the country's electronics future.

At the same time, however, hopes were raised in the Estonian electronics R&D community for the prospects of building a national profile in terms of the provision of *electronics design* activities, which would involve considerable creative innovation. In contrast to the extremely high barriers of entry faced by independent telecom equipment producers, the area of electronics design was argued to fit very well with Estonian conditions. However, this interesting strategy does not appear to have received any response in terms of active political support, and it thus took almost a decade before the first promising small electronics design firms began to emerge. Nevertheless, at the time of EU accession they did seem to provide an interesting and favorable – though unnecessarily delayed – direction for the Estonian electronics sector.

Innovative Performance in Services

Telecommunications services in the Soviet Union were subject to a peculiar neglect and serious underdevelopment in comparison to Western countries. Not only was innovation in this field severely hindered both by the well-known general shortcomings of the Soviet economic system, but also by institutional obstacles related to the interests of the totalitarian state and Marxist reasoning (as discussed in Chapter 4). Within these limits, Estonia performed surprisingly well during Soviet times, with the incumbent operator, in contrast to other Soviet operators, being involved not only in telephony, but also in data communications development. Despite the fact that Moscow decided to remove all telecommunications production from the republic after World War II, Estonian telecommunications engineers – thanks to their easy access to other Soviet high-level universities – were familiar with recent technological developments, and there were minor R&D activities in the Ministry of Communications and in academia. But the problems inherent in the Soviet-style system of innovation meant that the potential benefits of these developments could never be realized. Estonian engineers could have advanced knowledge of, for example, digital technologies, but in the actual telecommunications networks digital equipment was never introduced more than at a negligible level prior to 1992.

In the years around 1990, in comparison to other Central and Eastern European countries Estonia managed to attract foreign investment at a very early stage, which was used to modernize its telecom infrastructure as well as introduce more Western-oriented business and management practices. This was closely interrelated with important institutional changes. However, it was not in telecommunications in a traditional sense (i.e., fixed voice telephony) that innovation tended to concentrate during the following decade. Rather, it was the dramatic global shift in telecommunications innovation in the West, towards an increased emphasis on the new technological paradigms of digital mobile communications and the Internet, that became the main trend in Estonia. The mobile and Internet services subsectors evolved rapidly and aggressively in the country, which towards the end of the millennium had already emerged not only as an advanced *user* of new technologies, but also as a *generator* of creative service innovation.

In the mobile field, creative innovation spanned mobile commerce solutions, mobile positioning services, mobile telematics and a range of value-added, SMS- and WAP-based solutions. In the Internet subsector, Estonian banks were first-movers and aggressive innovators by the mid-1990s, and they were followed a few years later by world-leading developments in public Internet services in the field of e-government (where only a few – mostly Nordic – countries ranked higher than Estonia). Important developments were made especially in the applications and middleware layers, which also involved the difficult resolution of institutional bottlenecks and the creation of advanced security solutions. A number of smaller Estonian firms were accepted into Ericsson's exclusive global partner network, and Estonian mobile as well as Internet innovations won a series of prizes and awards at international fairs and conferences. An international 'E-Governance Academy' was established in Tallinn with the goal of spreading the unique Estonian achievements to the rest of Central and Eastern Europe, among former Soviet states and elsewhere – from Budapest to Ulan Bator – where developments were considerably slower and less creative. World-leading international telecommunications producers also chose Estonia as a test market for their new service platforms, and the use of new innovative services sometimes reached much higher levels of innovation not only compared to *Eastern*, but also to large *Western* countries such as Italy, France and Great Britain (not to mention some South European laggards).

It is important to point out that this does not mean the Estonian telecommunications sectoral system *as a whole* was more advanced than those of the above-mentioned Western countries. The material presented in this book does not allow us to draw any conclusions in this respect (and this was not the purpose). There is certainly a large number of segments in the field of telecommunications services in which Estonia's performance is not at all impressive. What *can* be said, however, is that the above results are striking in view of the fact there actually do exist segments in the telecommunications sector in which a small post-socialist country has managed to become a world leader in innovation. In this sense, it cannot be denied that the Estonian performance is eye-catching and thought-provoking.

2. GENERATING BUSINESS AND APPROPRIATING RETURNS

Apart from the performance of systems, one may ask what Estonia's creative innovativeness in telecommunications services has meant for the country's *economic performance*. This is an issue that goes beyond the performance of systems of innovation, as it addresses the *consequences* rather than the *determinants* of innovation. One may here refer to Figure 7.1, where it was found that telecommunications services now account for almost 6 per cent of Estonia's total GDP. This is in international comparison an extremely high figure, which most countries in the world could never dream of. One may here also point to the mobile operator EMT's uniquely high profitability (the profit margin has for many years been around 40 per cent, as shown in Chapter 7, despite fierce competition), a circumstance that has few – if any – parallels elsewhere in the telecom world. Similarly, it was shown that telecommunications equipment exports corresponded to nearly a third of Estonia's total exports in the year 2000, although this share declined after Ericsson's crisis. In this sense, there is no doubt about the fact that telecommunications is an important and central sector in the Estonian economy.

At the same time, it is difficult to find any direct links between creative innovation and sectoral growth. A painful problem for the Estonian telecommunications sector is undoubtedly the country's failure to transform its domestic success in creative service innovation into profitable service-related *exports*. Hopes have been raised high in this connection in almost every sub-field investigated in this book, but hardly ever have these expectations come true:

- *Regio*, the Estonian GIS firm and a key partner of EMT and Ericsson, got a unique opportunity to market its mobile positioning software through Ericsson's global network. It took two years, however, before the firm could use this situation to sell its solution to another country (Slovakia), and until 2004 this remained Regio's *only* export success. The company traveled around the world, marketing its positioning software with Ericsson, though largely in vain. Instead of choosing Regio, foreign operators often chose in-house competencies and solutions developed by domestic software firms. In addition, Regio enjoyed the bad luck of its breakthrough being followed by the global sector's general – and Ericsson's particular – downturn. The future thus remained uncertain.
- *EMT* sought to export its 'm-parking' solution, and in 2002 a variant of it was implemented in Norway. However, this export venture was

fraught with problems, technological and institutional. Most Western countries showed little interest, as most places already had car-parking infrastructures that functioned well, implying the value-added offered by EMT's innovations was much lower outside Estonia, where the innovation was stimulated by the lack of earlier, alternative systems. Moreover, EMT licensed the solution to Radiolinja domestically, but Tele2 as well as the majority of other Baltic operators – while recognizing the excellent idea underlying EMT's new product – preferred to develop their *own* versions of 'm-parking'.

- *Cybernetica* spun off the marketing and support functions of its information security department in 2000, with the explicit goal of using the new company, Privador, to penetrate foreign markets. High expectations were held for the export potential of advanced information security products such as 'Barrikaad', 'Privador VPN' and 'TrueSign', but while these products typically conquered the domestic market very successfully, foreign sales remained almost negligible.
- Many other new products created in the middleware layer of data communications, such as 'X-Road' (aimed at interconnecting public data bases over the Internet) and the 'e-Tax Board' (filing tax declarations over the web) were often customized solutions that were unlikely to have any economic impact beyond the original project itself. This was also true of the field of e-government in general, where services were often too closely related to the technological, organizational and institutional specificities of different countries.

Ironically, then, Estonia's export performance looked clearly much better with respect to the largely non-innovative equipment sector, as compared to the highly creative and innovative telecommunications services sector. In other words, Estonia has had huge problems appropriating returns from its striking advancements in service innovation. To understand why this is so, it is necessary to examine the nature of the sector-specific appropriability regime of the telecommunication sectoral system. Three key issues can be discerned:

1) First, in the case of customized innovations, these are, by definition, not directly transferable to other contexts. Success is therefore often difficult to reproduce elsewhere. This is particularly troublesome in a sector where innovations evolve in close interdependence with institutions, such as telecom service innovations in the Internet era, where new institutions such as digital signature acts, legislation regulating the use of databases, etc., are of crucial importance. As discussed in Chapter 7, the Estonian creative advancements in Internet services in the years preceding the

country's EU accession involved difficult processes aimed at resolving institutional problems. E-government innovations started to take off only after the approval of a digital signature act by the Estonian parliament, and other laws had to be amended as a result of the demand for further development of services. The mutual adaptation of innovations with the institutional landscape in Estonia meant that Estonian innovative solutions typically did not fit well with the legal situation in *other* countries. For example, Sweden's lagging behind in the introduction of electronic income tax declarations was hardly an issue of technological backwardness, but rather a result of the misfit of technologies, institutions and other non-technical factors in that country. As a result of unresolved legal issues, other countries also lagged behind Estonia in fields such as the interconnection of databases. On the mobile side, services related to mobile positioning was in Estonia never a problem from an institutional point of view, thus clearing the field for radically new developments. In other countries, however, services enabling the positioning of people with the help of a mobile phone did not necessarily appear as unproblematic as in Estonia from a legal perspective. In other fields, such as Internet banking and m-parking, other countries usually had already installed alternatives, and these older systems (bank checks, giro payments; pay-and-display machines, etc.) made it much more difficult to introduce the new developments as swiftly as in Estonia. This then reduced the prospects for appropriating returns from Estonian innovations.

2) Secondly, the very nature of Estonian product innovations often posed an obstacle to innovating firms in Estonia. The country seemed to specialize in creative innovations that were easy to imitate or that lay in service provision, rather than in any unique firm-specific accumulated technological competencies. Mobile parking turned out to be easy to imitate by other actors once the service had been created by EMT. Even with advanced solutions such as mobile positioning, other countries did not consider it necessary to rely on Regio's solutions, but typically preferred to develop their own variants, though based on Regio's idea. The only clear exception from this pattern was Cybernetica's deep competencies in information security, but here export performance was, as in the case of the Barrikaad firewall, impeded by severe competition with much more well-known brands on the global market.

3) Thirdly, Estonia faced a clear disadvantage from its standing as a 'former Soviet republic'. Other things being equal, Estonian firms were clearly disfavored on international markets in this respect. Even in the culturally close Nordic countries, suspicion and conservatism was substantial with

respect to the potential purchase of new solutions offered by Estonian firms, for the simple reason that Estonia was regarded as an 'eastern' country (rather than as a new member of the Nordic world, as many Estonians wished their country to be seen), with all the supposed attributes typically associated with that part of the world. For example, Estonian software developers had to fight hard in order to be able to get their Finnish or Swedish customers to seriously listen to technological expert advice from their unknown firm from the 'wrong' side of the Baltic Sea. However, Swedish and Finnish firms such as Ericsson, Cellnetwork, 4 Technology Solutions, Tele2, SEB and others that *did* partner with innovative Estonian firms, were as a rule highly satisfied with the arrangement and found their prejudices mostly ungrounded.

3. ORGANIZING INNOVATION: THE EVOLUTION OF NETWORKS OVER TECHNOLOGICAL AND INSTITUTIONAL DISCONTINUITIES

Despite the problems of appropriating returns from the development of new products, the actual and repeated emergence of creative innovation in the Estonian telecommunications system of innovation may be regarded as an interesting and surprising result in itself, as it contradicts the traditional view of catch-up countries being focused on more imitative types of innovation. A natural and crucial question is, then, the issue of *how* it has been possible for Estonia to generate creative innovation. Let us here first use the detailed empirical material presented in Chapters 4 to 7 to examine the issue of how creative innovation has been *organized* in Estonia, and how this relates to the relations between different kinds of organizations and institutions. This then corresponds to the system's structural evolution.

When the Estonian Republic achieved its re-independence in 1991, the telecommunications system of innovation, as described in Chapters 4 and 5, was characterized by *far-reaching organizational fragmentation* domestically. This was primarily a consequence of the main inter-organizational linkages being connections between Estonia and other Soviet republics (in particular Russia) rather than domestic connections within Estonia itself. Moreover, the institutional conditions in the Soviet Union tended to discourage most interactive linkages and interactive learning between organizations, thus spurring further fragmentation. Telecommunications operators, equipment producers, educational organizations, research institutes and other types of actors carrying out different activities in the system had little to do with each other. Although the small size of Estonia often meant that people from these various fields knew each other and tried to bridge the seemingly artificial fragmentation, it was seldom possible to transform such informal networks into actual collaboration for innovation. The situation was further complicated due to the virtual absence of small firms (including specialized suppliers).

In the years around 1990, in response to the institutional discontinuity associated with the collapse of socialism, a radical organizational reconfiguration followed. Initially, it appeared the fragmentation would continue, since the most important linkages still tended to transcend Estonia's national borders, though now they went West, to Sweden and Finland, instead of in Eastern directions. Telecommunications operators, equipment producers, educational organizations, research institutes and other types of actors carrying out different activities in the system built new and often very dynamic linkages to Nordic organizations, but within Estonia itself the same actors continued to have very little to do with each other. Thus Ericsson, for example, lived a kind of double-life in Estonia, as an important partner of operators on the one hand, and as a key customer of Estonian electronics subcontractors on the other, but the two different spheres, the operators and subcontractors themselves, had nothing to do with each other.

From around 1990, however, a strong trend towards new firm formation set in, including the emergence of such diverse actors as computer assemblers, small software firms and IT consultants, retailers of telecom equipment, private banks and new electronics subcontractors. Totally new government agencies were also formed, and the impoverished conditions of the old research institutes led to a number of interesting spin-offs. Even more central for the upcoming developments was the emergence of mobile operators and data communications providers. These were not the result of any specific *Estonian* developments, but rather belonged to the emergence of new technological trajectories in the global telecommunications sector at large. Mobile telecommunications and the Internet represented dramatic technological discontinuities which had the power to totally transform the telecommunications sectoral system of innovation, and Estonia was quickly drawn into these gales of radical change – which happened to coincide more or less precisely with the country's transition period from socialism to capitalism.

Building on these new technological trajectories, the organizational landscape in Estonian telecommunications underwent dramatic and decisive evolution in the late 1990s and the early 21st century. The opening up of a whole universe of potential innovation in digital mobile telecommunications encouraged the diversification of existing firms as well as the formation of collaborative relationships. As large operators gradually moved towards the field of data communications and new advanced services, demand for corresponding expertise increased dramatically, forcing Eesti Telekom and the other operators to look for new partners. Perhaps to their surprise, they discovered that an impressive domestic potential had developed in the field since the early 1990s.

Thus, for example, mobile operators and small domestic software and telematics firms, all of which had up to then developed in isolation from each other, discovered the potential for joining efforts in innovative projects, alliances and consortia. Mobile operators also collaborated in creative innovation with government agencies such as the Rescue Board and the City of Tallinn. Computer assemblers evolved to become both data communications providers and software subcontractors to government agencies in the latter's process of Internetization and development of e-government solutions. Government agencies built formal and informal linkages to banks in order to learn from the latter's experience in Internet applications, and creative innovation projects in the middleware layer brought together the Informatics Center with advanced domestic software firms and R&D organizations (such as the information security department of the Institute of Cybernetics/Cybernetica). Banks and telecom operators, which previously had nothing to do with each other, responded to the radical technological changes by forming innovation consortia, which except for joint innovative efforts were also intended to facilitate the establishment of new institutions in the form of national standards.

The role of foreign-owned firms in these networks varied. Equipment producers such as the Finnish-owned Elcoteq and the Swedish-owned Tarkon remained highly isolated from the rest of the Estonian system in terms of user-supplier relationships and innovation consortia. Their principal external links were almost totally dominated by connections to foreign systems of innovation. Others, such as EMT, Eesti Telefon and the Swedish-owned banks, were from the late 1990s much less organically integrated with their foreign (minority or majority) owners, i.e., Estonian subsidiaries enjoyed far-reaching organizational autonomy from their investors with respect to most activities. Yet others, notably Ericsson Eesti, functioned as key mediators between domestic and foreign actors. Although Ericsson Eesti was formally a *sales organization*, it was in reality deeply involved in key innovative projects such as the launch of mobile positioning and the introduction of mobile telematic services. Both Ericsson and Nokia also used Estonia as a test market for new technologies and solutions, which gave local firms the opportunity to get involved with the most modern developments at an earlier stage than in many other countries.

The legal absence of exclusive rights (i.e., the lack of monopolies) in mobile and data communications also made it possible for *competitive* relationships to emerge for the first time in telecommunications services, and these became crucial for the further evolution of creative innovation in the years around 2000. Competitive relationships played an important role not least for the evolution of institutions, which were constantly discussed, reinterpreted, and amended as a result of differing perspectives of the competitors. Competition was also key in boosting the number of users of mobile and data communications technologies, which in turn motivated other innovators – such as the banks – to engage at a very early stage in new developments.

Thus external relationships and networks for innovation in Estonian telecommunications developed in highly dynamic ways in response to technological discontinuities. The earlier far-reaching fragmentation turned into a constantly evolving and creatively innovative web of relationships. To the extent that existing relationships might have hindered this organizational evolution, the earlier fragmentation may be seen as an advantage. Another clear advantage was the country's size, where the 'everybody-knows-everybody' mentality from time to time recalled a 'regional' rather than 'national' system of innovation. The crucial difference that made Estonian telecommunications a clearly national rather than regional variant of a sectoral system, was the power to retain totally separate formal and informal institutional structures.

In conclusion, it is not possible to explain creative innovation in Estonia without taking into account a wide array of *linkages* and *interactions* between different kinds of organizations. Moreover, the patterns of interaction evolved in interesting ways over time and in close relationship to technological discontinuities. In view of the systemic and evolutionary view of innovation presented in Chapter 2, this is hardly surprising. Also, the focus in systems of innovation theorizing on a much wider array of actors than 'firms' is confirmed through the clear roles played by non-firm organizations such as public agencies and the Institute of Cybernetics. In addition, the close interdependencies between technological, organizational and institutional change, which in Chapter 2 were pointed out as key in understanding system evolution, cannot be abstracted from in the material presented here. This is discussed further in Chapter 9.

4. THE TRANSFORMATION OF THE ESTONIAN STYLE OF INNOVATION

How is the above structural evolution related to the transformation of the Estonian style of innovation, in terms of the various activities occurring in the system of innovation? In the concluding sections of Chapters 4 to 7, the character of the Estonian style was discussed with respect to the four historical periods under investigation. In this section, those results are summarized by linking the four periods to each other and identifying the major elements of transformation.

It is interesting to note that the *formulation of visions* in the Estonian system was an activity that was fairly well developed in the Soviet era. Many organizations felt they could have done much more in terms of innovation and technological upgrading, had it not been for the troublesome structural constraints within the Soviet system of innovation. Estonia saw it as natural to be a leading and pioneering republic within the USSR, a fact that together with the size of the republic probably contributed to Moscow's interest in using Estonia as a republic where novelties could be introduced on an experimental basis before being launched elsewhere in the USSR. This characteristic continued in the post-Soviet era, through the vision of Estonia as a 'test-site' for the capitalist world's advanced technological developments. More generally, after an uncertain transition period in the late 1980s and early 1990s, when several national strategies were formulated and discussed (including the idea of establishing a 'national champion' in equipment production), the visions that crystallized in the mid-1990s built to a great extent on deriving advantage in various ways from Estonia's small size and overall flexibility, rather than seeking to imitate the West.

The relative lack of inherited technologies and formal institutions also contributed to this vision. It was argued that many changes – not only technological, but also organizational and institutional – necessary for the mobile and Internet era, could be carried out less problematically in Estonia than in many Western countries, where relatively well-functioning but overly rigid technological, organizational and institutional arrangements hindered change. This perception was so strong that Estonia imagined that by building upon its special and unique characteristics it could become a leading country not only among Central and East European economies, but globally, at least in some segments of the telecommunications sector.

The *articulation of demand* for new, improved and less expensive products developed in an interesting way during the four periods. Both domestic and foreign demand played important roles in influencing the Estonian style of innovation. In the Soviet era, the character of demand was strongly shaped by central planning, and it was thus in response to Moscow directives that the electronics manufacturing sector expanded rapidly. This domestic industrial growth created a demand for related R&D in Estonia. Much more important, however, appears to have been the strong demand for Estonian R&D from other parts of the Soviet Union. It was such external demand that lay behind, for example, the phenomenal growth of the Institute of Cybernetics. In the end, this meant that both equipment production and R&D in the Estonian system were heavily dependent – directly or indirectly – upon external articulation of demand. This dependence became particularly obvious in the years around 1990, when large segments of Estonia's industrial enterprises and R&D organizations collapsed as a consequence of losing Soviet customers.

At the same time, however, there had during the Soviet era existed a strong latent, domestic demand in Estonia for new telecommunications services (above all in data communications) and equipment (digital equipment, computers, etc.). The Estonian Ministry of Communications was also interested in the rapid development of the field of mobile telephony worldwide. This latent demand did not play any role in the Soviet era, but it became important once the transformation of the innovation system got under way. The latent demand meant that Estonian firms, organizations and individuals seemed to know very early on what they wanted in terms of Western technologies, and after the re-establishment of national independence the demand for new telecommunications services sky-rocketed in an unprecedented way, as shown in telecommunications services representing a 5 or 6 per cent share of the country's GDP in the last period of investigation here. This was much higher than in Western countries.

In absolute numbers, however, the mere size of Estonia meant that aggregate domestic demand was still low. On the service side, this was not necessarily a problem, but in the scale-intensive field of equipment production, it meant that Estonia could not establish local manufacturers based on domestic demand. In the equipment subsector, therefore, it was – just as in Soviet times – *foreign* demand that proved decisive. However, in contrast to the Soviet era, there was no foreign demand for the re-independent country's vast inherited R&D capacities, only for labor-intensive manufacturing and related services. The integration of Estonian manufacturers into global corporate networks in combination with the neo-liberal economic environment also meant that local equipment producers did not articulate any demand for local R&D, as their production-related problems could be more efficiently solved within the networks of large multinationals. By 2004, this was still the predominant situation, and it was beginning to change only very slowly.

The *creation of new knowledge* in the Estonian system of innovation was in the Soviet period intertwined with the all-Soviet system in complex ways. R&D, where the strongest Estonian segments were those of radio engineering and computer-related research, was, as mentioned above, carried out in response to non-Estonian demand or to demand from production enterprises that served the entire Soviet market. After an upturn in the 'perestroika' period, Estonian R&D collapsed, primarily because of its strong previous integration with the Soviet system and its failure to attract new customers domestically or in the West. The positive effect of this was that valuable human resources were freed, enabling much of the knowledge-creating activities to move to the business sector or to government bodies. In the business sector, this contributed to impressive entrepreneurship in the Internet- and IT-related segments of telecommunications. In the government sector, it facilitated the launch of both IT policy and far-reaching institutional changes as well as the creation of modern governmental IT systems. Activities related to more basic research were generally less affected by transition, and they could often be successfully integrated into Western academic networks. This was facilitated by existing links from the Soviet era.

Most knowledge creation in business, at least up to the mid-1990s, addressed imitative rather than creative innovation, but in the years around 2000 Estonia saw a virtual explosion of activities related to the creation of new service products, many of which did not exist anywhere else in the world. In manufacturing, new knowledge creation was usually process-oriented and mostly imitative, but at Elcoteq Tallinn there was an interesting movement into more knowledge-intensive parts of the value chain. This was true also for the field of electronics design. In general, however, both creative and imitative innovation was largely based on knowledge that had little to do with the exploitation of deeper science or R&D.

In terms of *competence-building*, the Soviet-era Estonian educational infrastructure included a fairly strong set of facilities in radio engineering, electronics and the field of information and computer technology. The situation was worse with respect to fixed voice telephony, although this was partly counterbalanced by the generous opportunities to use the large Soviet universities and institutes that offered specializations in telecommunications. When the Soviet Union collapsed, it was regarded as necessary to strengthen higher telecommunications education locally, and this was supported financially directly by the incumbent Estonian operator.

The existing facilities in the growing fields of radio engineering and IT meant that Estonia was fairly well prepared for the mobile and Internet in terms of competence-building; it was rarely regarded necessary to carry out any radical changes in the existing structures, which thus largely remained in place (although competence-building, too, came to suffer from the Soviet collapse in terms of funding). A variety of foreign relations were developed from about 1988 on with Western (in particular Nordic) universities, a trend that was important for competence-building due to new opportunities for student exchange and for advice in the continuous adjustments in curricula along more Western-style trajectories.

A problematic issue was at the same time the role of the university within the domestic system of innovation; Estonia inherited from the Soviet period underdeveloped formal university-industry links, and this did not change much during the 1990s. On the other hand, the small size of Estonia and the concentration of telecommunications-related university education in one single technical university, meant that there were good prospects for building and exploiting informal links. Competence-building also appears to have profited from the fact that university professors and teachers were almost always active not only at the university, but also in government or business, preventing the isolation of curricula from 'real world' needs outside the university, which was so common elsewhere.

In general, however, competence-building continued to suffer from under-financing, and the lack of competent specialists proved a problem towards the late 1990s, when the IT boom set in. A partial solution to this was the creation of the new IT college in 2000, but it is clear that the low number of competent personnel inhibited much potential innovation in Estonian telecommunications, not least in collaboration with foreign firms.

The process of *forming new firms and other organizations* was one of the activities that had suffered the most from the Soviet institutional setting, as new firms and other organizations were not permitted to be created in any spontaneous way – let alone enter telecommunications-related markets as new competitors. However, Estonia had pre-war experience as a capitalist economy, and when the situation started to change again in the late 1980s, it seems probable that the collective memory of this experience was still alive. In any case, a variety of telecom-related firms were successfully established in the years around 1990. As the new institutional setting mirrored a strong belief in the power of market competition, new firms were allowed to be formed and enter sectoral segments of the market such as Internet and mobile services, and by 2001, Estonia had become the first among the Central and East European countries to allow competition in fixed services. A wave of new firm formations also followed in the equipment subsector, most of which remained small and focused on labor-intensive subcontracting. Apart from firms, a variety of other important organizations were also created. These were mainly government bodies (such as the Department of State Information Systems and the National Communications Board), but also educational organizations (such as the IT College), industry organizations (such as the Association of Computer Firms), etc.

While the process of new firm formation seldom seems to have been identified as a major problem in the Estonian system of innovation, a constant debate has concerned the lack of large Estonian firms with the power to become independent global players. Similarly, many small firms have managed to survive, but without engaging in any far-reaching innovative activities. Hence, *adapting firms and other organizations for innovation and growth* has often been a weak activity in the Estonian system of innovation. In the Soviet era, the situation was the opposite: most firms and other organizations were overly large, and there was a general neglect of small firms.

In the years around 1990, firm-internal adaptations were largely concentrated on the transition from the Soviet system to a capitalist market economy, as the establishment of totally new organizational cultures was necessary. Many overstaffed Soviet-era firms also had to decrease their number of employees radically and deal in other ways with the chaos following the lost Soviet demand for their products. In the mid-1990s, firm-internal adaptations became oriented towards responding to and exploiting rapid developments in the Internet and mobile subsectors. This was so both in the private and the public sectors, as well as in academia.

During the IT boom in the late 1990s, innovative opportunities increased, but the rapid market growth also allowed many Estonian firms in both manufacturing and IT services to survive and grow without necessarily engaging in innovation. At the same time, competition became tougher and firms had to learn to respond to this, which pushed organizational adaptations in terms of increased efficiency and, as in the case of EMT, new product developments. Competition also concerned employing new staff, which turned out to be a troublesome issue during the spectacular market growth in the years around 2000. From the mid-1990s, an interesting trend was also the expansion of firms' activities to foreign markets, particularly to the other Baltic states.

When it comes to inter-organizational *networking* activities, the point of departure for transformation and reorientation was a Soviet-style system where the build-up and exploitation of inter-organizational links was highly problematic. Networking was difficult from a formal point of view in the Soviet system, but it was informally practiced in Estonia. In the years around 1990, this situation partly remained in place, but rapid technological developments created opportunities to build links between firms and other organizations that had previously had very little to do with each other (see section 3). Most creative innovations occurring in the years around 2000 involved far-reaching networking and collaboration with domestic and foreign firms, and often private and public organizations. Once again, the small size of the Estonian telecommunications-related community of practitioners meant that it was easy for firms and other organizations to know who did what and thus where potential partners could be found. The relative absence of social hierarchies also probably encouraged these wide informal networks.

The *provision of finance* underwent an interesting evolution during the four periods. In the Soviet era, telecommunications suffered from an ideologically motivated neglect in the state budget. On the other hand, the Soviet system did provide finance for an impressive expansion of both radio engineering and information and computer technology in Estonia. The collapse of the Soviet system therefore severely hit these subsectors financially, but the investments that had already been made were not in vain. In particular, the existence of an infrastructure for research and education appears to have been decisive for the performance of the Estonian system in the post-Soviet era. For example, Soviet investments in research on information and computer technologies was the indirect basis for many new post-Soviet firms, government agencies and other organizations.

In the transition period, many firms managed to expand and grow through the provision of foreign investment, without which most fixed and mobile operators and manufacturers would hardly have been created. This was clearly crucial, as government expenditures on telecommunications and IT at the same time continued to be minimal. Education and research suffered greatly from the lack of finance in the entire post-socialist period, and this was clearly part of the reason for the serious lack of specialists. On the other hand, some private firms took the initiative and invested in innovative projects such as the Tiger Leap and Look@World. The preparedness of the population to spend almost 6 percent of GDP on telecommunications services indirectly provided an important source of finance for firms such as EMT, which made impressive profit margins that enabled reinvestment into new product development.

Many firms, however, did not invest in innovation at all. The very different financial situation in Eastern Europe as compared to the richer West, in general meant that firms and organizations had to learn to live with minimal budgets, and this forced those organizations that did innovate to do it efficiently. Estonia became an encouraging example in many cases, showing that it was possible even for a small and relatively poor country to actually generate new processes and products, though perhaps not always as fancy as Western ones. Foreign sources of finance played an important role in this, as exemplified by the Swedish Baltic Sea IT Fund, which provided assistance for many of the most interesting Estonian innovative projects.

In terms of *consultancy, advice and lobbying activities,* Estonia differed quite naturally from advanced Western countries. During the Soviet period, Estonia gathered considerable experience with respect to lobbying in Moscow, which was often absolutely necessary for things to get going in the telecommunications sector. In the years around 1990, lobbying activities lost much of their importance, although they were a significant factor influencing new legislation in connection with the re-establishment of national independence. More important was the advice and consultancy that took place in connection with projects involving foreign direct investment.

With surprising speed, however, the Estonians emerged as partners at a much more equal level with their Western investors, so that the initial consultancy was transformed into more regular communication and interaction within transnational firms, such as Tele2, Radiolinja and Elcoteq.

EMT and Eesti Telefon, as well as the large banks, appear to have developed in a more independent direction with respect to their foreign investors, although this situation was beginning to change towards the end of the period under investigation. Foreign consultancy became important in the government and educational subsectors. Lobbying was an important force in some parts of the sector, not least in the process of policy formation and institutional change.

In the Soviet era, it was difficult to *create, change and abolish formal institutions*, and Estonia had thus only limited experience in this field. This created some uncertainty in the years around 1990, when the new political situation made it necessary to carry out far-reaching and radical institutional reform. Much of the institutional activities thereby focused on learning from foreign institutional arrangements and adapting these to local conditions. In many instances, the new capitalist formal institutions that emerged had the advantage of being relatively transparent and clear, and above all they actually worked: the lack of exclusive rights in data and mobile communications, for example, led to real-world competition, which proved key for the evolution of the system of innovation. In the last period of investigation, Estonia also successfully launched competition in the fixed telephony subsector, and at the same time the system proved capable of resolving difficult institutional bottlenecks that were seen to hinder further innovation in data communications. The relatively uncomplicated legislation in the 'new' country – as compared to Western countries, where laws and rules had often evolved incrementally for decades or even centuries – was seen to be an advantage in the process of institutional change. However, there remained uncertainties and disagreement about how various actors and mechanisms could and should be involved in this process. This was, for example, one of the reasons behind the delayed modification of the telecommunications law in the late 1990s.

Similar to institutional change, the *formulation of public policy* was extremely difficult in the Soviet period. Many policies that concerned Estonian telecommunications were formulated outside Estonia, and they could seldom be molded to fit the particular Estonian conditions. Nevertheless, the idea of making the Estonian SSR a strong location in the field of electronics in a broad sense contributed to strengthening the sector locally. The situation changed radically in the years around 1990, when policy became dominated by the desire to build a liberal market economy in Estonia. Colorful debates ensued over introducing a temporary monopoly in fixed telephony, but the overall laissez-faire political climate meant that the government did not intervene directly to prevent the collapse of production enterprises and research institutes. This was also partly a result of the lack of financial means to support more complex policies.

In the mid-1990s, public policymaking evolved through the formulation of a number of priorities within the broad field of ICT, but government expenditures remained at such a low level that it was doubtful the priorities could be realized. An interesting project was the 'Tiger Leap' initiative, launched in 1996 and aimed at boosting the general awareness of and skills in computer technology and the Internet. The fact that no clear innovation policy within ICT emerged was probably due to the continuing neo-liberal political climate and the general lack of funds in this area. Estonia became more active in policymaking in the years around 2000, as reflected above all in the formulation of new visions, the passing of new important laws, and the carrying out of innovative projects coordinated by government agencies.

Having discussed the above activities and their transformation, it can be concluded that the evidence points to many clear elements of *increasing similarities* in the styles of innovation between Estonia and the advanced Western countries, and thus at the occurrence of a convergence process. For example, following the collapse of the Soviet Union and the introduction of capitalist economic institutions, decision-making became more decentralized and organization-centered, i.e., firms and other organizations became free to make their own decisions and were also responsible for the consequences of their activities. Competence-building shifted towards more Western-like arrangements, and the organizational landscape underwent a process of normalization in terms of an emerging variety of the earlier almost non-existent small firms. Consumer demand came to play a more important role, while the government gave up the task of centrally planning the evolution of systems.

At the same time, however, while most activities were certainly inspired by their counterparts in Western systems of innovation, one also discerns interesting peculiarities in the new Estonian style. These peculiarities give an idea about where the future characteristics of East European systems of innovation might be heading, and their emergence is therefore of great interest. From the above discussion, some Estonian peculiarities can be seen to lie in the following:

- With respect to the formulation of visions, Estonia differed from advanced Western countries, as it deliberately sought to take advantage of its status as a post-socialist country. It did so by emphasizing its lack of older Western technologies and complex institutional structures as a specific advantage. Visions were also formulated in close relation to the unique smallness of the country, its difficult financial situation, and a level of education that was much higher than one would expect for a country with Estonia's GDP level.

- The articulation of demand in Estonia similarly developed interesting deviations from other countries: Local articulation of demand came to play an unprecedented role in service innovation, as Estonian consumers turned out to be prepared to spend dramatically much more of their income on telecommunications services than their counterparts in Western countries – a tendency that encouraged many innovating firms to become involved in new developments at a very early stage, as compared to other countries.
- Competence-building in telecommunications-related fields was unusual in Estonia because higher education was concentrated in a single technical university (and, to a lesser extent, one other university). Higher education was seriously underfinanced, but this circumstance formed the basis of a further typical Estonian characteristic, namely that university professors and teachers were almost always active not only at the university, but also in business and/or government. Estonia also distinguished itself through initiatives such as the 'Tiger Leap', 'Look@World', the IT College and not least the 'E-Governance Academy', with its mission to engage in competence-building for *other* countries with a socialist past.
- With respect to the creation of new knowledge, Estonia moved towards a heavy concentration on service- and applications-oriented knowledge creation. It was these activities that paved the way for the introduction of a number of globally new products in the applications and middleware layers of telecommunications services. Estonian firms provided encouraging examples of what was possible to do in terms of new knowledge creation with very limited financial and human resources. Particularly in the field of e-government, Estonia came to show the world how creative innovations – though perhaps not as flamboyant as comparable Western ones – could be realized in an extremely efficient way.
- Estonia also demonstrated an unusual ability to engage in *resolving institutional bottlenecks*. This can, on the one hand, be seen in relation to many other post-socialist countries, which struggled much longer with basic formal institutions such as the implementation of liberalized telecommunications markets or digital signature laws. Estonia was the first among post-socialist countries not only in approving new legislation, but also in implementing the laws and adapting them to new circumstances. On the other hand, the country also demonstrated an ability to carry out institutional changes more efficiently than many *Western* countries. The still relatively uncomplicated legislation in the 'new' country – as compared to Western countries, where laws and rules had often evolved incrementally for decades or even centuries –

was here seen to be an advantage in the process of institutional change. Moreover, from an informal institutional perspective, there was – in comparison to Western countries – a uniquely broad consensus about the general path of future development in the country: the belief in (fair) competition as a driving force for innovation was genuine and broad, and although intense controversies and lobbying accompanied the working out of new legislation, this did not lead to any bigger scandals of the type that, for example, neighboring Latvia suffered from.

- Networking activities were also characterized by a special Estonian style. Once again, the small size of the Estonian telecommunications-related community of practitioners meant that it was easy for firms and other organizations to know who did what and thus where potential partners could be found. The relative lack in cultural terms of social hierarchies also probably encouraged these wide informal networks. Communication was easy in the small telecommunications-related community, which made it much simpler than in many other countries to exploit the dynamic economies of interaction that were so dearly needed to take advantage of the new emerging technological paradigms in the Internet and mobile era. This involved smooth communication between private and public actors (the latter being inspired by banks), and building external collaborative networks with local software firms and research organizations. For many Estonian firms, 'co-opetition' – a genuine combination of cooperation and competition – was a self-evident way of life.
- With respect to the adaptation of organizations, one of the most interesting aspects of Estonian innovating firms was the formation of a new organizational culture. This culture differed, of course, from the Soviet era, but it also differed strongly from neighboring Western countries such as Sweden and Finland. Swedish and Finnish firms were by many Estonian innovators regarded as overly careful and slow in their innovation-related decision-making. The Estonians were used to living in constant and severe scarcity of resources, and this forced them to become extremely efficient. From the Estonian perspective, nearly all Western countries appeared peculiarly defensive in their whole approach to innovation, and on the service side, Estonian actors found it hard to understand why things seemed to move so slowly in the West despite the enormous resources invested into their systems of innovation.

Admittedly, the above results are preliminary in the sense that they are not based upon any formal comparison between Estonia and other countries.

Nevertheless, they point in a thought-provoking way to the possibility for Central and East European countries to actually pursue developments along trajectories that do not necessarily coincide with Western ones. To the extent that Estonia can be said to have been successful in creative telecommunications innovation, this is based not on imitating the West, but rather learning from advanced countries and use that knowledge for creating their own specific styles of innovation.

Let us now turn to the issue of the underlying forces that have influenced the structural evolution discussed in the preceding section and the transformation of the Estonian style of innovation as addressed in this section. In the theoretical chapter, the forces responsible for system evolution were approached through the analysis of both internal and external dynamics. More precisely, the focus was on studying the influence of the momentum of the system's internal processes (i.e., history), and the impact of outside pushes and pulls (i.e., the influence of the foreign environment). The first of these issues, the influence of the system's internal momentum, is addressed in section 5 through the analysis of how Soviet heritage came to influence the post-Soviet Estonian system of innovation. The second issue, the influence of outside pushes and pulls, is addressed in section 6 by focusing on Estonia's international reorientation.

5. EXPLOITING THE SOVIET HERITAGE

This section first identifies the initial conditions for Estonia's transformation in the 1990s in terms of inherited competencies. After that, the section turns to the issue of how these competencies developed during this time, up until Estonia's EU accession in 2004.

At a first glance, the material does not appear to strengthen the proposition that inherited competencies play a crucial role for post-socialist systems of innovation. Thus in Chapter 4 it was concluded that telecommunications in a narrow sense, i.e., fixed voice telephony if we follow the view that dominated up to the 1980s, was not a prioritized or privileged sector in the Estonian SSR. Resources were minimal, existing factories had been closed down after the war, educational opportunities were highly troublesome (so that higher-level employees in the Ministry of Communications had to go to other parts of the Soviet Union to get their education), and the institutional landscape seemed to discourage all innovative developments in the field. At first sight, useful historical experiences and inherited competencies therefore seem to have been almost completely absent.

The problem with this view is that it does not take into account the technological discontinuities that later (from the 1990s on) came to completely transform the very concept of telecommunications, and almost totally came to dominate all innovative activities in the years preceding Estonia's EU accession. If these technological discontinuities – in particular the emergence of mobile communications and the Internet – are taken into account, then the role of past developments in Estonia has to be re-evaluated. It turns out that Estonia specialized in interesting fields.

First, the expansion of the electronics-related manufacturing sector in the Estonian SSR after World War II meant that post-Soviet Estonia inherited a large number of electronics production enterprises, which in some cases were attached to important research institutes, as well as a important electronics-related departments at the Technical University, which were involved in far-reaching competence-building as well as research. At the time of the Soviet breakdown, hundreds of electronics-related engineers had already graduated from these departments at the Technical University, and most of them played important roles in the economy, either in production enterprises, universities or research institutes. Although the specializations of these organizations did not necessarily have any strong ties to telecommunications-related fields at that time, their existence had a potential value in view of the increasing 'electronification' of telecommunications.

Secondly, and more specifically, the creation of the radio set producer RET in the 1930s became the starting point for a vast expansion of *radio engineering* in what became the Estonian SSR. This meant that production enterprises were expanded, R&D organizations were established in connection with the factories and education opportunities increased substantially through the creation of the Department of Radio Engineering at the Tallinn Polytechnical Institute, which also carried out R&D to a minor extent. In view of the fact that radio engineering came to lie at the technological heart of the mobile telecommunications sector, this inherited expertise in radio engineering was clearly an interesting feature of the Soviet-era Estonian system of innovation.

Thirdly, since the 1960s, advanced R&D and education had been carried out in Estonia in the field of *informatics* and *computer engineering*. The key organization here was the Institute of Cybernetics, which grew to become an extremely powerful research and development organization in Estonia, providing developments and expertise that were recognized all over the Soviet Union and even in the Western academic community. Up to the 1980s, informatics and computer engineering were not seen to have much in common with telecommunications, but with the appearance of the Internet discontinuity and the general focus on data communications in the 1990s, the activities of the Institute of Cybernetics became not only relevant, but entirely *central* for further developments. Estonia was thus, thanks to its Soviet heritage, in reality unusually well-prepared for the Internet era.

From this perspective, Estonia's Soviet-era specialization in electronics, radio engineering, informatics and computer engineering fitted in extremely well with the new, emerging technological trajectories of the 1990s. It is interesting to compare this situation with that of Latvia, a country that had an impressive focus on competencies in telecommunications in a traditional sense during the Soviet era, which may lead us to think that this country would seem better prepared for the post-socialist period. Due to the emergence of new technological trajectories, however, it is not clear that this was the case, as Estonia's already existing focus on those fields that would later stand at the center of innovative effort turned out to be a key advantage. (Certainly, this was not the result of a deliberate Soviet-era Estonian policy aimed at creating an advanced future telecommunications sector, but it nevertheless turned out to be a lucky coincidence.)

How, then, was this highly interesting Soviet heritage dealt with by actors in the post-Soviet Estonian telecommunications system of innovation?

Firstly, the empirical material presented in Chapter 4 to 7 pointed to a disastrous collapse of the electronics sector in Estonia during the years around 1990. A majority of the far-reaching inherited competencies that were embedded in Soviet-era production enterprises could thus hardly be exploited in the new capitalist context. However, those telecommunications-related electronics manufacturers that did survive the Soviet breakdown, as well as those electronics firms that were newly created, did build on competencies that had been built up in the past. At the level of organizations, this was the case with Tarkon, which in the post-Soviet period even regarded itself as the historical successor of the old Tartu Telephone Factory, which had been established in tsarist times. In contrast, the largest telecom producer, Elcoteq Tallinn, was a completely new company, but at the level of individuals it showed – similar to Tarkon – a clear continuity. Still in 2004, many employees at these firms had long experience with Soviet-era electronics enterprises, and the persons that acted as directors of both companies in the 1990s had trained at the Soviet-era Tallinn Polytechnical Institute (in electronics and machine construction, respectively). In addition, it was engineers with electronics-related training from the Soviet era who were the driving personalities behind creating in 2004 the important competence center ELIKO.

Secondly, with respect to the inherited competencies in radio engineering, the empirical evidence shows in a striking way that an impressive number of telecom-related business leaders and other key individuals in post-socialist Estonia's telecom system of innovation were actually graduates from the Department of Radio Engineering at the Polytechnical Institute. These individuals, i.e., expert radio engineers, contributed heavily to a surprisingly rapid build-up of an advanced mobile telecommunications sector in post-

Soviet Estonia. While the inherited Polytechnical Institute hardly participated in any *direct* way in those innovative projects discussed in the empirical part of the study, its Department of Radio Engineering was nevertheless a key supplier of the competence that actually made those innovative projects possible. Moreover, the availability of this remarkable competence also created the opportunity for key actors such as EMT to develop early on a far-reaching autonomy from their Nordic investors, with the build-up of highly creative Estonian R&D departments based exclusively on Estonian human resources.

Thirdly, from around 1990, researchers from the Institute of Cybernetics played key roles in advising government agencies in ICT-related issues, and it was representatives from the Institute who became the heads of the new ICT-centered public bodies. The collapse of research contracts from the East led to a radical reduction in employment at the Institute of Cybernetics, but many of its researchers responded by creating their own firms and therefore competencies were not lost. Several of these firms later took part in the creative innovation projects that were discussed in Chapters 6 and 7. Other researchers continued as teachers at the Technical University or were even appointed to chairs at Nordic and other foreign universities, where they became important channels of communication between the Western world and the Estonian sectoral system.

But the inherited Soviet system also explains many of the *problems* pointed out in sections 1 and 2, with respect to Estonia's innovative and economic performance in telecommunications. Technological competencies are, in capitalist systems of innovation, merely one of many necessary preconditions for generating and exploiting innovation. Innovations are, by the definition used in Chapter 1, new ways to create *economic* value, and mere technological achievements, however spectacular they may be, are therefore often totally uninteresting from an innovation perspective. New knowledge is, in this sense, interesting only in relation to its economic potential.

In the Soviet system, the understanding of this economic aspect of technology was persistently underdeveloped, and Estonia inherited from its Soviet past a perspective on innovation that from a Western perspective seemed peculiarly naïve in its understanding of technology and economics. The available evidence suggests that Estonia gradually learned to handle this domestically, but that with respect to export performance building on innovative developments, it faced severe difficulties due to an overestimation of the power of technology itself. By the EU entry in 2004, Estonia had not yet managed to transform its extensive technological competencies in telecommunications into any internationally well-known brands, and its far-reaching creative innovation in telecom services had hardly contributed to exports at all. It was symptomatic that the system's greatest successes were in fields in which marketing efforts were less important, such as subcontracting and customized solutions.

The problems faced in re-independent Estonia with respect to *university-industry relationships* also had a historical background. Universities and production enterprises had been strongly but separately integrated with the Soviet Union, and after 1990 they similarly developed strong but totally separate links with the West. The hopes raised in recent years for better utilizing the university potential in the innovation system has thereby been counteracted by an actual reinforced separation between universities and industries domestically, and in the creative innovation achievements that have been dealt with in this study one thus never sees any direct involvement from the side of universities. This does not contradict the conclusion drawn above that universities and academies have played an extremely important role as providers of advanced and highly relevant *education*. The interesting point is how close this pattern lies to the old roles that universities played in the Soviet Union.

In summary, although the Estonian system of innovation has been transformed in a very radical way, it is not possible to ignore the almost omnipresent traces of Soviet history in the making of the post-socialist system of innovation in telecommunications. Moreover, these traces are much more than minor curiosities; rather, they are completely central to understanding how the system of innovation has been transformed and how creative innovation has been enabled in the post-socialist context. The theoretical implications of this will be further discussed in Chapter 9.

6. THE NATURE AND IMPACT OF ESTONIA'S INTERNATIONAL REORIENTATION

This section first discusses the nature of the links between the Estonian system of innovation in telecommunications and its foreign environment. Building on that, the discussion then turns to the issue of how these links have influenced the transformation of the Estonian system.

Estonia's relationships with the West from the late 1980s to its EU accession in 2004 were both omnipresent and highly intense. Moreover, an interesting and striking aspect of Estonia's links to the foreign environment is that the relationships have been almost exclusively oriented in very specific directions: to Sweden and Finland. The linkages can be seen to take different forms:

- *International trade.* Estonia's foreign trade with the West increased dramatically in telecommunications from around 1990, and has been strongly focused on the Nordic countries. Foreign trade began in 1990 with the re-establishment of Ericsson as a trade partner, which thereby explicitly referred to its historical experience in the country from before 1940. Nordic equipment vendors became the most dominant suppliers in the years that followed, particularly in the mobile segment, while the old links to Russian and other East European suppliers were totally cut off. Conversely, Sweden and Finland became vastly dominant destinations for Estonia's telecommunications-related exports, while Estonia at the same time lost almost all its former customers in the East.
- *Ownership and investment.* Finland and Sweden have enjoyed an almost total dominance over other countries as external owners and investors in Estonian telecommunications-related firms. The trend is set by the Nordic ownership in Eesti Telefon and EMT and is confirmed through Elcoteq, Radiolinja, Tarkon, Tele2, Uninet and other firms in the mid-1990s. In the years around 2000 the trend is even further strengthened through the arrival of a second wave of subcontractors, IT acquisitions and above all the Swedish takeover of virtually the entire Estonian banking sector.
- *User-producer relationships.* After the collapse of the Soviet-era electronics sector, the new telecom-related equipment industry that emerged in the mid-1990s built very heavily on labor-intensive subcontracting, with customers almost exclusively from the Nordic countries. These relationships were economically powerful but largely non-innovative. Subcontracting also took place, though on a completely different scale, in the field of software (SEB IT Partner Estonia, Assert, Aprote, etc.).
- *Financial support.* Starting from minor development aid activities in the years around 1990, the Nordic countries emerged as the main bilateral partners in terms of various kinds of financial support. In student exchange, for example, Sweden became clearly the most dominant provider of scholarships for Estonian students. Somewhat later, the Swedish 'Baltic Sea IT Fund' played a role in an array of creative innovation projects in the telecommunications sector. In addition, Telia and Sonera – both directly and through their ownership of Eesti Telekom – emerged as donors in projects such as the new IT college.
- *Consulting.* At various stages of systems transformation in Estonia, one finds Western advisors and consultants involved. Again, the vast majority of these came from Finland and Sweden, and they were active in both the public and private sectors: as advisors in working out

telecom legislation, as temporary managers in operators and other firms, as trainers of public sector IT heads, etc.

- *Student and research exchange.* In Soviet times, Estonian researchers, to the extent that they were allowed to travel abroad, developed a habit of going for shorter guest research stays to Swedish and Finnish educational and research organizations. The scope for this was minimal, but the direction was followed up when better prospects for exchange began in the late 1980s. About half of all student exchange in the mid-1990s took place with Swedish and Finnish universities, and among the students who participated, many would later arrive at key positions in the Estonian sectoral system, particularly in business. As a matter of fact, it appears more difficult to find people in key positions who had *not* experienced some form of student, research or job exchange in Sweden or Finland. These individuals range from the directors of the Institute of Cybernetics, the Rector of Tallinn Technical University, the prime minister's key IT advisor, the R&D directors of EMT and Hanspank, the CEOs of Oskando, Ericsson Eesti, Elvior, Crebit, Eesti Telekom, Teleport, Meediamaa, etc. They can also be found among the driving personalities in projects such as the 'Tiger Leap' and 'Look@World'. In addition, several scientists and other people from Estonia found work in Sweden and Finland: they became professors and lecturers at technical universities in Stockholm, Gothenburg, Helsinki and other Nordic cities. Cooperation between universities also played a very direct role in connection to the birth of the Internet subsector in Estonia from 1988.
- *The Nordic world as a model.* In Estonian eyes, the Nordic countries became the 'natural' point of reference when laying out strategies for transformation. Through the success of Sweden and Finland as telecommunications countries, the Nordic world became a powerful source of inspiration for the Estonians, representing small but highly advanced European countries with which Estonia felt culturally related. A whole mythology of historical-cultural links exist between them, and most actors in the Estonian system acknowledge this 'feeling of belonging' with the Nordic world when asked to describe their relations to foreign partners. The enormous difference in the use of new telecom technologies in Estonia as compared to Latvia and Lithuania can hardly be understood without taking this argument into account. The Nordic inspiration can be seen very clearly already in the late 1980s, leading to the very early introduction of NMT in Estonia (a history that in the local telecom mythology is widely considered to have begun during an informal meeting 'in a Finnish sauna'), and little more than a decade later the same inspiration was used to formulate the national ICT

strategy in terms of 'overtaking Finland', with the prime minister speaking about Estonia potentially emerging as the 'no. 1 Internet country in the world'. Similarly, Tallinn Technical University has been heavily inspired by KTH in Stockholm and the Helsinki University of Technology.

- *Joint innovation.* However, extending beyond the Nordic countries as models and inspiration and instead participating in *concrete collaboration* for *concrete innovation projects* across the Baltic Sea, has been a relatively rare phenomenon. Ericsson's launch of mobile positioning in cooperation with a number of Estonian partners represents an unusual example. Joint activities in terms of Estonia as a test market for new developments from Ericsson and Nokia should also be mentioned. There are some signs that joint innovation activities are on the increase, and this appears to be the result of toughening international competition on telecommunications markets: with the consolidation of Nordic ownership in banking and telecom in the Baltic Sea region, there is a trend towards an efficient use of resources from different country branches of firms such as TeliaSonera, Tele2, SEB and Föreningssparbanken. In education, it is not at all unimaginable that TTÜ may merge with the Helsinki University of Technology. A similar trend can be seen in the IT sector, where Swedish and Finnish groups such as TietoEnator and Cellnetwork integrate their Baltic branches more deeply together. In addition, it can be seen that Elcoteq Tallinn is not at all regarded as a *subsidiary* of the Finnish-based corporation, but rather as an organic part of it, filling an important international function by playing key roles in connection to, for example, the establishment of new plants in Hungary and St. Petersburg in the late 1990s.

The general geographical, historical and cultural proximity to the Nordic countries has thus been accompanied from the late 1980s by the development of a very broad range of new and important channels of interaction between East and West. The loose historical and cultural feeling of 'belonging together' has the character of a self-fulfilling prophecy, as it, on the one hand, *provided a strong basis* for the new linkages, while at the same time was *reinforced* by the dramatically increasing interaction in terms of specific innovation-related linkages (as listed above). Through positive feedback, this has, as a matter of fact, gone so far that the Estonian prime minister in the years preceding the EU accession spoke explicitly about Estonia *being* a Nordic country.[1]

A somewhat different issue is the question of how the strong Estonian-Nordic linkages have shaped the Estonian system of innovation and its performance. Here it should first be pointed out that it is not self-evident that the strong Nordic contacts are associated with a *beneficial* impact on the Estonian telecommunications system of innovation. Doubts can be raised, in particular, with respect to the Nordic activities as customers of Estonian subcontractors in equipment production, where the strong – and economically successful – integration into the networks of Nordic equipment vendors are directly related to what appears to be an Estonian *lock-in* into labor-intensive and largely non-innovative manufacturing activities. With the anticipated increase of wage levels in the coming years, this arrangement for successful export performance will almost certainly be eroded within a decade or two. In order to survive, Estonian manufacturers may have to follow the paths of countries such as the East Asian 'tiger economies', where firms have survived by moving from labor-intensive manufacturing towards 'own-design manufacture' (ODM) and 'own-brand manufacture' (OBM). Whether this is possible within the networks of Nordic equipment vendors remains to be seen.

In services, on the other hand, many of the activities related to creative innovation discussed above are hardly imaginable without strong links to the Nordic world. The most clear-cut example is provided by Ericsson Eesti, a key mediator between Estonian and foreign actors, which directly participated in several mobile innovation projects in Estonia. Telematics innovation involved close contacts with Swedish and Finnish firms as both suppliers and customers. Several of the most innovative Estonian small firms – such as Crebit, Elvior, Voicecom, Oskando and others – were founded more or less directly following the return of Estonian entrepreneurs to Estonia after a time abroad in Finland or Sweden. In the public sector, key advisors had also a background of working and learning in the Nordic countries.

An interesting point is that this did not lead to pure imitations of Nordic developments. The Estonian circumstances appear to have been considered too different for that, demanding alternative and sometimes radically different solutions, and Estonian organizations thus responded by using the Nordic influence to create a different, Estonian style of innovation. Some of these specifically Estonian features were discussed in section 4. The key issue here is that the emergence of the Estonian style took place in parallel with omnipresent and highly intense bombardments of impressions from and interactions with the Nordic world. In other words, the efficient communication between the Estonian system of innovation and its environment did not prevent the Estonian system from developing a number of system-specific features.

NOTE

1. See, e.g., *Hufvudstadsbladet*, 4 July 2003.

9. The Dynamics of Innovation in Eastern Europe: Implications for Theory and Policy

In this chapter, the conclusions drawn from the case of the Estonian telecommunications system of innovation, as outlined in the preceding chapter, are taken a step further by discussing them in relation to further theory and policy development. The chapter begins by critically evaluating the relevance of a systems approach to the study of transformation and reorientation (section 1). This is followed by a discussion of how the results from this study can be used to enhance our understanding of interaction between a system of innovation and its environment (section 2), and how such interaction can be seen to influence systems transformation (section 3). Thereafter, the chapter turns to the question of the theoretical treatment of the socialist past in the analysis of post-socialist transformation and reorientation, focusing in particular on the role of inherited competencies from the socialist era (section 4). A further section explores the implications of the results of the study for public innovation policy (section 5). The chapter concludes by returning to the issue raised in the introductory chapter as to what roles the economies in Central and Eastern Europe may come to play in the European innovation-based economy of the future (section 6).

1. THE RELEVANCE OF A SYSTEMS APPROACH TO TRANSFORMATION AND REORIENTATION

In Chapter 2, a conceptual framework based on systems thinking was proposed for the study of innovation in the context of East European transformation and reorientation. It was suggested that in order to understand the mechanisms and processes of transformation and reorientation, the research should focus on organizations and institutions as forming the elements of systems, which together with the relations that connect the elements define the structure of the system. It was also argued that the study of the system's structure should be complemented by addressing the level of activities, which collectively define the particular style in which innovation is performed. Moreover, an important feature of the proposed systems approach was that a system of innovation should be treated holistically, including a wide array of organizations, institutions, relations and activities. Above all, the need for focusing on the interaction among various components and activities, and between the system and its environment, was emphasized.

Indeed, the empirical part of the study supports the argument that processes of innovation in the context of East European transformation and reorientation can be better understood through the application of a systems approach of the kind referred to above. In particular, one notes the following:

- As emphasized in Chapter 8, innovation in the post-socialist Estonian context involved *far-reaching networking activities*, linking various actors to each other in a variety of ways in the pursuit of innovation. Both collaborative and competitive links, as well as transactional relations, were found to play an important role for innovative outcomes, which are therefore difficult to grasp without taking the interactive dimension of innovation into account.
- Creative innovation in the Estonian telecommunications sector was the result not only of the activities of firms. *A variety of non-firm organizations* also contributed in decisive ways. These included actors such as research institutes, non-profit innovation consortia, financial organizations, sector-specific government agencies, as well as other public bodies. These acted both as demanding customers and initiative-taking developers of new products.
- A wide array of *institutions* played a decisive role for enabling creative innovation in Estonian telecommunications. This goes beyond the activity of networking among various actors, and is related to changes in legislation and other formal and informal rules and traditions (a number of concrete examples of such institutions were given in Chapter 8). The resolution of institutional bottlenecks, as a way for the system to respond to technological and institutional discontinuities, turned out to be key for the enabling of creative innovation.
- The *relations between organizations and institutions* proved crucial, as these two types of components were seen to strongly influence each other. In particular, the study highlighted the ways in which organizations took part in the process of institutional change in the difficult and highly fluid context of post-socialism. In this way, it could also be seen that institutions did not appear from nothing as a *deus ex machina*. Rather, they were the result of actors' deliberate efforts. Institutional changes could therefore hardly be understood without studying their relations to organizations (and individuals), as suggested

in the systems approach. Conversely, as noted above, organizations were dependent upon a variety of institutions in their pursuit of innovation.

- The focus on *a broad array of activities* turned out to be helpful when seeking to describe and explain the enabling of creative innovation in the post-socialist context. If the empirical investigation had limited itself to the study of, for example, the role of R&D activities or the business activities of specific types of firms, much of the dynamics of the system as a whole would have been missed, and creative innovation would have been hard to explain. By taking into account a wider number of activities – the formulation of visions, the articulation of demand, knowledge-creating activities, competence-building, the creation and adaptation of firms and other organizations, external networking, the provision of finance, advice, consultancy and lobbying, the creation and adaptations of formal institutions and policymaking – a much more satisfactory understanding of how creative innovation occurred in the Estonian telecommunications sector could be developed.

Apart from, but related to, the above aspects, the present study has addressed important questions relating to the issue of *system boundaries* and the relations between a system and its foreign environment. This issue is discussed in the following two sections. As we will see, it is associated with an even heavier stress on the need for a systems approach to innovation in the context of East European transformation and reorientation.

2. THE NEGLECTED ISSUE OF SYSTEM-ENVIRONMENT INTERACTION

In previous research on systems of innovation – both in Eastern Europe and more generally – one of the main interests has been in contrasting and comparing different national systems to each other. Freeman (1987) thus built his early delineation of the concept of 'systems of innovation' on contrasting the Japanese system of innovation with the national systems of innovation in the United States and Western Europe. The approach pursued in the seminal work by Nelson (1993) and his colleagues also focused on comparing different national systems,[1] and this was followed up by Patel and Pavitt (1994), who sought to develop more coherent methods for comparing advanced national systems.[2] More recently, Edquist and his colleagues have carried out far-reaching comparisons between national systems based on a framework with the concept of activities at the centre of analysis (Edquist and Hommen, 2004).[3]

In the context of Central and Eastern Europe, Hanson and Pavitt (1987) contrasted socialist and capitalist systems of innovation as two ideal types of systems (cf. Box 2.1). Applying a more quantitative methodology, Radosevic (2003) recently compared the 'national innovation capacities' in different countries (including a comparison with the Western EU), although he did not (in that work) explicitly apply a systems thinking.

A somewhat different issue, which has not gained as much interest as these comparative approaches, is the problem of how systems of innovation are *related* to each other and what this means for the dynamics and performance of systems. The evidence presented in this study suggests that the issue of how systems of innovation are related to each other is not merely a minor detail, but should be placed centrally in the analysis of systems. This is particularly so in an era of increasing internationalization and globalization. The study of East European systems of innovation here offers a fruitful empirical field for increasing our knowledge of such system-environment interactions in general. The transformation and evolution of the Estonian telecommunications system of innovation, as concluded in Chapter 8, is hardly possible to grasp to any satisfactory extent without taking into account Estonia's strong and dramatically shifting linkages to its foreign environment. The foreign links were found to be omnipresent in the activities taking place in the system, and they had a vast impact – directly or indirectly – on the Estonian system. Interaction and communication across system boundaries, i.e., between the system and its foreign environment, was thus of great importance.

As noted in Chapter 2, previous theorizing relating to the boundaries of systems lead to at least two ways of looking upon system-environment interaction. These can in turn be said to be associated with two types of inter-system integration processes. The first, which in Chapter 2 was labeled *Deutsch-type integration*, referred to a process of building relations between components in the system and organizations and institutions in the environment. The second, *Luhmann-type integration*, referred to a process of harmonization of one system's structures and styles with those of another in the foreign environment. Let me here first address the phenomenon of Deutsch-type integration. The next section then turns to Luhmann-type integration.

The radical reorientation of Estonia from its deep integration with the Soviet system of innovation, towards a new orientation towards the West, allowed us to investigate the dynamics of such links and how foreign environments have influenced the system and its performance. Building on the discussion in Chapter 8, an important implication from the perspective of further theory development along this path of enquiry, is the tendency for the foreign environment to strongly influence systems through a broad range of channels. Drawing on section 5 in Chapter 8, these channels (which reflect different types of system-environment relations) include at least the following:

- International trade;
- Foreign ownership and investment;
- Cross-border user-producer relationships (in particular subcontracting);
- Financial support from abroad (for various activities);
- Consulting as a way of transferring skills and knowledge from abroad (in business as well as in academia and government);
- International student and research exchange;
- Foreign systems of innovation as models;
- Cross-border joint innovation.

Some of these channels intersect with and influence each other. For example, foreign investment may be associated with consulting and user-producer relationships, which in turn may be linked to cross-border joint innovation. The above list is likely to be expanded through further empirical research, but the point is that the channels mentioned are not merely small details; rather, communication through these channels provide absolutely crucial pushes and pulls for the transformation and evolution of systems of innovation.

This becomes particularly clear if one accepts the conceptualization of systems in terms of comprising not only organizations and institutions, but also activities, as suggested in Chapter 2: for example, the role of foreign systems of innovation as a model for the domestic system is closely associated with the formulation of visions. Similarly, international student exchange might seem peripheral to innovation at first glance, but in reality it provides a powerful basis for the formulation of visions, for competence-building, for new firm formation based on insights gained abroad, for future networking with foreign partners, etc. With respect to financial support, the present study showed that such support was involved in a whole array of innovative projects. Developing the above typology further should be a goal of future research on system-environment interaction.

Furthermore, one of the most striking features of the Estonian case is that the linkages to the Nordic countries were virtually the *only* relevant linkages that emerged and mattered. In other words, Estonia's integration with the West in the field of telecommunications has been extremely focused in very specific directions. This may appear surprising from the perspective of traditional globalization theories, with the common expectation that an opening up of borders and liberalization of trade should lead to a general integration into the global economy, rather than to the formation of any new strongly connected groups of countries. If seen in relation to the systemic perspectives on innovation that were developed in Chapter 2, however, strong and focused links to a few specific countries may be seen as perfectly natural.

The reason is that innovation is a very different phenomenon from that of trade and other less complex processes of interaction and exchange. Innovation is heavily dependent upon difficult processes of learning and is, by definition, focused on developments which have never taken place before. Since learning in the context of innovation is often a highly interactive process, there is a strong need for efficient communication in order for the learning process to be effective. The problem in the context of learning from the foreign environment is that it is dependent upon 'resonance' at the system boundary. Such resonance, as argued in Chapter 2, is more likely to occur when the cross-border communication takes place with foreign environments that are located in geographical and cultural proximity of the system. The relative ease with which the system can understand and communicate with such geographically and culturally close environments, increases the prospects for the system to respond in fruitful ways to interaction with those environments. Therefore, they have a stronger influence on the system than geographically and culturally more distant environments.

Moreover, the relative success of communicating with certain foreign environments rather than with others, may stimulate further interaction and communication with those environments. This is because the practice of communication increases the efficiency of communication in a circular and cumulative way, and that networking activities across system boundaries are likely to create further opportunities for networking. Through processes of positive feedback, the resulting concentration of system-environment communication in highly specific directions is therefore likely to be self-reinforcing.

This seems to be a result that holds at least when culturally and geographically close environments coincide with highly advanced systems of innovation. This was the case with Estonia's links in the telecommunications sector to Sweden and Finland, i.e., countries that are among the absolute world leaders in that specific sector. An interesting question is how system-environment communication may evolve in cases where the culturally and geographically close environments are not world leaders. There is a possible risk that vicious circles may emerge, which reinforce a disadvantageous pattern of system-environment interaction. But at the same time, it cannot be excluded that the disadvantageous consequences of the established patterns may be counteracted by actors who deliberately seek interaction with other, more successful foreign systems of innovation. The empirical evidence presented in this book does not allow us to draw any further conclusions in this respect, but it remains an important topic for further research.

3. BEYOND CONVERGENCE: NATIONAL VARIANTS OF SECTORAL SYSTEMS

Having discussed the nature and variety of system-environment channels of interaction, let us now turn to the issue of what the process of Deutsch-type integration means for the system in terms of Luhmann-type integration. This type of inter-system integration, as it was defined in Chapter 2, relates to an interesting debate in previous research on Central and East European economies and systems of innovation, namely the quest for 'convergence' of Eastern with Western systems of innovation. The notion of convergence goes beyond 'catching-up' (in terms of GDP, productivity, patenting or other quantitative measures) with the more advanced countries in Western Europe and elsewhere, and rather, refers to a movement towards increasing similarities in the structure and style of Eastern and Western systems of innovation.

For example, Radosevic (1999) took as his point of departure Hanson and Pavitt's (1987) comparative study of socialist and capitalist systems of innovation in order to determine the extent to which post-socialist systems of innovation can be said to have achieved convergence with the West in terms of such factors as the character of financing; the emergence of links between large users and specialized suppliers; the importance of weapons R&D; the relative importance of process as opposed to product innovation; the degree of decentralization; pluralism in technological activities; ownership roles; etc. (Radosevic, 1999: 25). He found strong signs of convergence between Eastern and Western systems of innovation following the transition to market-based capitalist systems of innovation, but also remaining elements of divergence. The latter, however, were understood as temporary deviations (ibid.: 37).

The evidence presented in this book supports those earlier findings, as clear and strong elements of convergence can be found in both the structure and style of the Estonian system of innovation in telecommunications, while there are also a number of remaining elements from the Soviet-era system. At the same time, however, the results in this study go a step further, by addressing the question of where the East European styles of innovation may be heading after the transition from socialism to capitalism. It is well-recognized that systems of innovation differ strongly among advanced countries, and that it is not possible to speak of any 'optimal' style of innovation in the Western world. In the same way, there is no reason to assume that the new East European styles of innovation can – or should – converge fully with the style of any specific Western system. On the contrary, it is natural to expect that the unique circumstances of East European transformation, and the specific histories of the Eastern countries, will give rise to styles of innovation that differ in various respects – and to a greater or lesser extent – from those of Western countries.

In Chapter 8, it was concluded that the above perspective is supported by empirical evidence in the present study. It was found that the Estonian style of innovation had undergone a dramatic transformation since the collapse of the Soviet Union, but that this transformation was more than merely a convergence towards Western styles of innovation. Although the Estonian system had become much more Western than in the Soviet era, it had by the time of the country's accession to the European Union developed a style that also *differed* in interesting and perhaps surprising ways from the West. This was so despite the fact that Estonia was involved in very intense interaction and learning processes with its Western – mainly Nordic – neighbors. The foreign environment strongly influenced the evolution of the Estonian system of innovation, but this did not merely result in convergence.

To understand why this is the case, a systems thinking approach is absolutely necessary. As argued in Chapter 2, impressions from foreign systems of innovation cannot directly 'diffuse' into the domestic system, but have to be 'interpreted' by the system, which can only understand the foreign impressions through its own logic, rationality and history. The reason is that domestic organizations in the system do not carry out their various activities in isolation from each other. Through domestic relations – in terms of collaboration, transactions, competition and relations between organizations and institutions – they are dependent upon a variety of other components. More concretely, they find themselves forced to adapt to an ever-present local arena with which they are constantly interacting and which surrounds them from every possible angle, in terms of, for example, local politics, culture and traditions, as well as existing patterns of interaction with customers, suppliers, competitors, universities, government agencies, etc. In such a situation, organizations, institutions and activities in the system are not free to converge with the environment, but are forced to respond in more or less idiosyncratic ways to the process of system-environment interaction.

However, the above findings also point out the need to be more specific and precise when defining what we actually mean by a 'systems perspective'. Much of the previous use of the systems concept in innovation research has been surprisingly vague in this respect. Several authors have been inspired in their research on systems of innovation by a *comparative* thinking, in which the main interest has been in the structures and styles (although these terms are not necessarily explicitly used) of systems and how they differ among countries or among sectors. Examples of those studies were referred to in section 2. This comparative interest in systems can, in the terminology developed in this study, be associated with systems defined by Luhmann-type boundaries.

Other authors have associated systems of innovation more with the *interactive* and *collaborative* nature of the innovative process, taking as their point of departure the observation that firms do not innovate in isolation. Lundvall and his colleagues, in their seminal work on national systems of innovation, were thus mainly inspired by studies of user-producer relationships and the far-reaching domestic interactive learning processes that, according to these researchers, are much more pronounced than international relations (Lundvall, 1988, 1992; see also Hippel, 1988; Niosi et al., 1993: 102). Networking and the relations between organizations and institutions was also the main emphasis in Carlsson and Stankiewicz's early delineation of 'technological systems', leaving international comparisons as a relatively minor issue (Carlsson and Stankiewicz, 1991).[4] This intuitive understanding of what systems of innovation are about corresponds to systems being defined by Deutsch-type rather than by Luhmann-type boundaries.

The differences between these perspectives should not be exaggerated, and they are not necessarily contradictory. Several authors – notably Edquist (1997b: 15ff.) – have noted that both the comparative and interactive dimension are key in much of the research on systems of innovation. The evidence presented in this book supports the conclusion that *both* aspects of a systems thinking approach have to be taken into account. As a matter of fact, none of the two makes sense without the other. At the same time, however, there is a need to be more explicit about the two sides of the same coin, as they represent quite different aspects of what systems approaches are all about.

The stress on relations and interactions is clearly crucial for a systems thinking approach. However, if 'systems of innovation' are only about relations and interactions, it should be enough to address the concept of systems in terms of networks of organizations, or more generally in terms of organizational and institutional structures. If such concepts as 'networks' or 'structures' are enough to cover the general intent of the argument, the introduction of the term 'system' does not obviously add any value and becomes only confusing. This is so even more because of the more specific uses of the term 'system' in other fields of research, including 'general systems theory', 'social systems', etc. Thus if we use the notion of system of innovation, it should relate to something more than simply a network or a structure.

Those researchers who apply a comparative perspective are, arguably, on the right path towards relating the notion of system of innovation to more than networks and structures. Several studies have demonstrated how systems differ from each other, thus indicating differences in structure and style across countries, regions and sectors (see e.g. Nelson, 1992; Edquist and Lundvall, 1993; Carlsson, 1995b; Breschi and Malerba, 1997; Malerba, 2002, 2004). Such differences mean, by definition, that Luhmann-type boundaries exist and that national, regional and sectoral systems exist in a quite different sense than suggested by the interactive definition of systems. However, as discussed above, it is exactly the intensity and variety of interactive relations within the system that, through processes of circular and cumulative causation, makes it possible for nationally specific structures and styles to emerge and be sustained. Hence, the existence of Luhmann-type boundaries cannot be understood without taking into account Deutsch-type boundaries.

In conclusion, any approach that does not take into account both Luhmann-type boundaries *and* Deutsch-type boundaries should not be labeled a 'systems approach'. Strictly speaking, it is only Luhmann-type boundaries that motivate the use of the concept of 'system', because in the absence of Luhmann-type boundaries, a 'system' is reduced to a network or a structure. At the same time, however, in order to explain the dynamics of systems, as defined by Luhmann-type boundaries, the analysis has to address interactions within the system and between the system and its environment – an issue that is closely related to Deutsch-type boundaries.

4. EVOLUTION AND DISRUPTION: THE UNDERESTIMATED ROLE OF INHERITED COMPETENCIES

In Chapter 2, it was suggested that East European transformation can be interpreted as a problem of how systems of innovation respond to institutional discontinuities. One of the most interesting questions in that process is what role the socialist past plays. Drawing on the evidence presented in this book, Eastern Europe seems to offer an interesting empirical field for developing a better understanding of this problem.

In earlier research on innovation in Central and Eastern Europe, there has been considerable disagreement and confusion with respect to the role that inherited competencies from socialist systems of innovation play in the post-socialist context. Some authors have argued that inherited socialist strengths in science and engineering were largely obsolescent, and that Central and East European firms have therefore been compelled to start from scratch in their accumulation of relevant capabilities. For example, Pavitt argued that 'the competencies in R&D generated under central planning . . . turned out to be obsolete when brought face to face with the standards of the rest of the world' (Pavitt, 1997: 52). Hirschhausen and Bitzer (2000: 13) went even further and suggested that 'the socialist knowledge-base and the socialist technological trajectories did not significantly influence the development of post-socialist enterprises and sectors'. Other authors, however, have convincingly pointed to a number of cases where Soviet-era competencies have been successfully exploited in the post-Soviet period (e.g., Dyker, 1996, 2001; Shaw, 1996).

A major problem with the studies that raise the argument that competencies have become obsolete is that they often deal much more with *production* rather than *innovation*; with the *use* of innovation rather than the *generation* of innovation; and with the *failure* to innovate rather than with the actual *occurrence* of innovation. The clearly dominant interest has been in explaining East European failure to innovate, whereby it has been tempting to associate this with a very low value of the inherited knowledge base. The collapse of inherited production enterprises, the enormous inflow of foreign products, the radical downsizing of R&D, the emergence of non-innovative East European subcontractors, and the extremely low levels of international patenting are all examples of trends that make it tempting to conclude that inherited strengths have little role in the post-socialist context.

However, there is a major methodological problem with this: if one wants to show that inherited competencies do not matter for post-socialist innovation, one would have to focus on cases of successful innovation and show that these cases do not involve any far-reaching exploitation of inherited competencies. Instead of carrying out such investigations, studies arguing that inherited competencies do not matter base their argument on empirical observations of non-successful innovation. Studies that actually focus on the relationship between inherited competencies and creative innovation have been sadly rare, and those that do exist tend to reject the proposition that inherited competencies do not matter for creative innovation (e.g., Dyker, 1996, 2001; Shaw, 1996).

In Chapter 8, it was concluded that innovation in the Estonian telecommunications sector has been influenced by its Soviet heritage in a strong and very evident way. The material presented in this book thus stands in sharp contrast to the idea that inherited socialist competencies have become obsolete in the post-socialist context. Applying the conceptualization of systems of innovation and their transformation as proposed in Chapter 2, this result is hardly surprising. Innovation was there viewed as a systemic and evolutionary process, whereby the effectiveness with which structural tensions can be resolved and styles of innovation be transformed after an institutional discontinuity is likely to depend upon the availability of inherited competencies. Due to limitations in human cognitive capacity, learning tends to be incremental, and it often also depends upon the availability of complex organizational structures, such as universities, R&D institutes and large corporations. From this perspective, it is hardly surprising that knowledge and competencies that are needed for various activities in the system of innovation cannot be accumulated from nothing overnight. Conversely, the removal or change of inherited traditions and patterns of behavior requires explicit and time-consuming efforts. Systems that already possess relevant competencies – embedded in individuals as well as in system components and activities – are therefore likely to have an advantage. This is so at least to the extent that the competencies fit reasonably well with the new requirements.

The evidence in this study leads to the conclusion that the importance of inherited competencies is strongest at the level of *individuals*. While many organizations – in particular production enterprises and applications-oriented R&D organizations – collapsed in Eastern Europe in the years around 1990, a clear pattern in the present study is that individuals who have accumulated far-reaching competencies within Soviet-era organizations, often come to play important roles in the post-socialist system of innovation. A major source of such key individuals appear to be R&D organizations of various kinds – both institutes focused on deep fundamental research and more applications-oriented design bureaus. Even more important, however, is the availability of Soviet-era graduates from technical and other universities; these people show up nearly everywhere where creative innovation occurs. Inherited competencies are in that case not necessarily associated with the inherited R&D system, but rather with the Soviet educational infrastructure.

It is somewhat surprising that individuals have gained so little attention in previous research on innovation in Central and Eastern Europe, as well as in research on systems of innovation more generally. The main units of analysis have usually been firms and other organizations. Judging from the evidence in this study, future research has much to gain from an emphasis not only on organizational but also on personal histories, and not only on inter-organizational relations in systems of innovation, but also on inter-personal relations in such systems. The focus on *activities* as a unit of analysis, which complements the interest in organizations and institutions, appears to be a promising approach in this context, as the investigation of a particular activity does not point *ex ante* to any specific organizational or institutional arrangement, but rather broadens the existing focus to encompass alternative arrangements and explanations, including the role of individuals and their histories.

At the level of organizations, the power of inherited competencies is not as obvious in the post-socialist context as at the individual level. Old production enterprises and applications-oriented R&D organizations, in particular, often collapsed completely as an effect of the break-up of the old socialist systems of innovation, losing, in the process, a great deal of competencies. These lost competencies may in many cases have been obsolete, while in other cases they may have been potentially useful in the post-socialist era. However, inherited production enterprises and applications-oriented R&D organizations are hardly the key types of organizations that matter for creative innovation in the post-socialist context. Instead, it is one other type of organization that seems to be crucial, and which has almost always managed to survive surprisingly well in Eastern Europe: universities.

Although post-socialist universities have suffered from a loss of teachers and researchers, mainly as a consequence of under-financing and deteriorated opportunities within Soviet and East European educational networks, they undoubtedly provide a highly valuable asset for post-socialist systems of innovation. The value of inherited universities becomes particularly obvious when looking at their organizational histories, with the gradual emergence of complex structures in terms of their various departments and curricula. It takes time to establish such structures and build educational strengths, and if they had to be created out of nothing starting from the years around 1990, Central and East European systems of innovation would have faced an immensely more difficult situation. Although the inherited universities came to face extremely important processes of adaptation to the radically altered systems of innovation of which they are part, the departments and curricula that exist today have often been able to successfully exploit their past in the new capitalist context. Judging from the evidence in this book, this is so at least in fields where the underlying principles of science and engineering that were taught under socialism did not differ radically from what was taught in the capitalist world (in contrast to, for example, social sciences and the humanities, where this was much more difficult).

A somewhat different issue is the role of universities in the *creation of new knowledge*. While the role of universities in competence-building and the creation of absorptive capacities can hardly be underestimated, it is not clear to what extent universities have been important in knowledge creation. The evidence presented in this book suggest that universities, although they are active as R&D performers, are often strikingly absent as actors when it comes to concrete innovative projects. Of the innovations discussed, there were hardly any cases in which universities took part as knowledge creators, and creative innovation was thus almost always achieved without the participation of universities.

However, this result may be biased by the specific focus in this book on the telecommunications sector. Estonia was oriented towards innovation in services rather than in equipment, and most concrete innovations in telecommunications services were by the very nature of this sectoral segment rarely science-based. In the manufacturing subsector, which was not the main focus of innovation in Estonian telecommunications, there *was* a trend, albeit slow, towards increasing involvement of universities in the innovative process, in particular related to electronics design. In order to draw any further conclusions here, future research may benefit from a more explicit focus on the dynamics of creative innovation in East European *science-based* sectors.

Returning to the argument about the role of inherited universities for competence-building, this line of reasoning can certainly be generalized to include the entire educational infrastructure, not least with respect to primary and secondary schools. In particular secondary schools are probably an underestimated source of competencies and absorptive capacities in East European systems of innovation. Compared to Western EU countries, Central and East European countries are placed much more favorably with respect to the number of graduates from secondary schools,[5] and it would be highly interesting to examine more specifically what this means in the context of systems of innovation.

Moreover, the strong educational traditions in Eastern Europe, as inherited from the socialist past, can arguably be seen as part of a broader picture, in which one discerns the importance of being part of European culture, with its general interest in science, engineering, education and business. Central and Eastern Europe inherited from their socialist pasts a European tradition of seeing science and technology as integral parts of society at large, and the post-socialist era has added new dimensions to this: nowadays new technologies are not only found all around in everyday life, but they are also under constant scrutiny in the media, giving rise to intense debates and criticisms, etc. Such an omnipresence of science and technology in today's Central and Eastern Europe can hardly be overlooked.

An interesting question is also which inherited competencies stem from the socialist period and which from the period before that. Here, it is important to remember the fact that most Central and East European countries experienced a period of independence between the world wars, in which organizations and individuals learned to act in the environment of Western-style capitalism. The orientation towards the West in that era was particularly strong due to the difficult circumstances in the young Soviet Union during that period. It was also in the interwar period that many of Central and Eastern Europe's most important businesses were founded.

On the other hand, by the outbreak of World War II most Central and East European countries were still largely agricultural economies, while the real drive to industrialize occurred only during the socialist era. In the Estonian telecommunications sector, one notes that only around 2 per cent of the population had a telephone by 1940, while this figure grew ten-fold during the Soviet period. It was also during the Soviet period that the Institute of Cybernetics and most relevant departments at the Technical University were established. In this perspective, it is difficult to argue that the Soviet period was merely a short interregnum in the history of East European systems of innovation. Rather, it was decisive. However, as the present study has focused primarily on developments after World War II and up to today, rather than on the role of the interwar period, further research would be welcome to clarify this interesting issue in further detail.

In summary, even though the far-reaching changes in Central and Eastern Europe in the years around 1990 belong to the most radical political and economic disruptions of our time, their systems of innovation in the post-socialist period continue to show strong traces of their previous history. Understanding this – more or less strong – power of the past, is absolutely key when analyzing processes of systems transformation and reorientation.

5. IMPLICATIONS FOR POLICY

If the importance of system-environment interaction and inherited competencies is recognized, the appropriate way to describe the process of systems change would be to see it as a combination of domestic transformation and international reorientation. As argued above, both these processes are extremely important and have to be seen in relation to each other. From the perspective of public innovation policy, a general point of departure for the process of policy formation should therefore be to influence and shape the complex interplay between transformation and reorientation.

The study has pointed out the crucial importance of interacting with and learning from foreign systems of innovation. Such learning should be encompassed in a broad sense, seeking possibilities to build on all the channels of system-environment interaction that were listed in section 3. However, while recognizing the importance of foreign systems, policy initiatives will also have to be firmly rooted in a deep understanding of the nature and dynamics of the specific domestic system itself. This is because when, for example, new technological opportunities emerge globally, the ways in which policymaking can best respond depends crucially on how the new technology is 'interpreted' by the domestic system, i.e., on how different components and activities can be expected to respond to the new technology and on their ability to adapt to it. Therefore, policy measures cannot simply be copied from experiences in other countries.

An understanding of this is crucial not least when it comes to the formulation of policy priorities. For example, a system with weakly developed traditions of university-industry interaction may require enormous efforts with respect to strengthening such interaction in order to make the system respond in a fruitful way to a new science-based technology. In another system, however, with a different style, a bigger problem may be to make legislation fit with the new technology. In general, it would be dangerous to define innovation policy in terms of more narrow policy fields – R&D policy, education policy, competition policy, etc. – because innovation, as indicated by the results of this study, is something much broader, covering a wide array of activities.

Turning to the more specific context of this book, an interesting issue is how the exploitation of increased technological opportunities may be supported in periods of radical political change. In the Estonian telecommunications sector, at least two important technological discontinuities – the emergence of the Internet and the take-off of mobile telecommunications – coincided with Estonia's political and economic transition period from socialism to capitalism. It appears that such a system of innovation, i.e., a system that is simultaneously subject to both institutional and technological discontinuities, may, somewhat paradoxically, have an advantage over institutionally more stable systems. The technological landscapes of more smoothly developing countries are more likely to be characterized by the strong presence of well-established firms that seek to protect older technological paradigms as well as organizational and institutional orders from the 'invasion of alien technologies' (e.g. Utterback, 1994). In countries undergoing far-reaching political and economic changes, in contrast, such well-established firms are usually much weaker and are often totally swept away by systems transformation. In short, the structure and style of the system – in terms of organizations, institutions, relations and activities – are in a stage of high turbulence in connection to the collapse of the old institutional regime (such as the old Soviet institutions). Such a system may therefore be less likely to resist the pushes for new arrangements, and thus have greater chances to quickly adapt to the new technological paradigms and trajectories.

This is perhaps one of the most fascinating aspects of the relationship between political disruptions and technological change, and it has been pointed out previously in research on catch-up in technology more generally (see in particular Perez and Soete, 1988: 462f.). However, the actual circumstances under which politically disruptive systems of innovation can actually use this situation to their advantage are far from obvious. The evidence presented in this book suggests that the issue is closely related to the question of inherited competencies. Clearly, the collapse of an existing system and the removal of problematic momentum does not in itself lead to higher-level innovative outcomes. Instead, it is necessary that the system controls substantial competencies in the relevant field for it to be able to take advantage of the fluid situation. From a policy perspective, this means that in times of radical institutional change, a system of innovation may profit from deliberately seeking to exploit technological discontinuities in fields which fit its existing competencies. If the system has competencies that fit with the emerging technology, it may have an advantage over systems in other countries with similar-level competencies but that lack the more favorable, institutionally fluid situation. Much of Estonia's innovative success built, as we have seen, exactly upon such exploitation of inherited competencies in technologically turbulent fields.[6]

At the same time, policy initiatives have to link the exploitation of its already existing strengths to the build-up of new (and destruction of old) linkages to foreign systems of innovation. In the absence of such linking and delinking, the transformation of the domestic system, and in particular the exploitation of inherited competencies, is likely to be without result. Innovation policy should under such circumstances, therefore, devote considerable effort to system-environment relations. But the opposite relation also holds: Efficient system-environment relations cannot *compensate* for weak domestic dynamics. It is only through far-reaching domestic processes that system-environment interaction can make an influence on the domestic structure and style and therefore on innovation. National systems in Eastern Europe would therefore face an enormous loss, from the perspective of innovation, if the domestic R&D system totally disappeared, or if innovating firms lost their domestic links and became too integrated with their foreign owners.

An important issue is also with *which* foreign countries Central and Eastern Europe should build system-environment links. The present study has pointed to the good prospects for success when links are built to geographically and culturally close countries that at the same time belong to the world's most advanced systems. Although the book has dealt with only one sector, it seems plausible that different countries may constitute the ideal partners depending on the sector with which we are dealing. For example, although Sweden and Finland belong to the world leaders in telecommunications, this is not necessarily the case in other sectors, in which they might therefore be less suitable partners. Obviously, East European countries face a much more difficult situation when the geographically and culturally closest neighbors do not belong to the more successful systems of innovation.

A more general policy issue is to what extent Central and East European systems of innovation should attempt to imitate Western systems, or rather deliberately seek to build their innovative strengths on system trajectories that differ, more or less strongly, from their Western neighbors. The present study shows that this is a largely hypothetical issue, as the very nature and dynamics of systems of innovation make it impossible to imitate another system. This is in line with, for example, Nelson's observation that although national systems have often tried to 'copy' various features of more advanced systems, such attempts have almost always had very different outcomes from those that were expected (Nelson, 1992: 387).[7] Again, systems may take inspiration and learn from foreign systems, but even if far-reaching imitation was possible, it appears that a system of innovation has more to gain from taking into account its own historical specificities and inherited strengths, rather than spending enormous efforts on imitating foreign systems in a 'blind' way. However, the extent to which such more country-specific policies may actually be developed, depends on the availability of necessary policymaking competencies in the system.

Drawing on the more specific problems experienced in Estonia, there are also several more concrete issues that Central and East European countries have to tackle. The balance between innovation policy and other policy fields is one issue. In electronics, as pointed out in section 1 of Chapter 8, Estonia seems for a long time to have been oriented much more towards strengthening production capacities while largely neglecting the prospects regarding the build-up of innovative strength. Policies were aimed at attracting foreign investors, but do not appear to have been linked to any strategy for accumulating capabilities in the generation of new products and processes. To some extent, this was a result of mere necessity, since the financial means by which the government could have backed more innovation-oriented policies were very meager. But at the same time, Estonia possessed advanced competencies that could clearly have been used to a greater extent, by integrating them much more firmly into the policy process. Far-reaching ideas based on advanced knowledge were raised in the years around 1990 by, for example, the academic community, but were only to a minor extent allowed to influence actual government policies.

That it actually *is* possible for East European countries to pursue advanced innovation policy is shown in the Estonian case by the ability to create a variety of new institutions and organizations in response to the technological challenges of the mobile telecom and Internet services era. The Estonian policy developments with respect to, for example, the digital signature act, or more generally in the field of e-government, were not mere imitations of Western policies. Rather, they usually built upon critical evaluations of Western experiences by largely using domestic competencies that had accumulated for decades, and this made it possible to pursue somewhat alternative and interestingly creative paths that fitted better with the local economy and history. This does not mean that the Estonian developments were in any way ideal. But it did demonstrate that it is actually possible for a small post-socialist country to formulate and implement advanced innovation-related policies even under severe financial constraints.

A more fundamental policy-related problem, which goes back to the Soviet era, was the somewhat artificial separation between, on the one hand, science and technology, and on the other, economic and industrial development. As shown in this study, the relations between universities, academies and industry in the Estonian telecommunications sector were not always weak, but it seems that they could have contributed much more to actual processes of innovation. At the same time, the relations between technology and economy were troublesome *within* industry. A linear model of innovation seems to have been persistent in many Estonian firms, which were surprised and disappointed when their often excellent technological solutions found it hard to conquer markets. As discussed in section 2 of Chapter 8, this was especially a problem in relation to exports. This points to an obvious need for public policies to focus more on the commercial and market-oriented aspects of science and technology in Central and Eastern Europe. However, this is certainly a difficult task, as the post-socialist countries hardly possess any inherited competencies in the commercialization of science and technology.

6. INNOVATION AND EAST-WEST INTEGRATION: TOWARDS A NEW EUROPE

In Chapter 1 it was stated as a background to the study that it is not at all clear what roles the former socialist economies in Central and Eastern Europe will play as members of – and neighbors to – the European Union. What, then, does the making of the Estonian telecommunications system of innovation tell us about the future prospects in this respect?

First of all, the material presented in this book confirms the important point that East-West integration must not be thought of primarily as a 'transfer' of Western technologies, practices and structures to the East. Instead, the case of Estonian telecommunications points to the encouraging opportunity – or necessity – for Central and East European countries to use inspirations from and interactions with culturally and geographically close neighbors in the West to boost domestic innovation in accordance with local conditions and inherited competencies, i.e., by taking into account the logic of the domestic system itself. However, this conclusion leads to radically different scenarios. Let me here briefly discuss two extremes:

First, Eastern systems of innovation may very well choose to continue building on the availability of low-wage labor, using largely non-innovative activities in order to boost economic growth and exports. Eastern Europe would thus continue to be a conveniently located reservoir of relatively simple manufacturing services for large Western multinationals. This corresponds to the present directions set out by many Eastern countries, with close inter-organizational East-West integration in terms of user-producer networks, but reinforcing a non-creative style of innovation that considerably differs from the Western EU systems of innovation.

When wages rise, this style risks being eroded through relocation of production to other low-wage countries, and this is likely to result in increasing unemployment and dramatic declines in exports. The only thing which could rescue this development path would be to imitate the East Asian experience of climbing the ladder towards increasing involvement in innovation, as exemplified by Elcoteq in the Estonian case.

But there is considerable uncertainty with respect to the viability of such a strategy in the longer term: when the East Asian 'tiger' economies achieved their remarkable success, the world looked different. Their success built on exploiting technological discontinuities with a potential that may be of a different character today as compared to earlier. Much tougher global competition and vastly increasing technological complexity in electronics-related sectors also reduce the prospects for small East European countries to become world leaders on their own, without integrating closely into global networks. Even enormous political efforts of the kind that were commonplace in East Asia can today hardly compensate for this. Under these circumstances, it is difficult to know to what extent it will be possible for East European producers to move towards 'own-brand manufacture' (OBM) and frontier R&D.

A radically different scenario is one in which the Eastern EU gradually leaves its traditional style of economy behind in a politically conscious way. Eastern Europe would carefully use its geographical and cultural proximity to advanced West European countries in combination with its long tradition of higher education and other potential advantages in brave and clever ways, backed by a broad national belief in the actual possibility to proudly fight its way into the global economy. Eastern Europe becomes a region famous for its courageous new developments, which often surpass Western countries, the latter eagerly following the first-movers in the East. This is made possible through the still much more loosely defined institutional landscapes in the Eastern countries, their lack of path-dependence-inducing inherited technical systems and their far-reaching experience of managing radical systems change.

The Eastern countries become the ones that show the West how institutional bottlenecks can be resolved. They become the sites where new technologies of powerful Western firms are tried out in highly dynamic and rapidly evolving post-socialist markets (which lack the conservatism and inertias that become more and more characteristic of Western Europe). Eastern Europe, in this scenario, follows a completely different path of development compared to countries such as China, as they understand how to use their small size to move rapidly and cleverly and to retain a continuous dynamism, perhaps with a specialization in service sectors.

Global competition also encourages the formation of East-West joint corporations and conglomerates – such as TeliaSonera and the Swedish-based banks – which gradually learn to use the continuous flows of knowledge between East and West and their still vastly differing systems of innovation in order to boost the whole process of innovative development. The very gap between East and West thus becomes an engine of progress. While it remains difficult to create 'Estonian Nokias' in the small Eastern countries, many large Western brands soon become closely associated with the East European dynamism. Western EU support helps Eastern countries revive their advanced but heavily under-financed educational sectors, and the huge unemployment in countries such as Poland is reduced in connection to the appearance of new technological trajectories, with the formation of new industries that pull the Eastern countries fearlessly into the gales of creative destruction. While Western EU countries tend to stagnate in their conservatism and overly careful political arenas, Eastern Europe, where a younger generation already rules in politics as well as business, moves ahead and becomes the new symbol of European progress.

How unrealistic is the latter scenario? One might argue that the most unrealistic aspect is the envisaged continuing division line between an 'Eastern' and a 'Western' half of Europe. The general trend, after all, is that the differences in development levels *among* Western and *among* Eastern countries are becoming more pronounced than the differences between 'East' and 'West'. But again, when it comes to 'style' rather than 'level', the Eastern countries will for decades be heavily influenced by their socialist histories. Turning these difficult and traumatic pasts into new and flourishing systems of innovation continues to be an extremely difficult task.

To judge from Estonia, success has followed when the key process of building linkages to geographically and culturally close countries has been combined with well-focused interactive learning between a wide range of actors – foreign and domestic, public and private – while both collaboration and competition have been allowed to function as driving forces. What lies behind this, however, is not merely the establishment of formal institutions; rather, the ability to uphold smoothly functioning relationships between components in a system is heavily dependent upon a genuine will to change and a belief in the power of their nations' capacities. The case of Estonia also points, again, to the key role of the education system, a circumstance that is illustrated by the fact that educational activities have become both a crucial enabler of and an obstacle to further progress.

Above all, the need for a *holistic* approach – both to analyzing innovation and to elaborating innovation policy – cannot be exaggerated. The various activities discussed in this study cover a very broad array of fields, and they all turn out to be necessary to incorporate if one wants to understand and explain the forces that enable the emergence of creative styles of innovation. Separations between seemingly unrelated areas such as foreign direct investment, telecommunications regulation and public education policy are totally artificial from the perspective developed in this book. If technologies and sciences are getting increasingly interdisciplinary, then there is probably an even greater need for public policies to merge and intermingle and be seen as a whole.

*

It is fully imaginable that, as levels of development cease to point to a division between East and West, a wider Nordic world – including the Baltic countries and possibly further areas – emerges as a powerful economic bloc within a new Europe. There is no longer an Iron Curtain, dividing up this part of the globe, with the power to keep the Eastern and Western shores of the Baltic Sea apart. On the other hand, while it remains unclear how the new border between the Baltic countries and Russia – now identical to the EU border – will influence the Baltic Sea region, it is undeniable that the great tensions of the world, the real conflicts and fights for the future of the planet, now take place somewhere else. Once upon a time, the Baltic Sea region was the scene for radical movements of historical importance: it hosted the introduction of socialism as a completely new societal system, and it was on the shores of the Baltic that World War II broke out. In the cold war era, the region was in many people's views a major potential battlefield in a future nuclear war. Later, it was here that the East European revolutions of 1989 were initiated.

All that has changed now. Even EU enlargement and NATO expansion are processes that in a worldwide perspective fall completely in the shadow of much more crucial global events: the terrorist attacks in the United States in 2001, the Iraq war in 2003, the enormous economic development in Asia, etc. It is getting increasingly difficult to see Europe as the center of the world, and Northern Europe has in the 21st century perhaps – at last – become a true outskirt of the world, a periphery of civilization. In a sense, this could be seen as a largely positive development: the Baltic Sea region as a peaceful place, where it is possible to live and work without great disruptions, where there is plenty of time to concentrate on more essential things, to build a life that rests firmly on a sense of security, stability and careful progress through harmonic, daily work.

But this is in reality only a serene surface that conceals relentless, streaming waters. Northern Europe is more than ever drawn into the gales of global competition and complex technological revolutions, we are more interdependent than ever with other parts of the world, more sensitive than ever to developments elsewhere and to the world as a whole. And capitalism is by its very nature a process of change, of disruption, uncertainty and destruction, a process of 'constant revolution', as Marx was so clear to point out. In our own age, this is closely related to enormous tensions between the economic and other spheres of life that are getting increasingly troublesome. If these tensions are not to transform the visions of sustainable social progress and welfare, built on knowledge- and innovation-based economic development, into empty and merely cynical catch-phrases, then they first of all demand a thorough – and broad – understanding of the actual mechanisms of change and development, in order to respond creatively to the present threats and opportunities. This book has taken a small step in that direction.

NOTES

1. The countries included in the study were the United States, Japan, Germany, the United Kingdom, France, Italy, Denmark, Sweden, Canada, Australia, South Korea, Taiwan, Brazil, Argentina and Israel.
2. They focused mainly on OECD countries and on comparisons between Japan, the United States and Western Europe.
3. The focus is here on developing a comparative framework for ten small national systems of innovation. The participating countries are Denmark, Finland, Hong Kong, Ireland, the Netherlands, Norway, Singapore, South Korea, Sweden, South Korea and Taiwan.
4. See also Carlsson (1995a, 1997), where detailed studies are carried out on the dynamics of various technological systems specifically in Sweden, although a concluding chapter relates the Swedish system to systems in other countries. See also Carlsson et al. (2002), whose understanding of what systems of innovation are actually about builds on similar ideas.
5. Thus the share of adult population with at least secondary education was 64.6 per cent for the Western part of the EU, while the corresponding figure for Estonia was 87.5 per cent, for Latvia 82.6 per cent, for Lithuania 84.8 per cent, for Poland 80.8 per cent, for Slovakia 85.8 per cent, for Slovenia 76.8 per cent, for the Czech Republic 87.8 per cent and for Hungary 71.4 per cent. See, e.g., *Ny Teknik*, 28 April 2004.
6. Note that this argument goes beyond the debate about whether post-socialist countries should focus on high- or low-tech fields in their build-up of post-socialist systems of innovation. There is no perfect strategy here; rather, every country has to evaluate its own inherited competencies in order to find out what fields or sectors are going to be prioritized.
7. For example, Japan tried to 'pick and choose from European and American experience, and came out with something quite different'. Similarly, the United States tried to 'adopt the German university system, and actually built a very different one'.

Appendix A: Interviewed Organizations, by Sectoral Segment

Institutional and policy developments:

- Ministry of Communications
- Ministry of Foreign Affairs
- Department of State Information Systems (RISO)
- Estonian Informatics Center
- Estonian Informatics Council
- Prime Minister's ICT advisor
- Institute of Cybernetics

Fixed voice telephony:

- Ministry of Communications/Eesti Telefon
- Uninet
- Tele2
- Telefant
- EsData

Mobile telecommunications:

- Eesti Mobiiltelefon (EMT)
- Radiolinja Eesti
- Tele2/Ritabell
- Estonian Rescue Board
- Mobi
- Oskando
- Regio
- Voicecom
- Ericsson Eesti
- ADM Communications
- Aprote
- Delfi

- Abobase Systems
- ProEkspert

Data communications:

- Uninet
- Eesti Telefon
- Tele2/Levicom
- MicroLink
- EsData
- Privador
- Institute of Cybernetics/Cybernetica
- Estonian Tax Board
- Delfi
- Ühispank
- Hansapank
- Crebit
- Abobase Systems
- Department of State Information Systems (RISO)
- Estonian Informatics Center
- Assert/Cellnetwork
- SEB IT Partner Estonia
- Absolutions Eesti
- ProEkspert

Equipment production and retail:

- Elcoteq
- Tarkon
- Keila Kaabel
- Oskando
- Elvior
- Reveko
- Ericsson Eesti

Education and research:

- Institute of Cybernetics/Cybernetica
- Department of Radio and Communications Engineering at TTÜ
- Department of Electronics at TTÜ
- Tartu Technological Park
- IT College

- Tiger Leap Foundation
- Baltic Computer Systems

References

Printed material

Adirim I. (1991), 'Current development and dissemination of computer technology in the Soviet economy', *Soviet Studies*, **43** (4), 651-67.

Agur, U. (1993), 'Kaks aastat informaatikafondi', *Arvutimaailm*, **1** (1).

Agur, U. (1996a), 'Public procurement of information technology in Estonia', *Baltic IT Review*, **1** (1), 41-42.

Agur, U. (1996b), 'Estonia on the road toward the information society: an overview', *Baltic IT Review*, **1** (2), 6-9.

Amann, R. and Cooper, J. (eds) (1982), *Industrial Innovation in the Soviet Union*, New Haven: Yale University Press.

Amann, R. and Cooper, J. (eds) (1986), *Technical Progress and Soviet Economic Development*, Oxford: Basil Blackwell.

Anderson, P. and Tushman, M. (1990), 'Technological discontinuities and dominant designs: a cyclical model of technical change', *Administrative Science Quarterly*, **35**, 604-33.

Balzat, M. and Hanusch, H. (2004), 'Recent trends in the research on national innovation systems', *Journal of Evolutionary Economics*, **14**, 197-210.

Bell, M. (1997), 'Technology Transfer to Transition Countries: Are there lessons from the experience of the post-war industrializing countries?', in Dyker, D. (ed.), *The Technology of Transition. Science and Technology Policies for Transition Countries*, Budapest: Central European University Press, pp. 63-93.

Bell, M. and Pavitt, K. (1993), 'Technological Accumulation and Industrial Growth: Contrasts Between Developed and Developing Countries', *Industrial and Corporate Change*, **2** (2), 157-210.

Berg-Andersson, B. (1998), 'Den estniska industrins framtidsutsikter', *Ekonomiska samfundets tidskrift*, (1), 15-26.

Bergek, A. (2002), *Shaping and Exploiting Technological Opportunities: The Case of Renewable Energy Technology in Sweden*, Gothenburg: Chalmers University of Technology (PhD dissertation).

Berner, B. (1999), *Perpetuum mobile? Teknikens utmaningar och historiens gang*, Lund: Arkiv förlag.

Bijker, W., Hughes, T. and Pinch, T. (eds) (1987), *The Social Construction of Technological Systems: New Directions in the Sociology and History of Technology*, Cambridge, MA: MIT Press.

Bitzer, J. (2000), 'The Eastern European Computer Industry: National Champions with a Screwdriver', in Hirschhausen, C. von and Bitzer, J. (eds), *The Globalization of Industry and Innovation in Eastern Europe. From Post-Socialist Restructuring to International Competitiveness*, Cheltenham, UK and Northampton, MA, USA: Edward Elgar, pp. 257-281.

Breschi, S. (1995), *Spatial patterns of innovation. Evidence from patent data*, CESPRI Working Paper no. 80, Milan: CESPRI.

Breschi, S. and Malerba, F. (1997), 'Sectoral Innovation Systems: Technological Regimes, Schumpeterian Dynamics and Spatial Boundaries', in Edquist, C. (ed.), *Systems of Innovation. Technologies, Institutions and Organizations*, London: Pinter, pp. 130-156.

Bresnahan, T., Gambardella, A. and Saxenian, A. (2001), 'Old economy inputs for new economy outcomes: cluster formation in the new silicon valleys', *Industrial and Corporate Change*, **10** (4), 835-860.

Campbell, R. (1995), *Soviet and Post-Soviet Telecommunications*, Boulder: Westview Press.

Caracostas, P. and Soete, L. (1997), 'The Building of Cross-Border Institutions in Europe: Toward a European System of Innovation?', in Edquist, C. (ed.), *Systems of Innovation: Technologies, Institutions, and Organizations*, London: Pinter, pp. 395-419.

Carlsson, B. (ed.) (1995a), *Technological Systems and Economic Performance. The Case of Factory Automation*, Dordrecht, Boston and London: Kluwer.

Carlsson, B. (1995b), 'The Technological System for Factory Automation: An International Comparison', in Carlsson, B. (ed.), *Technological Systems and Economic Performance. The Case of Factory Automation*, Dordrecht, Boston and London: Kluwer, pp. 441-75.

Carlsson, B. (ed.) (1997), *Technological Systems and Industrial Dynamics*, Boston, Dordrecht and London: Kluwer.

Carlsson, B. and Stankiewicz, R. (1991), 'On the nature, function and composition of technological systems', *Journal of Evolutionary Economics*, **1** (2), 93-118.

Carlsson, B., Jacobsson, S., Holmén, M. and Rickne, A. (2002), 'Innovation systems: analytical and methodological issues', *Research Policy*, **31**, 233-245.

Casper, S. and Glimstedt, H. (2001), 'Economic organization, innovation systems, and the internet', *Oxford Review of Economic Policy*, **17** (2), 265-281.

Certification Center (2001), *EMT, Hansapank, Union Bank of Estonia and Eesti Telefon founded certification joint venture* (press-release, 16 February).

Certification Center (2003), *Finland and Estonia start harmonizing digital signature and document practices* (press-release, 4 June).

CIESIN (2003), *Internet in Estonia*, Tallinn: CIESIN.

Cooper, J. (1982), 'Innovation for Innovation in Soviet Industry', in Amann, R. and Cooper, J. (eds), *Industrial Innovation in the Soviet Union*, New Haven: Yale University Press, pp. 453-512.

Corrocher, N. (2003), 'The Internet Services Industry. Sectoral dynamics of innovation and production', in Edquist, C. (ed.), *The Internet and Mobile Telephony Sectoral System of Innovation: Equipment Production, Access Provision and Content Provision*, Cheltenham, UK and Northampton, MA, USA: Edward Elgar.

Cunningham, G. (1998), 'Telecommunications regulation in Estonia', *Computer and Telecommunications Law Review*, **4** (7), 234-37.

Dahmén, E. (1988), 'Development blocks in industrial economics', *Scandinavian Economic History Review*, **36** (1), 13-14.

David, P. (1985), 'Clio and the economics of QWERTY', *American Economic Review*, **75** (2), 332-37.

Department of Radio and Communications Engineering (2001), *Täna avati Tallinna Tehnikaülikoolis mobiilrakenduste arenduskeskus* (press-release, 7 May).

Deutsch, K. (1949), 'Innovation, Entrepreneurship and the Learning Process', in *Change and the Entrepreneur. Postulates and Patterns for Entrepreneurial History.* Prepared by the Research Center in Entrepreneurial History, Cambridge, MA: Harvard University Press.

Deutsch, K. (1963), *The Nerves of Government. Models of Political Communication and Control*, New York: Free Press.

Digitaalallkirja seadus (2000), *Digitaalallkirja seadus, vastu võetud 8. märtsil 2000. a.* [Digital Signatures Act, adopted on 8 March 2000]

Dosi, G. (1988), 'The Nature of the Innovative Process', in Dosi, G., Freeman, C., Nelson, R., Silverberg, G. and Soete, L. (eds), *Technical Change and Economic Theory*, London: Pinter, pp. 221-38.

Dyker, D. (1996), 'The computer and software industries in the East European economies – a bridgehead to the global economy?', *Europe-Asia Studies*, **48** (6), 915-30.

Dyker, D. (2001), 'Technology exchange and the foreign business sector in Russia', *Research Policy*, **30**, 851-68.

Dyker, D. and Radosevic, S. (1999), *Building the Knowledge-Based Economy in Countries in Transition – From Concepts to Policies*, SPRU Electronic Working Paper no. 36, Brighton: University of Sussex.

Edquist, C. (ed.) (1997a), *Systems of Innovation: Technologies, Institutions, and Organizations*, London: Pinter.

Edquist, C. (1997b), 'Systems of Innovation Approaches – Their Emergence and Characteristics', in Edquist, C. (ed.), *Systems of Innovation: Technologies, Institutions, and Organizations*, London: Pinter, pp. 1-35.

Edquist, C. (1999), *Forms of learning and categories of innovation.* Lecture held in San Sebastian, Spain, 1 October.

Edquist, C. (ed.) (2003), *The Internet and Mobile Telephony Sectoral System of Innovation: Equipment Production, Access Provision and Content Provision*, Cheltenham, UK and Northampton, MA, USA: Edward Elgar.

Edquist, C. (2004), 'Systems of Innovation – Perspectives and Challenges', in Fagerberg, J., Mowery, D. and Nelson, R. (eds), *The Oxford Handbook of Innovation*, Oxford University Press.

Edquist, C. and Hommen, L. (2004), *Comparative Framework for, and Proposed Structure of, the National Innovation Systems in Ten Small Countries*, working paper, Lund: Division of Innovation, Lund Institute of Technology.

Edquist, C. and Johnson, B. (1997), 'Institutions and Organizations in Systems of Innovation', in Edquist, C. (ed.), *Systems of Innovation: Technologies, Institutions, and Organizations*, London: Pinter, pp. 41-63.

Edquist, C. and Lundvall, B.-Å. (1993), 'Comparing the Danish and Swedish Systems of Innovation', in Nelson, R. (ed.), *National Systems of Innovation – A Comparative Analysis*, Oxford: Oxford University Press.

Edquist, C., Hommen, L. and McKelvey, M. (2001), *Innovation and Employment. Process versus Product Innovation*, Cheltenham, UK and Northampton, MA, USA: Edward Elgar.

Edquist, C., Hommen, L. and Tsipouri, L. (eds) (2000), *Public Technology Procurement and Innovation*, Boston, Dordrecht and London: Kluwer.

Eesti Telefon (1994), *Annual Report 1994*, Tallinn: Eesti Telefon.

Eesti Telefon (1995), *Annual Report 1995*, Tallinn: Eesti Telefon.

Eesti Telefon (1996a), *Olulised hetked ja arvud AS Eesti Telefonis 1995. aastal* (press-release, 25 April).

Eesti Telefon (1996b), *Mõksi RAS 1000 sai eluõiguse* (press-release, 13 May).

Eesti Telefon (1996c), *Esimesed RASi abonendid hakkavad tööle sel nädalal* (press-release, 20 May).

Eesti Telefon (1997a), *Annual Report 1997*, Tallinn: Eesti Telefon.

Eesti Telefon (1997b), *Eesti side uus baasvõrk on valmis* (press-release, 23 December).

Eesti Telefon (1997c), *Eesti Telefon teeb statistilisi kokkuvõtteid* (press-release, 14 March).

Eesti Telefon (1997d), *Interneti sissehelisatmiskeskused jõuavad maakonnakeskustesse* (press-release, 30 January).

Eesti Telefon (1998a), *Annual Report 1998*, Tallinn: Eesti Telefon.

Eesti Telefon (1998b), *Eesti Telefon avas Tallinna kesklinnas uue digitaaljaama* (press-release, 24 March).

Eesti Telefon (1998c), *Eesti Telefon ja Ericsson sõlmisid 300 miljoni kroonise sidelepingu* (press-release, 29 January).

Eesti Telefon (1999a), *Annual Report 1999*, Tallinn: Eesti Telefon.

Eesti Telefon (1999b), *Eesti Telefon tõi esimesena Eestis turule interneti-telefoni* (press-release, 4 November).

Eesti Telefon (2000a), *Annual Report 2000*, Tallinn: Eesti Telefon.

Eesti Telefon (2000b), *Uue ajastu teenuste pakkumiseks luuakse avaliku võtme infrastruktuur* (press-release, 15 March).

Eesti Telefon (2001a), *Annual Report 2001*, Tallinn: Eesti Telefon.

Eesti Telefon (2001b), *Estonian Telephone Company and Wicom Communications start to provide corporate IP telephony solutions* (press-release, 30 January).

Eesti Telefon (2001c), *Eesti Telefon and Tallink to launch pilot project of IP-based call handling system* (press-release, 4 June).

Eesti Telefon (2002), *Annual Report 2002*, Tallinn: Eesti Telefon.

Ehandi, A. (2001), 'The "Look@World" Project: an initiative from Estonia's private sector to boost Internet use', *Baltic IT&T Review*, **6**, (21), 36-39.

Elcoteq (1996), *Presidents Lennart Meri and Martti Ahtisaari to lay the foundation stone of the factory expansion of AS Elcoteq Tallinn* (press-release, 15 June).

Elcoteq (1997a), *Elcoteq's Tallinn factory to produce new GSM phones* (press-release, 17 June).

Elcoteq (1997b), *Estonian Foreign Investor '96 Award goes to Elcoteq* (press-release, 13 March).

Elcoteq (1999), *No basis for news about Elcoteq's investments in new plant in Russia* (press-release, 19 November).

Elcoteq (2000), *Elcoteq to start up new plant in Tallinn making telecommunication network products* (press-release, 31 July).

Elcoteq (2001), *Elcoteq to strengthen customer relationship and new product introduction services in Europe* (press-release, 11 July).

Elcoteq (2002a), *Elcoteq and Benefon into a deal to jump-start Elcoteq Design Center* (press-release, 1 July).

Elcoteq (2002b), *Elcoteq reduces personnel in Estonia* (press-release, 15 April).

Elcoteq (2002c), *Elcoteq to manufacture combined GSM/GPRS and WLAN terminals for Wireless House* (press-release, 22 May).

Elcoteq (2002d), *Elcoteq Tallinn wins the first Estonian Quality Award* (press-release, 15 November).

EMT (1993), *Annual Report 1993*, Tallinn: EMT.

EMT (1995), *Annual Report 1995*, Tallinn: EMT.

EMT (1996), *Annual Report 1996*, Tallinn: EMT.

EMT (1999), *Annual Report 1999*, Tallinn: EMT.

EMT (2000), 'EMT joins Mgine as strategic partner' (press-release, 27 October).
EMT (2001a), *EMT siis ja praegu*, Tallinn: EMT.
EMT (2001b), 'AS EMT ostis osaluse Voicecom OÜ-s' (press-release, 23 January).
EMT (2003a), 'The first 3G call in Estonia was made in the EMT trial network' (press-release, 17 September).
EMT (2003b), 'GPRS in LMT and Omnitel networks for EMT customers at the domestic rate' (press-release, 29 October).
Estades, J. and Ramani, S. (1998), 'Technological competence and the influence of networks: a comparative analysis of new biotechnology firms in France and Britain', *Technology Analysis & Strategic Management*, **10** (4), 483-95.
Estonian Tax Board (2000), *Annual Report 2000*, Tallinn: Estonian Tax Board.
Farkas, P. (1997), *The Effect of Foreign Direct Investment on Research, Development and Innovation in Hungary*, Working Paper no. 81, Institute of World Economics of the Hungarian Academy of Sciences, Budapest.
Freeman, C. (1987), *Technology Policy and Economic Performance: Lessons from Japan*, London: Pinter.
Freeman, C. (1992), 'Networks of Innovators: A Synthesis of Research Issues', in *The Economics of Hope. Essays on technical change, economic growth and the environment*, London: Pinter, pp. 93-120.
Freeman, C. (1994), 'Technological Revolutions and Catching-Up: ICT and the NICs', in Fagerberg, J., Verspagen, B. and Tunzelmann, N. von (eds), *The Dynamics of Technology, Trade and Growth*, Aldershot, UK and Brookfield, US: Edward Elgar, pp. 198-221.
Freeman, C. (2002), 'Continental, national and sub-national innovation systems – complementarity and economic growth', *Research Policy*, **31**, 191-211.
Freeman, C. and Soete, L. (1997), *The Economics of Industrial Innovation*, Third Edition, London: Pinter.
Gavroglu, K. (1994), 'Types of Discourse and the Reading of the History of the Physical Sciences', in Gavroglu, K., Christianidis, J. and Nicolaidis, E. (eds), *Trends in the Historiography of Science*, Dordrecht: Kluwer, pp. 65-86.
Geenhuizen, M. van (2001), 'Which Role for Foreign Direct Investment? Active Space Development in Central and Eastern Europe', in Geenhuizen, M. van and Ratti, R. (eds), *Gaining Advantage from Open Borders. An active space approach to regional development*, Aldershot: Ashgate.
GITR (2002), *The Global Information Technology Report 2001-2002*, New York and Oxford: Oxford University Press.
Hacking, I. (1994), 'Styles of Scientific Thinking or Reasoning: A New Analytical Tool for Historians and Philosophers of the Sciences', in Gavroglu, K., Christianidis, J. and Nicolaidis, E. (eds), *Trends in the Historiography of Science*, Dordrecht: Kluwer, pp. 31-48.
Hansapank (1999), *As of 1 January 2000, Tõnis Sildmäe, IT Director and Board Member of Hansapank, will step down* (press-release, 6 October).
Hanson, P. and Pavitt, K. (1987), *The Comparative Economics of Research Development and Innovation in East and West: A Survey*, Chur: Harwood Academic Publishers.
Hendre, E. (2003), *The Early History of the Legendary Mini-Camera Minox in Estonia*, Paper presented at the ICOHTEC Conference, St. Petersburg, August.
Hippel, E. von (1988), *The Sources of Innovation*, New York: Oxford University Press.

Hirschhausen, C. von and Bitzer, J. (eds) (2000), *The Globalization of Industry and Innovation in Eastern Europe. From Post-Socialist Restructuring to International Competitiveness*, Cheltenham, UK and Northampton, MA, USA: Edward Elgar.

Hobday, M. (1995), *Innovation in East Asia: The Challenge to Japan*, Alershot, UK and Brookfield, US: Edward Elgar.

Högselius, P. (2001), *East-West Interfirm Collaboration in Telecommunications and Electricity. A Comparative Study across Sectors and Countries, 1980-2000*, Working Paper No. 228, Linköping: Department of Technology and Social Change.

Hollingsworth, R. (2002), *Some Reflections on How Institutions Influence Styles of Innovation*, Paper presented to the Swedish Collegium for Advanded Study in the Social Sciences (SCASSS), 26 September.

Hond, F. (1998), 'The "similarity" and "heterogeneity" theses in studying innovation: evidence from the end-of-life vehicle case', *Technology Analysis & Strategic Management*, **10** (4), 529-43.

Hughes, T. (1983), *Networks of Power. Electrification in Western Society*, Baltimore: John Hopkins University Press.

Hughes, T. (1987), 'The Evolution of Large Technological Systems', in Bijker, W., Hughes, T. and Pinch, T. (eds) (1987), *The Social Construction of Technological Systems: New Directions in the Sociology and History of Technology*, Cambridge, MA: MIT Press, pp. 52-82.

Hultén, S. and Mölleryd, B. (1997), 'Estonia, Latvia and Lithuania: Balancing the demands for short term profitability and network expansion', in Schenk, K.-E., Kruse, J. and Müller, J. (eds), *Telecommunications Take-Off in Transition Countries*, Aldershot: Avebury.

Ideon, A. (2001), *Kapitali sünd*, Tallinn: Hansapank.

Institute of Cybernetics (1987a), *Personal computers and their applications: abstracts of the papers presented at Soviet-Finnish international symposium on mini-, micro- and personal computers and their applications, Tallinn, September, 16-20, 1985*, Tallinn: Academy of Science of the Estonian S.S.R., Institute of Cybernetics.

Institute of Cybernetics (1987b), *Computers in education: proceedings of Soviet-Finnish Symposium, Tallinn, 12-14 October, 1987*, Tallinn: Academy of Science of the ESSR, Institute of Cybernetics.

Institute of Cybernetics (1989), *Informatics '89: proceedings of the Soviet-French symposium in Tallinn, May 29-June 2, 1989*, Tallinn: Estonian Academy of Sciences, Institute of Cybernetics.

Institute of Cybernetics (1990), *Institute of Cybernetics of the Estonian Academy of Sciences 1990*, Tallinn: Valgus.

Institute of Cybernetics (2000), *In Memoriam: Arnold Reitsakas*, Tallinn: Institute of Cybernetics.

Irwin, H., More, E. and McGrath, M. (1998), 'Relationship management for innovation: the central role of communication in Australia's participation in two high-tech industries', *Technology Analysis & Strategic Management*, **10** (4), 467-81.

Jaaksoo, Ü. (1993), 'Jätkugu meil tahtmist ja tarkust', *Arvutimaailm*, **1** (1).

Jacobsson, S. (2002), 'Universities and industrial transformation: an interpretative and selective literature study with special emphasis on Sweden', *Science and Public Policy*, **29** (5), 345-65.

Jacobsson, S., Sandén, B. and Bångens, L. (2004), 'Transforming the energy system – the evolution of the German technological system for solar cells', *Technology Analysis & Strategic Management*, **16** (1), 3-30.

Jangfeldt, B (1998), *Svenska vägar till S:t Petersburg. Kapitel ur historien om svenskarna vid Nevans stränder*, Stockholm: Wahlström & Widstrand.

Johnson, B. (1992), 'Institutional Learning', in Lundvall, B.-Å. (ed.), *National Systems of Innovation. Toward a Theory of Innovation and Interactive Learning*, London: Pinter, pp. 23-44.

Kaal, M. (1996), 'Estonian Forexbank – seeking its own way', *Baltic IT Review*, **1** (2), 38-39.

Kaasik-Aaslav, V. and Maripuu, H. (1999), *Tallinna Telefonivõrk 110 aastat 1886-1996*, Tallinn.

Kaijser, A. (1994), 'Motstånd och maktkamp när Sverige fick telefon och elektricitet', *Forskning och framsteg*, (2), 18-24.

Kalvet, T. (1998), 'Electronic services in public administration: forms on the Internet', *Baltic IT Review*, **3** (11), 38-39.

Kalvet, T. (2001), *ICT, Innovations and Innovation Policy: The Case of Estonia*, Working Paper within the project 'Establishment of the Virtual Centre of Excellence for IST RTD in Estonia' (Estonian eVikings), Tartu: Archimedes Foundation.

Karlsson, M. (1998), *The Liberalization of Telecommunications in Sweden. Technology and Regime Change from the 1960s to 1993*, Linköping: Department of Technology and Social Change (PhD Dissertation).

Kase, J. (2003), 'The real-life situation with digital signatures in Estonia', *Baltic IT&T Review*, **8** (29), 42-46.

Kauber, R. (1996), 'Data communications networks and services in Estonia', *Baltic IT Review*, **1** (1), 57-60.

Kirme, J. (1993), 'Suurte infosüsteemide loomine ja eksport firmas Baltic Computer Systems', *Arvutimaailm*, **1** (2).

Kline, S. (1991), 'Styles of innovation and their cultural basis', *Chemtech*, August-November, 472-80, 525-35, and 654-59 (three parts).

Knowledge-Based Estonia (2002), *Knowledge-Based Estonia. Estonian Research and Development Strategy 2002-2006*, Tallinn 2002.

Kontsessioonileping (1992), *Kontsessioonileping Eesti Vabariigi valitsuse ja kontsessionääri vahel*, 16 December 1992, Tallinn.

Kõstner, O. (ed.) (1996), *Test-Site Estonia. Ideenäituse prospekt*, Tallinn.

Küberneetika Instituut (2000), *Küberneetika Instituut muutuvas ajas*, Tallinn: TTÜ.

Kurik, S., Lummiste, R., Terk, E. and Heinlo, A. (2002), *Innovation in Estonian Enterprises 1998-2000*, Tallinn: Enterprise Estonia.

Küttner, R. (1998), *Teadus- ja arendustegevus ülikoolis*, lecture at the conference Tallinna Tehnikaülikooli teadus, 18 September.

Lall, S. (1992), 'Technological capabilities and industrialization', *World Development*, **20** (2), 165-186.

Liik, T. (1996), 'Estonian Savings Bank – built on the ruins of a Soviet bank', *Baltic IT Review*, **1** (2), 37-38.

Liu, X. and White, S. (2001), 'Comparing innovation systems: a framework and application to China's transitional context', *Research Policy*, **30**, 1091-1114.

Luhmann, N. (1984), *Social Systems*, Stanford: Stanford University Press (English translation 1995; German original 1984).

Luhmann, N. (1986), *Ecological Communication*, Cambridge: Polity Press (English translation 1989; German original 1986).

Lundvall, B.-Å. (1988), 'Innovation as an Interactive Process: From User-Producer Interaction to the National System of Innovation', in Dosi, G., Freeman, C., Nelson, R., Silverberg, G. and Soete, L. (eds), *Technical Change and Economic Theory*, London: Pinter, pp. 349-369.

Lundvall, B.-Å. (1992), 'Introduction', in Lundvall, B.-Å. (ed.), *National Systems of Innovation. Toward a Theory of Innovation and Interactive Learning*, London: Pinter, pp. 1-19.

Lundvall, B.-Å., Johnson, B., Andersen, E. and Dalum, B. (2002), 'National systems of production, innovation and competence-building', *Research Policy*, **31**, 213-231.

Luštšik, O. (2003), 'E-pangandus Eestis: kiire kasve põhjused ja tegurid', *Kroon & majandus*, (3).

Malerba, F. (2002), 'Sectoral systems of innovation and production', *Research Policy*, **31**, 247-64.

Malerba, F. (ed.) (2004), *Sectoral Systems of Innovation. Concepts, Issues and Analyses of Six Major Sectors in Europe*, Cambridge: Cambridge University Press.

Malerba, F. and Orsenigo, L. (1996), 'The dynamics and evolution of industries', *Industrial and Corporate Change*, **5** (1), 51-87.

Mandel, M. (1999), *Eesti majandus 1989-1999*, Tallinn: Äripäeva kirjastus.

Martens, T. (1994), 'Interneti protokollistik Eestis', *Arvutimaailm*, **2** (6).

Martens, T. (1997a), 'Data communications in Estonia: a status report', *Baltic IT Review*, **2** (5), 67-69.

Martens, T. (1997b), 'Tulemüür Barrikaad', *Arvutimaailm*, **5** (2).

Martens, T. (1998), 'Security in data communication in Estonia: theory and practice', *Baltic IT Review*, **3** (8), 46-48.

Marx, K. and Engels, F. (1848), *Manifest der Kommunistischen Partei*, London: Office der Bildungsgesellschaft für Arbeiter.

McKelvey, M. (1996), *Evolutionary Innovation. The Business of Biotechnology*, Oxford: Oxford University Press.

McKelvey, M. (1997), 'Using Evolutionary Theory to Define Systems of Innovation', in Edquist, C. (ed.), *Systems of Innovation. Technologies, Institutions and Organizations*, London: Pinter, pp. 200-222.

Merilaid, A. (1938), *Sidetehnika ajalooline areng. Telegraaf, telefon, raadio, kaugnägemine, pilditelegraaf*, Tallinn.

Meriste, M. and Kalvet, T. (2001), 'E-Estonia and the information society', *Baltic IT&T Review*, **6** (20).

Min, M. (2000a), 'Eletroonikatööstus ja -haridus Eestis: olukord ja väljavaateid', in *Tallinna Tehnikaülikooli 80. aastapäev*, Tallinn: TTÜ.

Min, M. (2000b), 'Elektroonikud ettevõtluses: teadus, arendus, äri', lecture held at the conference Elektroonika 2000, 19 May, published in *Tallinna Tehnikaülikooli Aastaraamat 2000*, Tallinn: TTÜ.

Minka, R. (1996), 'Information technology helps rationalize banking services', *Baltic IT Review*, **1** (2), 30-33.

Muchie, M., Gammeltoft, P. and Lundvall, B.-Å. (eds) (2003), *Putting Africa First. The Making of African Innovation Systems*, Aalborg: Aalborg University Press.

Müller, J. (2000), 'Restructuring of the Telecommunications Sector in the West and the East and the Role of Science and Technology', in Hirschhausen, C. von and Bitzer, J. (eds), *The Globalization of Industry and Innovation in Eastern Europe. From Post-socialist Restructuring to International Competitiveness*, Cheltenham, UK and Northampton, MA, USA: Edward Elgar.

Murmann, J. (2003), *Knowledge and Competitive Advantage. The Coevolution of Firms, Technology, and National Institutions*, New York: Cambridge University Press.

Myrdal, G. (1957), *Economic Theory and Underdeveloped Regions*, Methuen and London: University Paperbacks.

NEBI (2003), *The NEBI Yearbook 2003. North European and Baltic Sea Integration*, edited by Lars Hedegaard and Bjarne Lindström, Berlin: Springer.

Nelson, R. (1992), 'National innovation systems: a retrospective on a study', *Industrial and Corporate Change*, **1** (2), 347-74.

Nelson, R. (ed.) (1993), *National Innovation Systems – A Comparative Analysis*, New York: Oxford University Press.

Nelson, R. (1995), 'Recent evolutionary theorizing about economic change', *Journal of Economic Literature*, **33**, 48-90.

Niosi, J., Saviotti, P., Bellon, B. and Crow, M. (1993), 'National systems of innovation: in search of a workable concept', *Technology in Society*, **15**, 207-227.

Noorem, T. (2002), Presentation at the Eesti Eduka Arengu Foorum, 27 November.

North, D. (1990), *Institutions, Institutional Change and Economic Performance*, Cambridge: Cambridge University Press.

Odrats, I. (1998), 'Information technologies in Estonian public administration agencies', *Baltic IT Review*, **3** (10), 24-29.

OECD (2003), *Main Science and Technology Indicators*, Volume 2003/2, Paris: OECD.

Oit, M. and Praust, V. (1998), 'Information Security in the Republic of Estonia: whence it started, and whereto it tends', in *IT in Public Administration. Yearbook 1998*, Tallinn: RISO, section 7.1.

Oskando (2002), *New m-lotto as a result of development work by Oskando* (press-release, 5 December).

Ott, A. (1993), 'Infotehnoloogia vahendite riigihangetest', *Arvutimaailm*, **1** (4).

Ott, A. (1996a), 'Government bodies and management structures of information technology in public administration in Estonia', *Baltic IT Review*, **1** (1), 9-11.

Ott, A. (1996b), 'Foreign aid for projects involving the development of state information systems in Estonia', *Baltic IT Review*, **1** (2), 10-11.

Ott, A. (1998), 'Information systems in Estonia's public administration: the role of information policy', *Baltic IT Review*, **3** (11), 34-36.

Ott, A. (2001), 'Changes in the coordination of state information systems in Estonia', *Baltic IT&T Review*, **6** (20), 23-24.

Patel, P. and Pavitt, K. (1994), 'National innovation systems: why they are important, and how they might be measured and compared', *Economics of Innovation and New Technology*, **3**, 77-95.

Pavitt, K. (1984), 'Sectoral patterns of technical change: Toward a taxonomy and a theory', *Research Policy*, **13**, 343-73.

Pavitt, K. (1997), 'Transforming Centrally Planned Systems of Science and Technology: The problem of obsolete competencies', in Dyker, D. (ed.), *The Technology of Transition. Science and Technology Policies for Transition Countries*, Budapest: Central European University Press.

Penjam, J. (2000a), 'Infotehnoloogia', in *Tehnoloogiaseire II. Eesti majanduse tehnoloogilise taseme analüüs*, Tallinn: Sihtasutus Eesti Innovatsioonifond.

Penjam, J. (2000b), 'Infotehnoloogia õpetamine Eesti ülikoolis – mida oskame ja tahame teha, mida peaksime tegema?', in *Tallinna Tehnikaülikooli Aastaraamat 2000*, Tallinn: TTÜ.

Perez, C. and Soete, L. (1988), 'Catching-up in Technology: Entry Barriers and Windows of Opportunity', in Dosi, G., Freeman, C., Nelson, R., Silverberg, G. and Soete, L. (eds), *Technical Change and Economic Theory*, London: Pinter, pp. 458-479.

Pihl, T. (2001), *Analysis of Estonian IT Sector Innovation System. Estonian ICT Cluster: Present State and Future Outlooks*, Working Paper within the project 'Establishment of the Virtual Centre of Excellence for IST RTD in Estonia' (Estonian eVikings), Tartu: Archimedes Foundation.

Pihlamägi, M. (1999), *Eesti industrializeerimine 1870-1940*, Tallinn: Teaduste Akadeemia Kirjastus.

Punch, K. (1998), *Introduction to Social Research. Quantitative and Qualitative Approaches*, London: Sage.

Raagmaa, G., Laan, M. and Raamat, A. (2003), 'Transformation of Estonia's industrial structure', *Baltic Review*, **22**.

Radiolinja (2002), *New positioning service: Find the Closest* (press-release, 20 November).

Radosevic, S. (1997), 'Systems of Innovation in Transformation. From Socialism to Post-Socialism', in Edquist, C. (ed.), *Systems of Innovation: Technologies, Institutions, and Organizations*, London: Pinter, pp. 371-394.

Radosevic, S. (1999), 'Divergence or Convergence in Research and Development and Innovation between "East" and "West"', in Fritsch, M. and Brezinski, H. (eds), *Innovation and Technological Change in Eastern Europe. Pathways to Industrial Recovery*, Cheltenham, UK and Northampton, MA, USA: Edward Elgar.

Radosevic, S. (2003), *A Two-Tier or Multi-Tier Europe: Assessing the Innovation Capacities of Central and East European Countries in the Enlarged EU*, Working Paper No. 31, London: Centre for the Study of Economic and Social Change in Europe, School of Slavonic and East European Studies, UCL.

Radosevic, S. (2004), 'The Electronics Industry in Central and Eastern Europe: An Emerging Production Location in the Alignment of Networks Perspective', in McGowan, F., Radosevic, S. and Tunzelmann, N. von (eds), *The Emerging Industrial Structure of the Wider Europe*, London: Routledge.

Rang, T. (1996), *Elektroonika Eestis – tööstus, haridus, teadus*, lecture held at the Baltic Electronics Conference at the Tallinn Technical University, 7 October 1996.

Rang, T. (2000), *Elektroonikainseneride harimisest meil ja mujal*, lecture held at the conference Elektroonika 2000, Tallinn, 19 May.

Rang, T. (2001), *Elektroonikaspetsialistide koolitus tööstuse tarbeks – Elektroonikainstituudi visioon*, lecture held at the conference Raadiotehnika 2001.

Rannu, R. (2003), *Mobile Services in Estonia. Policy Analysis No. 2*, Tallinn: PRAXIS Centre for Policy Studies.

Regio (2001), *Regios väljatöötatud tarkvara Ericssoni müügivõrgu kaudu maailma* (press-release, 25 September).

Regio (2003), *Reach-U location-based services (LBSs) and middleware were launched by Orange, the first operator in Slovakia to introduce mobile positioning* (press-release, 17 February).

Rickne (2000), *New Technology-Based Firms and Industrial Dynamics. Evidence from the Technological System of Biomaterials in Sweden, Ohio and Massachusetts*, Gothenburg: Chalmers University of Technology (PhD dissertation).

Riigi haldusinfosüsteemide arendamise ja riikliku juhtimise alused (1992), Document prepared for the Estonian government, 3 December.

Riigikantselei (1998), *Principles of Estonian Information Policy*, Tallinn: Riigikantselei.

Ritabell (1998), *Aastalõpu üllatus Q GSM-ilt: uus toode ja uued hinnad*, (press-release, 7 December).

Ritabell (1999), *NetCom Systems AB omandas 90% Q GSM omanikfirmast* (press-release, 29 January).

Roland, G. (2000), *Transition and Economics. Politics, Markets, and Firms*, Cambridge, MA: MIT Press.

Rosenberg, N. (1982), *Inside the Black Box. Technology and Economics*, Cambridge: Cambridge University Press.

Rosenberg, N. and Birdzell, L. (1986), *How the West Grew Rich. The Economic Transformation of the Industrial World*, New York: Basic Books.

Ruut, J. (1996), 'Estonia is planning a "Tiger Leap" into the information society', *Baltic IT Review*, **1** (3), 61-64.

Saarma, E. (1999), 'Mälestused 1989-1993', in Kaasik-Aaslav, V. and Maripuu, H. (1999), *Tallinna Telefonivõrk 110 aastat 1886-1996*, Tallinn.

Sadowski, B. (2001), 'Toward market repositioning in Central and Eastern Europe: International cooperative ventures in Hungary, Poland and the Czech Republic', *Research Policy*, **30**, 711-724.

Sandberg, M. (1989), *Learning from Capitalists. A Study of Soviet Assimilation of Western Technology*, Stockholm: Almqvist & Wiksell International.

Sandberg, M. (1999), *Green Post-Socialism? Environmental Aid, Polish Innovation and Evolutionary Political Economics*, London: Routledge.

Schumpeter, J. (1943), *Capitalism, Socialism and Democracy*, London: Allen & Unwin.

Schumpeter, J. (1947), 'The Creative Response in Economic History', *Journal of Economic History*, **7** (2), 149-159.

Shaw, B. (1996), 'Networking in the Russian arospace industry', *R&D Management*, **26** (3), 255-265.

Sideamet (2001a), *Sideamet 10 aastat*, Tallinn: Sideamet.

Sideamet (2001b), *Aastaraamat 2001*, Tallinn: Sideamet.

Siil, I. (1997), 'Evolution of Estonia's information policy', *Baltic IT Review*, **2** (7), October 1997, 9-12.

Siil, I. (2000), 'Estonia probes the information society', *Baltic IT&T Review*, **5** (3).

Sirkel, E. (1999), 'Eesti telefoniside 1877-1917', *Kõnetraat*, (4).

Sõmera, T. (2003), *Eesti Telefoni asutamislugu* (unpublished document).

Statistical Yearbook of Estonia, selected years, Tallinn: Statistikaamet.

Statistika aastaraamat (1985), *Eesti NSV rahvamajandus 1985. aastal. Statistika aastaraamat*, Tallinn: Eesti NSV Riiklik Statistikakomitee.

Statistiline aastaraamat (1975), *Eesti NSV rahvamajandus 1975. aastal. statistiline aastaraamat*, Tallinn: Eesti NSV Riiklik Statistikakomitee.

Statistiline kogumik (1968), *Eesti NSV rahvamajandus 1967. aastal. Statistiline kogumik*, Tallinn.

Tamm, U. (2002), '40 years of the Department of Electronics at Tallinn Technical University', in *Telekommunikatsioon 2002: IX. rahvusvahelise telekommunikatsioonipäeva materjalid*, Tallinn.

Tammet, T. (1996), *Teejuht võrgumaailma*, Tartu: Ilmamaa.

Tani, H. (1993), 'Kas Eestil on šansse tehnoloogias?', *Arvutimaailm*, **1** (1).

Tani, H. (1999), 'Elektri- ja elektroonikatööstus', in *Tehnoloogiaseire I. Eesti majanduse tehnoloogilise taseme võrdlev analüüs*, Tallinn: Sihtasutus Eesti Innovatsioonifond.

Tani, H. (2000), 'Meenutusi EKTA tekkeloost ja arengust', in Küberneetika Instituut (2000), *Küberneetika Instituut muutuvas ajas*, Tallinn: TTÜ.

Tarkon (2001), *History of Enterprise* (unpublished document).

Tarkon (2003), *Ettevõtte tutvustus* (unpublished document).

Teemant, J. (2003), 'The telecommunications market in the Baltic states', *Baltic IT&T Review*, **8** (28), 29-30.

Telekommunikatsiooniseadus (2000), *Telekommunikatsiooniseadus, vastu võetud 9. veebruaril 2000. a.* [Telecommunications Law, accepted on 9 February 2000].

Televerket (1991), *Nytt baltiskt telebolag – Televerket blir delägare i estniska televerket* (press-release, 10 October).

Telia (1999), *Annual report 1999*, Stockholm: Telia.

Telia (2002), *Telia och Sonera går samman – skapar nytt ledande nordiskt telekommunikationsföretag* (press-release, 26 March).

Tidd, J., Bessant, J. and Pavitt, K (2001), *Managing Innovation. Integrating Technological, Market and Organizational Change*, Second edition, New York: Wiley.

TPI (1986a), *Raadiotehnika kateeder*, Tallinn: TPI.

TPI (1986b), *Tallinna Polütehniline Instituut 1936-1986*, Tallinn: TPI.

Trapido, G. (1994), 'Side areng projekteerija pilguga', *Kõnetraat*, (4).

TTÜ (1994), *Tallinna Tehnikaülikooli Aastaraamat 1994*, Tallinn: TTÜ.

TTÜ (1995), *Tallinna Tehnikaülikooli Aastaraamat 1995*, Tallinn: TTÜ.

TTÜ (1996), *Tallinna Tehnikaülikooli Aastaraamat 1996*, Tallinn: TTÜ.

TTÜ (1997), *Tallinna Tehnikaülikooli Aastaraamat 1997*, Tallinn: TTÜ.

TTÜ (1998), *Tallinna Tehnikaülikooli Aastaraamat 1998*, Tallinn: TTÜ.

TTÜ (2000), *Tallinna Tehnikaülikooli Aastaraamat 2000*, Tallinn: TTÜ.

TTÜ (2002), *Tallinna Tehnikaülikooli Aastaraamat 2002*, Tallinn: TTÜ.

TTÜ Nõukogu (1996), *Tallinna Tehnikaülikooli Nõukogu avalik pöördumine Eesti Vabariigi haridusministri poole*, vastu võetud TTÜ nõukogu istungil 13. veebruaril 1996. aastal.

TTÜ professorid (1996), *Tallinna Tehnikaülikooli professorite avalik pöördumine Eesti Vabariigi Riigikogu ja valitsuse poole*, Tallinnas, 21. märtsil 1996.

Tylecote, A., Cho, Y. and Zhang, W. (1998), 'National technological styles explained in terms of stakeholding patterns, enfranchisement and cultural differences', *Technology Analysis & Strategic Management*, **10** (4), 423-35.

Uba, T. (1996), 'Bank cards in Estonia', *Baltic IT Review*, **1** (2), 33-35.

Ubar, R. (1994), 'Elektroonika ja arvutustehnika vabariiklikust sihtprogrammist', *Arvutustehnika & Andmetöötlus*, **8** (1).

Utterback, J. (1994), *Mastering the Dynamics of Innovation. How Companies Can Seize Opportunities in the Face of Technological Change*, Boston: Harvard Business School Press.

Vallner, U. (1998), 'Governmental Institutions and the Internet', in *IT in Public Administration. Yearbook 1998*, Tallinn: RISO, section 5.1.

Vallner, U. (2002), 'An e-community, e-citizens and e-services – a new infrastructure and reengineering of institutions', *Baltic IT&T Review*, **7** (27), 39-44.

Veblen, T. (1925), *The Theory of the Leisure Class: An Economic Study of Institutions*, London: Allen & Unwin.

Vertova, G. (1998), 'Technological similarity in national styles of innovation in a historical perspective', *Technology Analysis & Strategic Management*, **10** (4), 437-49.

Viik, L. (2002), 'Kuldne aasta mobiilsides', *Äripäev*, 21 August 2002.

Wõrk, A. (1993), 'Arvutivõrgud Internet, NORDUnet ja FUNET', *Arvutimaailm*, **1** (1).

Wõrk, A. (1997), 'Data communication and Internet services in Estonia: history and the present situation', *Baltic IT Review*, **2** (7), 15-19.

Yin, R. (1989), *Case Study Research. Design and Methods*, London: Sage.

Interviewees

Aaviksoo, Peep	CEO, EMT, 1999-2003
Anton, Jaak	CEO, Voicecom; advisor, Ministry of Education and Research; former chairman of the Supervisory Board, Estonian Information Technology Foundation; CEO, Astrodata 1989-1998; former researcher at the Institute of Astrophysics and Atmospheric Physics
Carlsson, Ove	CEO of Tarkon, 2003-; former marketing director, Hallberg-Sekrom; formerly at Ericsson
Dane, Davids	Press secretary, LMT
Ejdemark, Anders	Formerly at Televerket
Elsérius, Jan-Erik	CFO, Lietuvos Telekomas; formerly at Telia
Enno, Jaak	Manager of Technical Development, Keila Kaabel
Erme, Enn	Director, Tartu Technological Park, -2002
Haljand, Holger	Business Development Manager at the Development Department, EMT
Härm, Heikki	Reveko Telekom
Jaaksoo, Ülo	CEO, Cybernetica, 1997-; Director, Institute of Cybernetics, 1989-1997; researcher at the Institute of Cybernetics since 1961; Member of the Estonian Academy of Sciences
Jagomägi, Teet	Chairman of the Board, Regio
Kalja, Ahto	Head of Department and Professor, Institute of Cybernetics, TTÜ; project manager, Department of State Information Systems
Klanberg, Peeter	Managing Director, SEB IT Partner Estonia
Kõuhkna, Tarmo	Former General Manager, Oskando
Kull, Andres	CEO, Elvior
Lapõnin, Aare	E-Tax Board Director, Estonian Tax Board, 1994-2003
Lehtimäki, Matti	Ericsson Finland; formerly CEO, Ericsson Eesti
Leimann, Jaak	Professor of Organization and Management, TTÜ; former Minister of Foreign Affairs, 1990-92 and 1996-99
Makkonen, Matti	Managing Director, Finnet; former Chairman of the Board, EMT
Mälk, Meelis	CEO, ADM Group
Männik, Jaan	CEO, Eesti Telekom, 2000-; CEO, Eesti Telefon, 1997-2000; former Head of Tallinn Telephone Network
Martinson, Allan	Chairman of the Board, MicroLink, 1998-2004; Chairman of the Board, Baltic News Service, 1991-98
Mets, Rein	Eesti Telefon; former Head of Tallinn Telephone Network
Nielsen, Mart	IT Director for the Baltic Region, Tele2
Nisula, Markus	Chairman of the Board, Uninet
Noorem, Toomas	Director, Tartu Technological Park, 2003-; CEO, Tarkon,

	1993-2002
Nurk, Mart	Former Head of Telia's representation in Estonia; formerly at Telia Infomedia
Ots, Avo	Lecturer, Department of Radio and Communications Engineering, TTÜ; former member of the Supervisory Board, EMT
Ott, Arvo	Director, Department of State Information Systems, 1993-; CEO, Univoice Estonia 1990-93; former researcher at the Institute of Cybernetics
Peets, Andres	CEO, Reveko Telekom
Petersen, Ilmar	General Manager, New Production Introduction Services, Elcoteq Tallinn, 2002-; project director at Elcoteq's headquarters in Helsinki, 2000-2002; plant director at Elcoteq Tallinn, 1994-2000; formerly at ANet Eesti
Rang, Toomas	Director and Professor, Department of Electronics, TTÜ; Director of ELIKO
Raudsalu, Andrus	Development Director, Delfi, 1999-; former purchasing and product manager at MicroLink; former lecturer at KBFI;
Reimo, Tõnis	CEO, Privador
Saard, Niilo	CEO, CellNetwork Estonia; Member of the Board, Association of Estonian Information Technology and Telecommunications Companies (ITL)
Saarma, Edvard	Head of the Communications Department, Ministry of Economic Affairs and Communication; former Head of Tallinn Telephone Network
Saarma, Kaido	Managing Director, Abobase Systems; former President, Computer Firms Association
Saarsen, Toomas	CEO, Aprote
Sepp, Veiko	CEO, Ericsson Eesti
Siil, Imre	Director of the Development Center, Department of State Information Systems/Estonian Informatics Center, -2003
Sild, Ants	CEO, Baltic Computer Systems
Sildmäe, Tõnis	Former IT Director, Hansapank; formerly CEO, Crebit
Sõmera, Toomas	Chairman of the Board, Estonian Information Technology Foundation, 2000-, CEO, Eesti Telekom, 1991-2000; former Minister of Communications
Suitsu, Andrus	Proekspert; former researcher, KBFI
Tamm, Alo	Project Manager, the Development Department, EMT, 2000-; formerly at Radiolinja Eesti
Tamm, Jaan	Director, e-Technology and Operations Division, Eesti Ühispank
Tammemäe, Kalle	Rector, Estonian Information Technology College, 2000-; former researcher at the Institute of Cybernetics
Tee, Tõnu	CEO, EMT, 1992-1999
Tiltsu, Taavi	Data Exchange Senior Specialist, Estonian Rescue Board
Tüksammel, Ove	Chairman of the Board, Proekspert
Ulman, Jaak	Chairman of the Board, Telefant; formerly Deputy Minister of Communications; formerly Head of Tallinn Telephone Network

Vallner, Uuno	Head of Development Division, Department of State Information Systems; former researcher at the Institute of Cybernetics
Viik, Linnar	Member of the Board, Estonian Information Technology College; former IT advisor to the Prime Minister;
Viira, Ants	Director of Development and External Affairs of Radiolinja Eesti; former Development Director of EMT
Wõrk, Ants	Managing Director, EsData; former Vice-Director, Institute of Cybernetics

Index